VIRUSES IN FOODS

FOOD MICROBIOLOGY AND FOOD SAFETY SERIES

Food Microbiology and Food Safety publishes valuable, practical, and timely resources for professionals and researchers working on microbiological topics associated with foods, as well as food safety issues and problems.

Titles

PCR Methods in Foods, John Maurer (Ed.) (2006)
Viruses in Foods, Sagar M. Goyal (Ed.) (2006)
Foodborne Parasites, Ynes R. Ortega (Ed.) (2006)

VIRUSES IN FOODS

Edited by

Sagar M. Goyal
University of Minnesota
St. Paul, Minnesota, USA

Springer

Sagar M. Goyal
Department of Veterinary Population Medicine
1333 Gortner Avenue
University of Minnesota
St Paul, MN 55108

Library of Congress Control Number: 2006921351

ISBN-10: 0-387-28935-6
ISBN-13: 978-0-387-28935-9

Printed on acid-free paper.

(BS/EB)

9 8 7 6 5 4 3 2 1

springer.com

To my mother

Bimia Devi Goyal

Contributors

F. Xavier Abad, Enteric Virus Laboratory, Department of Microbiology, University of Barcelona, 08028 Barcelona, Spain

Robert L. Atmar, Departments of Medicine and Molecular Virology and Microbiology, Baylor College of Medicine, Houston, Texas 77030

Sabah Bidawid, Microbiology Research Division, Bureau of Microbial Hazards, Health Canada, Ottawa, Ontario, Canada K1A 0L2

Albert Bosch, Enteric Virus Laboratory, Department of Microbiology, University of Barcelona, 08028 Barcelona, Spain

Javier Buesa, Department of Microbiology, School of Medicine, University of Valencia, 46010 Valencia, Spain

Christopher Y. Choi, Department of Agricultural and Biosystems Engineering, University of Arizona, Tucson, Arizona 85721

Doris H. D'Souza, Food Science Department, North Carolina State University, Raleigh, North Carolina 27695

Samuel R. Farrah, Department of Microbiology, University of Florida, Gainesville, Florida 32611

Charles P. Gerba, Department of Soil, Water, and Environmental Science, University of Arizona, Tucson, Arizona 85721

Sagar M. Goyal, Department of Veterinary Poplulation Medicine, University of Minnesota, St. Paul, Minnesota 55108

Gail E. Greening, Institute of Environmental Science and Research, Kenepuru Science Center, Porirua, New Zealand

Craig W. Hedberg, Division of Environmental Health Sciences, School of Public Health, University of Minnesota, Minneapolis, Minnesota 55455

Lee-Ann Jaykus, Food Science Department, North Carolina State University, Raleigh, North Carolina 27695

Efstathia Papafragkou, Food Science Department, North Carolina State University, Raleigh, North Carolina 27695

Suresh D. Pillai, Poultry Science Department, Institute of Food Science and Engineering, Texas A&M University, College Station, Texas 77843

Rosa M. Pinto, Enteric Virus Laboratory, Department of Microbiology, University of Barcelona, 08028 Barcelona, Spain

Gary P. Richards, United States Department of Agriculture—Agricultural Research Service, Microbial Food Safety Research Unit, Delaware State University, Dover, Delaware 19901

Jesús Rodríguez-Díaz, Department of Microbiology, School of Medicine, University of Valencia, 46010 Valencia, Spain

Syed A. Sattar, Centre for Research on Environmental Microbiology (CREM), Faculty of Medicine, University of Ottawa, Ottawa, Ontario, Canada K1H 8M5

Preface

Viral contamination of food and water represents a significant threat to human health. The cases of viral food borne outbreaks are on the rise partly because of increases in population, scarcity of clean water, and changes in eating habits. Outbreaks attributed to toxic, fungal, parasitic, and bacterial agents are very well known and characterized because we have known about these diseases for a long time and have developed appropriate methods to investigate and track them. Methods to investigate viral food borne diseases, on the other hand, have only recently begun to be developed. One reason for the lack of these methods is that the number of viruses present in food is too small to be detected by methods used in clinical virology, although low levels of viral contamination can still cause infection in a susceptible host. Another problem is that two of the most important food borne viruses either do not grow in cell cultures (norovirus) or grow poorly in primary isolation (hepatitis A virus). However, with the advent of molecular diagnostic methods, the role of viruses in food borne disease outbreaks is beginning to be understood.

Shellfish, fresh produce, and ready-to-eat foods are especially vulnerable to viral contamination. Although viral disease outbreaks associated with shellfish have been known to occur for decades, non-shellfish foods have only recently been implicated in several large outbreaks. In fact, the incidence of produce-associated outbreaks has increased in recent years because the consumption of such foods has increased due to health reasons and because produce is often imported from areas lacking in strict hygienic measures. Because of their very nature, fresh produce and ready-to-eat foods are more likely to contribute to the disease burden because they are often eaten uncooked, thereby eliminating the added safety factor provided by cooking and because they often come in contact with potentially contaminated water, ice, human hands, and surfaces from farm-to-table continuum. Even a single contamination event can result in widespread outbreaks as was demonstrated by the raspberry-associated outbreaks that occurred simultaneously in several countries. In addition, food is also subject to intentional contamination with highly infectious pathogens including viruses such as smallpox virus, filoviruses, arenaviruses, and alphaviruses.

A number of books are available on food borne disease outbreaks but none on the role of viruses in such outbreaks. *Viruses in Foods* was written to fill that gap. A team of international scientists has contributed material for this volume. We hope that the book serves a useful purpose, howsoever small, in the prevention and control of viral food borne outbreaks.

SAGAR M. GOYAL

Contents

CHAPTER 1

Food Virology: Past, Present, and Future

Charles P. Gerba

Food was first recognized as a vehicle for the transmission of viruses in 1914 when a raw milk–associated outbreak of poliomyelitis was reported (Jubb, 1915). Additional milk-borne outbreaks were recognized after this time, but with the development of a vaccine for poliovirus, no outbreaks were reported in the developed world after the early 1950s (Sattar and Tetro, 2001). However, in the mid-1950s, hepatitis A transmission by shellfish was first reported in Sweden (Roos, 1956) and then in the United States (Mason and McLean, 1962). Later, food-borne outbreaks of nonbacterial gastroenteritis were recognized, although a specific viral agent could not be isolated *in vitro*. For the next two decades, the study of food-borne viruses centered on virus transmission by shellfish and potential risks from the use of reclaimed wastewater for irrigation of food crops.

In the 1990s, molecular methods became available for the detection of difficult to cultivable or noncultivable viruses, which led to the realization that viruses are the leading cause of food-borne illness in the developed world (Bresee et al., 2002; Koopmans and Duizer, 2004). Although any virus capable of transmission by the fecal-oral route can be transmitted by foods, human caliciviruses (noroviruses and sapoviruses) are believed to be the major cause. Mead et al. (1999) estimated that 76 million cases of food-borne illness occur in the United States each year and that viruses cause an estimated 67% of these. They also estimated number of cases per year of calicivirus gastroenteritis at 23,000,000, or 40% of all cases (Mead et al., 1999). The number of documented food-borne virus outbreaks is believed to be on the increase worldwide (Koopmans et al., 2002). This trend will likely continue for a number of reasons as discussed in the next paragraph.

Persons that will experience more serious illness and greater chance of mortality from infectious disease are on the increase in the United States (Gerba et al., 1996). These include young children, pregnant individuals, older people, and immunocompromised individuals (cancer and organ transplant patients). This group currently represents almost 25% of the population in the United States and is expected to grow as the population ages. Persons in nursing homes are 100 times more likely to die of a rotavirus infection than the general population (Gerba et al., 1996). Enteric adenovirus 40 and 41, which appear to rarely cause disease in adults, can result in mortalities of 50% to 69% in immunosuppressed cancer and organ transplant patients (Hierholzer, 1992).

Increased consumption of foods traditionally eaten raw and globalization of international trade have increased the risks of viral contamination of foods. Much of the produce consumed in the developed world now originates from less developed countries where sanitation and hygiene are not adequate. Recent outbreaks of hepatitis A and norovirus from foods imported to the United States and Europe demonstrate the importance of international trade in aiding the transmission of viral diseases (Dentinger et al., 2001). An increase in the reported number of produce-associated foodborne outbreaks corresponds with the increased consumption of fresh fruit and vegetables and the expanded geographical sources and distribution of these products during the past two decades (Sivapalasingam et al., 2004). Produce-associated outbreaks have increased from 0.7% of all outbreaks in the 1970s to more than 6% in the 1990s in the United States. From 1990 to 2002, produce was second only to seafood in the total number of outbreaks documented, and in the same period the number of cases of produce-associated illnesses was almost equal to those reported for all of beef, poultry, and seafood combined (Center for Science in the Public Interest, 2002). The development of molecular methods has shown us how important food is in the transmission of viruses. We are now able to detect viruses, which cannot be grown in cell culture and can track their origin with molecular fingerprinting.

Until recently, it was thought that food-borne enteric viruses could only originate from humans and hence their transmission was limited to contaminated food handlers, cross-contamination of food, and contamination by water. However, several recent outbreaks of hepatitis E virus have demonstrated that it is a zoonotic virus capable of transmission by consumption of raw or lightly cooked meat products (Mishiro, 2004; Tei et al., 2004). The close relationship between human norovirus and some animal caliciviruses, such as those found in calves, suggests that interspecies transmission to humans may be possible (Koopmans et al., 2002). In addition, there are several other animal viruses that have demonstrated an ability to cross species barriers and have the potential to become involved in human disease. These include coronavirus of severe acute respiratory syndrome (SARS), influenza virus, picobirnaviruses, toroviruses, parvoviruses, bornaviruses, and pestivirus.

The fundamental problem with regard to the detection of viruses in food is the presence of low numbers of highly infectious viruses, which requires sampling of large volumes of food to assess the risk of infection. Ingestion of even one virus particle has a significant probability of causing an infection. In contrast, bacteria require ingestion of thousands of cells to have the same probability of infection. In addition, viruses have to be extracted or removed from food before amplification in cell culture or detection by molecular methods. This is necessary to reduce the assay volume and remove substances that are toxic to cell cultures or interfere with detection by molecular methods. This limits the sensitivity of viral assays and makes it a more difficult and costly process as compared with the detection of food-borne bacteria. Fortunately, in the case of produce contamination, it is most likely to occur on the surface by contaminated irrigation water, improper handling, or cross-contamination. Uptake of human viruses by roots and into the ingested part of produce seems an unlikely or rare event. However, in the case of shellfish, internal structures of the animal become contaminated creating additional difficulties in the recovery of the virus.

The application of polymerase chain reaction (PCR) assay has been a major advance in our ability to detect viruses in foods. However, removal of interfering substances, small assay volumes, and determination of infectivity are areas that still require much improvement before full advantage of this technology can be realized by food virologists. With the development of this technology comes the question: What do we do with it? We can use it to investigate outbreaks, demonstrate the virus in the implicated food(s), and track sources of contamination. In addition, we can test the effectiveness of the treatment processes or evaluate strategies to reduce exposure of foods to viral contamination. Beyond this, we may be limited by the sensitivity of the methods and what they can tell us about the safety of the food and the risk of viral illness.

Because low numbers of viruses are usually present in a food supply, we cannot sample large volumes of food like we have been able to do with water. Even with drinking water, the sampling of untreated water is done to estimate the risk of infection after treatment, because even low concentrations of virus (1 infectious virus in 100,000 L) are capable of causing a risk of infection of 1:10,000 per year if the water is consumed (Regli et al., 1991). This is where hazard analysis of critical control points (HACCP) comes into play. As our knowledge expands on viral contamination of foods, we need to identify how and where virus contamination of foods occur.

We obviously need a much greater understanding of viral contamination by water used in irrigation and processing. Controlling contamination by food handlers may be very difficult because of asymptomatic infections. It is currently unclear what proportion of food-borne infections can be attributed to workers in different parts of the food chain. It is important that HACCP systems be used to identify risks and help identify gaps in knowledge (Koopmans et al., 2002). Koopmans and Duizer (2004) noted that although infected food handlers at the end of the food chain are implicated in most outbreaks, contamination can occur anywhere (e.g., seasonal workers picking berries, sick individuals harvesting oysters, and recreational activities on lakes used for irrigation of crops, etc.).

Given the recent recognition of the significance of viruses in food-borne disease and the development of methods for virus detection, it appears that food virology as a field is poised to rapidly grow in the coming years.

REFERENCES

Bresee, J. S., Widdowson, M. A., Monroe, S. S., and Glass, R. I., 2002, Foodborne viral gastroenteritis: challenges and opportunities. *Clin. Infect. Dis.* 35:748–753.

Center for Science in the Public Interest, 2002, *Outbreak Alert.* CSPI, Washington, DC.

Dentinger C., Bower, W. A., Nainan, O. V., Cotter, S. M., Myers, G., Dubusky, L. M., Fowler, S., Salehi, E. D., and Bell, B. P., 2001, An outbreak of hepatitis A associated with green onions. *J. Infect. Dis.* 183:1273–1276.

Gerba, C. P., Rose, J. B., and Haas, C. N., 1996, Sensitive populations: who is at the greatest risk? *Int. J. Food Microbiol.* 30:113–123.

Hierholzer, J. C., 1992, Adenoviruses in the immuncompromised host. *Clin. Microbiol. Rev.* 5:262–274.

Jubb, G., 1915, The outbreak of epidemic poliomyelitis at West Kirby. *Lancet* 167.

Koopmans, M., and Duizer, E., 2004, Foodborne viruses: an emerging problem. *Int. J. Food Microbiol.* 90:23–41.

Koopmans, M., von Bonsdorff, C. H., Vinje, J., de Medici, D., and Monroe, S., 2002, Foodborne viruses. *FEMS Microbiol. Rev.* 26:187–205.

Mason, J. O., and McLean, W. R., 1962, Infectious hepatitis traced to the consumption of raw oysters: an epidemiologic study. *Am. J. Hyg.* 75:90–111.

Mead, P. S., Slutsker, L., Dietz, V., McCaig, L. F., Bresee, J. S., Shapiro, C., Griffin, P. M., and Tauxe, R. V., 1999, Food-related illness and death in the United States. *Emerg. Infect. Dis.* 5:607–625.

Mishiro, S., 2004, Recent topics on hepatitis E virus: emerging, zoonotic, animal-to-human transmission in Japan. *Uirusu* 54:243–248.

Regli, S., Rose, J. B., Haas, C. N., and Gerba, C. P., 1991, Modeling the risk from *Giardia* and viruses in drinking water. *J. Am. Water Works Assoc.* 83:473–479.

Roos, B., 1956, Hepatitis epidemic conveyed by oysters. *Svensk Lakartidn* 53:989–1003.

Sattar, S. A., and Tetro, J. A., 2001, Other foodborne viruses, in: *Foodborne Disease Handbook*, 2nd ed., vol. 2 (Y. H. Hui, S. A. Sattar, K. D. Murrell, W. K. Nip, and P. S. Stantfield, eds.), Marcel Dekker, New York, pp. 127–163.

Sivapalasingam, S., Friedman, C. R., Cohen, L., and Tauxe, R. V., 2004, Fresh produce: a growing cause of outbreaks of foodborne illness in the United States, 1973 through 1997. *J. Food Protect.* 67:2342–2353.

Tei, S., Kitajima, N., Ohara, S., Inoue, Y., Miki, M., Yamatani, T., Yamabe, H., Mishiro, S., and Kinoshita, Y., 2004, Consumption of uncooked deer meat as a risk factor for hepatitis E virus infection: an age- and sex-matched case-control study. *J. Med. Virol.* 74:67–70.

CHAPTER 2

Human and Animal Viruses in Food (Including Taxonomy of Enteric Viruses)

Gail E. Greening

1.0. INTRODUCTION

In recent years, there has been an increase in the incidence of food-borne diseases worldwide, with viruses now recognized as a major cause of these illnesses. The viruses implicated in food-borne disease are the enteric viruses, which are found in the human gut, excreted in human feces, and transmitted by the fecal-oral route. Many different viruses are found in the gut, but not all are recognized as food-borne pathogens. The enteric viral pathogens found in human feces include noroviruses (previously known as Norwalk-like viruses), enteroviruses, adenoviruses, hepatitis A virus (HAV), hepatitis E virus (HEV), rotaviruses, and astroviruses, most of which have been associated with food-borne disease outbreaks. Noroviruses are the major group identified in food-borne outbreaks of gastroenteritis, but other human-derived and possibly animal-derived viruses can also be transmitted via food.

The diseases caused by enteric viruses fall into three main types: gastroenteritis, enterically transmitted hepatitis, and illnesses that can affect other parts of the body such as the eye, the respiratory system, and the central nervous system including conjunctivitis, poliomyelitis, meningitis, and encephalitis. Four of the enteric viruses—noroviruses, HAV, rotaviruses, and astroviruses—are included in the thirteen major food-borne pathogens identified by the Centers for Disease Control and Prevention (CDC) (Mead et al., 1999). These four viruses are reported to comprise 80% of all food-borne illnesses in the United States, with noroviruses by far the greatest contributor at an estimated 23 million cases per year (Mead et al., 1999).

All enteric viruses except the adenoviruses contain RNA rather than DNA, have a protein capsid protecting the nucleic acid, and are nonenveloped. In the environment and in food, the enteric viruses are inert particles and do not replicate or metabolize because, like all viruses, they are obligate pathogens and require living cells to multiply. Many of the enteric viruses such as astroviruses, enteric adenoviruses, HAV, and rotaviruses are fastidious in their *in vitro* growth requirements but can still be grown in cell cultures. Noroviruses, on the other hand, do not grow *in vitro*, and no animal model exists for the human noroviruses yet. For many years, the lack of a culture system limited investigations focusing on the role of noroviruses in food-borne disease, although progress is now being made after the *in vitro* culture of a mouse norovirus (Wobus et al., 2004). Cell cultures are generally used for the analysis of culturable viruses. Using culture methods, infectious viruses can be identified through their ability to produce changes in inoculated cells (cytopathic effects or CPE) or through expression of viral antigens that may be detected serologically. The advantage of culture-based methodology is that it can be either quantitative or qualitative and produces unambiguous results with respect to virus presence and infectivity.

Until the introduction of molecular methods, enteric viruses were mainly identified by electron microscopy (EM) including solid-phase immune electron microscopy (SPIEM). The SPIEM is more sensitive than direct EM because, in the presence of specific antibodies, the virus particles are coated with specific antibody and aggregated together, making them more easily distinguishable from the background matrix. Many of the "small round viruses," which include astroviruses, noroviruses, sapoviruses, and parvoviruses, were first discovered through the use of EM.

Molecular methods are now the most commonly used techniques for the identification of enteric viruses in foods, but other methods are also available for virus detection in human specimens. Identification of enteric viruses can also be carried out by enzyme-linked immunosorbent assay (ELISA), radioimmunoassay (RIA), and, for the culturable viruses, culture-PCR, which is a combination of cell culture and polymerase chain reaction (PCR) methods. The latter technique detects only the infectious virus and is preferable to direct PCR, which currently detects both infectious and noninfectious viruses.

Enteric viruses are generally resistant to environmental stressors, including heat and acid. Most resist freezing and drying and are stable in the presence of lipid solvents. It is not clear whether pasteurization at 60°C for 30 min inactivates all enteric viruses. Many enteric viruses show resistance to ultrahigh hydrostatic pressure, which is now being widely used as a novel food-processing treatment for shellfish, jams, jellies, and dairy products (Wilkinson et al., 2001; Kingsley et al., 2002). The resistance of enteric viruses to environmental stressors allows them to resist both the acidic environment of the mammalian gut and also the proteolytic and alkaline activity of the duodenum so that they are able to pass through these regions and colonize the lower digestive tract. These properties also allow survival of enteric viruses in acidic, marinated, and pickled foods; frozen foods; and lightly cooked foods such as shellfish. Most enteric viruses are believed to have a low infectious dose of 10–100 particles or possibly even less. Hence, although they do not multiply in food, enough infectious virions may survive in food, be consumed, and cause disease.

Enteric viruses have been shown to retain infectivity in shellfish and in fresh, estuarine, and marine waters for several weeks at 4°C (Jaykus et al., 1994; Scientific Committee on Veterinary Measures relating to Public Health, 2002). The length of virus survival appears to be temperature dependent and is inversely related to increased temperature. The enteric viruses may survive longer if attached to particulate matter or sediments, where they can present a greater potential risk to human health (Jaykus et al., 1994).

Most viruses causing food-borne disease are of human origin, and the source of viral contamination generally originates from human fecal material. Viral contamination of foods can occur pre- or postharvest at any stage in the food harvesting, processing, and distribution chain. The key factors influencing the risk of contamination of fresh produce are water quality, field-worker hygiene, and food-handler hygiene. Thus, sewage contamination and poor hygiene practices play a major role in the contamination process.

The globalization of the food supply means that the source of fresh produce may not always be known and the quality may not always be controlled. Although it is presumed that fresh produce is "clean, green, and healthy," it may not be so, especially when it is imported from countries where general hygiene practices do not meet international standards. This knowledge, combined with a number of outbreaks associated with contaminated fresh produce, has led to consumer suspicion of imported foods in many countries.

The opportunities for both pre- and postharvest viral contamination are numerous. The quality of the growing waters is important for shellfish quality. Preharvest virus contamination occurs when filter-feeding bivalve shellfish grow in waters contaminated with sewage or fecal material. Shellfish filter between 4 and 20 L of water every hour, sieving out and accumulating food particles, including bacteria, viruses, and heavy metals. Feeding rates depend on water temperature and salinity and availability of food and particulate matter. Bacteria and viruses become trapped in the mucus of the gills, which is then pushed into the digestive gland where viruses appear to concentrate. Shellfish can accumulate high concentrations of viruses within a few hours when surrounding waters contain sufficient levels of virus, so that concentrations in shellfish may be 100 to 1,000 times greater than the surrounding waters.

Virus uptake varies between shellfish species and also between individuals. In winter, the shellfish are physiologically less active and so do not accumulate or remove viruses as fast as in the warmer seasons. In clean waters, shellfish depurate or cleanse themselves of bacteria and particulate matter. However, some studies have shown that depuration does not remove viruses efficiently, and there is no correlation between the removal of bacteria and viruses (Lees, 2000). This was demonstrated in a large hepatitis A outbreak in Australia where oysters were depurated for 36 hr before consumption but still retained infectious HAV (Conaty et al., 2000).

Fresh produce may have been irrigated or washed in water containing human fecal material or handled by field workers or food handlers with poor hygiene practices. In such situations, the produce may be contaminated with disease-causing enteric viruses. Foods at the greatest risk of virus contamination at the preharvest stage are shellfish, soft berry fruits, herbs, and salad greens. Foods at risk from contamination by food handlers include a wide range of foods that are subjected to much handling and are subsequently consumed cold or uncooked. These include bread and bakery goods, lightly cooked or raw shellfish, sandwiches, salads, herbs, fresh fruits, cold meats, and cold desserts. It is probable that the current trend for the consumption of raw or lightly cooked ready-to-eat (RTE) foods, especially salads and sandwiches, has increased the risk of food-borne viral disease. Poor food handling was shown to be a key risk factor in the transmission of noroviruses and rotaviruses in The Netherlands (de Wit et al., 2003).

All food-borne viruses are transmitted by the fecal-oral route and are generally host specific for humans, although animal strains of the same virus may also exist. Viruses are frequently host specific, preferring to grow in the tissue of one species rather than a range of species. Both animal and human strains exist in all of the enteric viral genera. A key question still to be answered is whether animal viruses can infect humans and vice versa. The pathogenic strains of astrovirus, adenovirus, and enterovirus that infect animals appear to be distinct from those that infect humans. Thus, although noroviruses have been isolated from animal feces, so far they have not been implicated in human disease (Sugeida et al., 1998; van der Poel et al., 2000; Oliver et al., 2003).

Zoonotic infections are generally not transmitted by food. However, the risk of zoonotic viral disease from meat products contaminated with animal viruses has been identified in some countries; tick-borne encephalitis virus (TBE) and hepatitis E virus (HEV) being two examples. HEV is possibly the first virus reported to cause zoonotic food-borne viral disease (Tei et al., 2003). Nonviral infectious proteinaceous agents, or prions, that cause diseases such as bovine spongiform encephalopathy (BSE), scrapie, and Creutzfeldt-Jakob disease, also transmit disease from animals to man via the food-borne route but are not discussed in this chapter.

As a result of the advances in methodology for detection of viruses in foods, the extent and role of viruses in food borne viruses have been clarified in recent years. The development of new molecular methods, including real-time PCR–based methods, for the detection of nonculturable or difficult to culture viruses has shown their frequent presence in the environment and in foods, especially shellfish. These methods have also allowed investigation of virus responses to environmental stressors and have contributed to increased knowledge of enteric virus behavior in foods and in the environment.

2.0. HEPATITIS A VIRUS

2.1. Distribution and Transmission

Several different viruses cause hepatitis but only two, HAV and HEV, are transmitted by the fecal-oral route and are listed as "Severe Hazards" in Appendix V of the U.S. Food and Drug Administration's Food Code (Cliver, 1997). The hepatitis viruses are so named because they infect the liver, rather than sharing phylogenetic or morphological similarities, and each of the five different hepatitis viruses is classified in a distinct viral family. HAV causes hepatitis A, a severe food and waterborne disease that was formerly known as infectious hepatitis or jaundice. The virus is primarily transmitted by the fecal-oral route but can also be transmitted by person-to-person contact. Hepatitis A infection occurs worldwide and is especially common in developing countries where more than 90% of children have been reported to be infected by 6 years of age (Cliver, 1997; Cromeans et al., 2001). The infection is often asymptomatic in children.

In recent years, the incidence of hepatitis A infection in many countries has decreased as sewage treatment and hygiene practices have improved, but this has also led to an overall lowering of immunity in these populations with consequent increase in susceptibility to the disease. As a result, there is an increasing risk of contracting hepatitis A infection from fresh foods imported from regions of the world where HAV is endemic and general hygiene standards are poor. Hepatitis A is a serious food-borne infection and hence is a notifiable disease in most of the developed countries. This means that accurate data on its occurrence are recorded in these countries. In the United States, hepatitis A is reported as the most common cause of hepatitis with a reported death rate of 0.3%. However, the actual incidence of hepatitis A is assumed to be 10 times that of the reported cases. Between 1980 and 2001, the CDC was notified of an average of 25,000 cases/year, but when corrections were made to the data, the average case numbers were estimated to be approximately 260,000 per year (Fiore, 2004).

No seasonal distribution of HAV has been observed, with infection occurring throughout the year, but the disease has been reported to have a cyclic occurrence in endemic areas. This cyclic pattern has been observed in the United States, particularly among low socioeconomic, Native American, and Hispanic populations, with large increases in hepatitis A infections occurring approximately every 10 years. However, the main transmission route is probably from person to person rather than being food-borne (Cromeans et al., 2001; Fiore, 2004).

2.2. Taxonomy and Morphology

HAV is a 27- to 32-nm, nonenveloped, positive-sense, single-stranded RNA virus with a 7.5-kb genome, icosahedral capsid symmetry, and a buoyant density in cesium chloride of 1.33–1.34 g/ml. The virus is classified in the Picornaviridae family in its own distinct genus, Hepatovirus (Table 2.1) but was formerly classified in the Enterovirus genus as Enterovirus 72. It has a structure similar to that of other picornaviruses. There is one species, HAV, with two strains or biotypes: human HAV and simian HAV. These two distinct strains are phylogenetically distinct and have different preferred hosts. Human HAV infects all species of primates including humans, chimpanzees, owl monkeys, and marmosets, whereas simian HAV infects green monkeys and cynomolgus monkeys. Seven genotypes have been recognized, of which four infect humans and the remaining three infect nonhuman primates.

Unlike many RNA viruses, the genome of HAV is highly conserved, with an average variation of only 1–4%, but there are two groups within the genus

Table 2.1 Characteristics of Food-borne Viruses

Virus Genus or Species	*Family*	*Nucleic Acid Type*	*Envelope*	*Morphology/ Symmetry*	*Size of Virion (nm)*	*Culturable*[a]	*Genome Size (kb)*	*Disease*
Adenovirus	Adenoviridae	dsDNA	N	Icosahedral	70–90	Y[a]	28–45	Respiratory, eye, and gastroenteritis infection
Astrovirus	Astroviridae	(+) ssRNA	N	Icosahedral	28–30	Y[a]	7–8	Gastroenteritis
Norovirus	Caliciviridae	(+) ssRNA	N	Icosahedral	28–35	N	7.4–7.7	Epidemic gastroenteritis
Sapovirus	Caliciviridae	(+) ssRNA	N	Icosahedral	28–35	N	7.4–7.7	Gastroenteritis
Hepatovirus: hepatitis A virus	Picornaviridae	(+) ssRNA	N	Icosahedral	27–32	Y[a]	7.5	Inflammation of liver, hepatitis
Hepevirus: hepatitis E virus	Hepeviridae	(+) ssRNA	N	Icosahedral	32–34	N	7.2	Inflammation of liver, hepatitis
Rotavirus	Reoviridae	dsRNA	N	Icosahedral	60–80	Y	16–27	Gastroenteritis
Enterovirus	Picornaviridae	(+) ssRNA	N	Icosahedral	28–30	Y[a]	7.2–8.4	Poliomyelitis, meningitis, encephalitis
Parvovirus	Parvoviridae	ssDNA	N	Icosahedral	20–30	N	5	Gastroenteritis
Tick-borne encephalitis virus	Flaviviridae	(+) ssRNA	N	Icosahedral	45–60		9.5–12.5	Tick-borne encephalitis via milk
Coronavirus	Coronaviridae	(+) ssRNA	Y	Helical	80–220	Y[a]	20–30	Gastroenteritis, respiratory infections
Torovirus	Toroviridae	(+) ssRNA	Y	Helical	100–150	Y	20–25	Gastroenteritis in animals and? humans
Picobirnavirus	Birnaviridae	(+) ssRNA	N	Icosahedral	35			Gastroenteritis in animals and? humans

[a] Not all strains within the genus are culturable; wild-type strains are often difficult to culture.

that show diversity of 10% and up to 25% (Cromeans et al., 2001). HAV has been classified into seven genotypes based on sequence analysis of the VP1 and VP3 genes that code for surface proteins (Robertson et al., 1991, 1992). Characterization of these genotypes has been useful in outbreak investigations for tracing infection sources, and strains within these genotypes have shown more than 85% genetic similarity (Niu et al., 1992; Cromeans et al., 2001).

2.3. Growth and Biological Properties

HAV can be cultured in several different primate cell lines including African green monkey kidney cells (BSC-1), fetal rhesus monkey kidney cells (FRhK-4 and FRhK-6), and human fibroblasts (HF), but wild-type strains are difficult to culture and generally do not produce CPE in cell cultures. Immunofluorescence is often used for detection of HAV antigen in infected cells because of the lack of CPE. The virus is usually slow-growing and the yield in cell cultures is lower as compared with most other picornaviruses. Consequently, it is difficult to identify the virus in clinical, food, or environmental sources by culture alone. Under normal conditions, the virus requires 3 weeks for *in vitro* growth. Laboratory-adapted strains such as HM 175 are able to produce CPE and so have been used extensively in research studies. These viruses require less time for *in vitro* growth and produce visible CPE or plaques. Molecular techniques, including culture-PCR, have become the method of choice for detection of virus in nonhuman samples, whereas clinical diagnosis is usually based on the patient's immune response. HAV antigens are conserved and antibodies are generated against a single antigenic site composed of amino acid residues of VP3 and VP1 proteins on the virus surface.

HAV is very stable, showing high resistance to chemical and physical agents such as drying, heat, low pH, and solvents, and has been shown to survive in the environment, including seawater and marine sediments, for more than 3 months (Sobsey et al., 1988). The virus retains integrity and infectivity after 60-min incubation at 60°C and is only partially inactivated after 10–12 hr at 56°C. The heat resistance of HAV is reported to be greater in foods and shellfish. After heating in a can for 19 min at 60°C, HAV inoculated into oysters was not fully inactivated. Under refrigeration and freezing conditions, the virus remains intact and infectious for several years. It is also resistant to drying, remaining infectious for more than 1 month at 25°C and 42% humidity, and shows even greater resistance at low humidity and low temperatures.

Although HAV infectivity decreased by 2 to 5 $\log_{10}$ after exposure to 70% alcohol for 3 min and 60 min at 25°C, it was resistant to several preservatives and solvents including chloroform, Freon, Arklone, and 20% ether and was not inactivated by 300 mg/L perchloroacetic acid or 1 g/L chloramine at 20°C for 15 min (Hollinger and Emerson, 2001) . The virus is stable at pH 1.0 and survives acid marination at pH 3.75 in mussels for at least 4 weeks (Hollinger and Emerson, 2001; Hewitt and Greening, 2004). Gamma irradiation is not effective for inactivation of HAV on fresh fruits and vegetables, but the virus does appear to be inactivated by high hydrostatic pressure. Hydrostatic pressure, now used as an isothermal preservation method for perishable foods, inactivated HAV after 5-min exposure at 450 MPa (Kingsley et al., 2002). Overall HAV exhibits greater resistance to stressors than other picornaviruses.

2.4. Infection and Disease

HAV infects the epithelial cells of small intestine and hepatocytes, causing elevation of liver enzymes and inflammation of the liver. The cytotoxic T-cell immune response destroys infected liver cells, releasing the virus particles into the bile duct from where they are excreted in the feces. The virus is believed to initially enter the liver via the bloodstream, and it is not clear if intestinal replication occurs.

The virus has an incubation period of 2 to 6 weeks with an average of 28 days. Initially the symptoms are nonspecific and include fever, headache, fatigue, anorexia, dark urine, light stools, and nausea and vomiting with occasional diarrhea. One to 2 weeks later, characteristic symptoms of hepatitis such as viremia and jaundice appear. Peak infectivity occurs in 2 weeks preceding the onset of jaundice, and the virus is present in the blood at 2 to 4 weeks. The HAV is shed in large numbers ($>10^6$ particles/g) in feces from the latter 2 weeks of the incubation period for up to 5 weeks. Jaundice is usually evident from week 4 to 7 and virus shedding generally continues throughout this period. Diagnosis is based on the detection of anti-HAV IgM antibody, which can be detected before the onset of symptoms and becomes undetectable within 6 months of recovery. Acute hepatitis is usually self-limiting, but overall debility lasting several weeks is common and relapses may occur.

The HAV has not been associated with development of chronic liver disease, but on rare occasions fulminant disease that results in death may occur. Because the onset of symptoms occurs several weeks after infection, it is rare to have the suspected food available for analysis. A killed vaccine that provides long-lasting immunity has been commercially available since 1995 and is commonly given to travelers at high risk. This vaccine could be used in the food industry to immunize food workers to reduce the risk of food contamination by workers.

2.5. Food-borne Disease

HAV has been associated with many outbreaks of food-borne disease. Contamination generally occurs either preharvest or during food handling. There are a number of documented outbreaks of disease resulting from consumption of HAV-contaminated shellfish, the largest of which occurred in China in 1988 when approximately 300,000 people were infected after consumption of partially cooked, HAV-contaminated clams harvested from a growing area impacted by raw sewage (Halliday et al., 1991). A few of the shellfish-associated outbreaks include oysters in Australia (Conaty et al., 2000), oysters in Brazil (Coelho et al., 2003), mussels in Italy (Croci et al., 2000), and clams in Spain (Bosch et al., 2001). In most of these outbreaks, sewage was generally the source of pollution.

Contamination of shellfish with HAV is still common in Italy, Spain, and other European countries. Preharvest contamination of fruits and vegetables, including strawberries (Niu et al., 1992), raspberries (Reid and Robinson, 1987; Ramsay and Upton, 1989), blueberries (Calder et al., 2003), lettuce (Pebody et al., 1998), and green onions (CDC, 2003) has also been reported and has resulted in outbreaks of disease in countries such as Finland and New Zealand, where populations have low or no immunity to the disease (Pebody et al., 1998; Calder et al., 2003). The source of contamination in these outbreaks was reported to be either infected fruit-pickers or contaminated irrigation waters.

The other main source of HAV infection is from food handlers and food processors. Because HAV is shed before symptoms become apparent and $>10^6$ infectious virus particles can be excreted per gram of feces, HAV-infected produce harvesters and food handlers, without knowing, can become a source of contamination. In areas with poor hygiene practices, this can present a risk to human health. Food-borne outbreaks of HAV are relatively uncommon in developing countries where there are high levels of immunity in the local population, but tourists in these regions can be susceptible if they are not vaccinated.

3.0. HEPATITIS E VIRUS

3.1. Distribution and Transmission

HEV is believed to be a major etiologic agent of enterically transmitted non-A, non-B hepatitis in humans worldwide (Emerson and Purcell, 2003). The virus is transmitted by the fecal-oral route and occurs widely in Asia, northern Africa, and Latin America, including Mexico, where waterborne outbreaks are common. Although originally it was believed that HEV did not occur in industrialized countries, in recent years it has been identified in Europe, Australasia, and the United States. However, it rarely is a cause of overt disease in these countries (Clemente-Casares et al., 2003; Emerson and Purcell, 2003). The virus has been isolated from raw sewage in Spain, France, Greece, Italy, Austria, and the United States (Jothikumar et al., 1993; Pina et al., 1998). Transmission is generally via fecally contaminated water, and evidence for food-borne transmission has not been definitively documented. Epidemics and sporadic cases of HEV are responsible for the majority of enterically transmitted acute hepatitis in regions where HEV is considered endemic. Antibodies to HEV have been detected in many animal species, which has led to a discussion on the possible zoonotic aspect of HEV.

3.2. Taxonomy and Morphology

HEV was first isolated and identified by Balayan et al. (1983) in acute and convalescent specimens collected from a case of non-A, non-B hepatitis. It is a 32- to 34-nm, nonenveloped, positive-sense, single-stranded RNA virus with a linear genome of 7.2kb (Table 2.1). The capsid symmetry is icosahedral, and the buoyant density in potassium tartrate–glycerol gradient is 1.29g/ml. HEV was originally classified in the Caliciviridae because of similarities in structural morphology and genome organization. The virus was then reclassified in the family Togaviridae because of similarities between the replicative enzymes of HEV and the togaviruses. However, the current International Committee on Taxonomy of Viruses (ICTV) classification places HEV under a new family, Hepeviridae, genus Hepevirus (van Regenmortel et al., 2000). Two HEV serotypes and four major HEV genotypes have been identified based on nucleotide and protein sequencing. Genotype 1 includes Asian and African strains, genotype 2 includes a Mexican strain, genotype 3 includes United States swine and human strains, and genotype 4 includes strains from China, Japan, and Taiwan (Emerson and Purcell, 2003).

3.3. Growth and Biological Properties

Although there are reports describing culture of HEV, none have shown sustained replication with production of virus particles, and there is no recognized culture system for the virus. HEV is generally identified by molecular methods. The inability to grow the virus has hampered research on the ability of this virus to survive in the environment.

3.4. Infection and Disease

As is the case with HAV, HEV produces an acute disease with generally mild symptoms. Although the disease can be quite severe in some cases, it is usually self-limiting and does not progress to a carrier or chronic state. The virus infects the liver and produces symptoms of hepatitis after a 22–60 day incubation period. Symptoms may include viremia, nausea, dark urine, and general malaise. The virus is excreted in bile and feces from 2 weeks before the elevation of liver enzymes and continues until the enzyme levels return to normal. Identification and diagnosis is generally by detection of IgM and IgG responses in patients' sera to recombinant HEV protein antigens or by molecular assays to identify the virus in feces or sera. In general, the mortality rate from hepatitis E infections is about 1% but may reach as high as 17–30% in pregnant women (Cromeans et al., 2001; Emerson and Purcell, 2003). The major mode of transmission appears to be contaminated water. Secondary person-to-person transmission has been estimated at 0.7–8.0% and is relatively uncommon (Cromeans et al., 2001).

3.5. Food-borne Disease

Food-borne outbreaks of HEV are the most common in developing countries with inadequate environmental sanitation. Large waterborne outbreaks have also been reported in Asian countries (Cromeans et al., 2001; Emerson and Purcell, 2003). HEV was not thought to be endemic in developed countries, and the first reported human cases of acute hepatitis E in the United

States were attributed to travel in HEV-endemic countries. In 1997, however, HEV was isolated from a U.S. resident with hepatitis with no history of travel. Simultaneously, the virus was also identified in domestic swine (Meng et al., 1997; Schlauder et al., 1998) and has now been documented in humans and swine in many other countries including Argentina, Australia, Austria, Canada, Germany, Greece, Japan, Korea, The Netherlands, New Zealand, Spain, and Taiwan (Clemente-Casares et al., 2003). The waterborne transmission route has been proved, and there have been reports of possible foodborne outbreaks in China, but corroborating evidence is lacking. There has been no evidence of HEV transmission via seafood.

3.6. Zoonotic Transmission

The HEV has been isolated from swine in several countries where hepatitis E in humans is rare, including Spain, New Zealand, The Netherlands, Japan, and Canada (Emerson and Purcell, 2003). The reservoirs of infection for HEV are unknown, but the virus has been isolated from the feces of a wide range of domestic animals. Recent reports from Japan show that the virus may be transmitted to humans by close contact with infected swine or from the consumption of contaminated raw or undercooked pork, wild boar liver, and deer meat (Tei et al., 2003; Yazaki et al., 2003). The most convincing evidence of zoonotic transmission to date is a recent report in which consumption of raw deer meat by a Japanese family was implicated in the transmission of HEV (Tei et al., 2003). In another study, Tei et al. (2004) investigated the risks associated with consumption of uncooked deer meat in a case control study and found that, in the area studied, eating uncooked deer meat was a risk factor.

Meng et al. (2002) found that veterinarians and people working with pigs were more likely to have antibodies to HEV. The isolation of a swine HEV that cross-reacts with an antibody to the capsid antigen of human HEV provides additional evidence for a zoonotic transmission route (Meng et al., 1997). Nonhuman primates can also be infected by swine HEV; inoculation of rhesus monkeys with swine HEV led to seroconversion, fecal shedding of virus, viremia, and development of a mild acute hepatitis with slight elevation in liver enzymes. Similar studies in chimpanzees also resulted in seroconversion and fecal shedding of virus but not viremia or hepatitis (Meng et al., 1998). These findings suggest that swine HEV may infect humans and that swine could be a zoonotic reservoir for HEV.

HEV is shed in the feces and bile of swine for 3–5 weeks after infection (Halbur et al., 2001). The excretion of HEV in feces of infected pigs could lead to the spread of HEV in the environment and increase the potential for zoonotic transmission. Similarly, fecal contamination of runoff waters from pig farms or from lands on which untreated pig manure has been spread could contaminate irrigation and surface waters with subsequent HEV contamination of fruits, vegetables, and shellfish. Although there is increasing evidence of the zoonotic transmission of HEV, the risk factors are still largely unknown.

4.0. NOROVIRUS AND SAPOVIRUS

4.1. Distribution and Transmission

Noroviruses, previously known as small round structured viruses (SRSVs) and Norwalk-like viruses (NLVs), are now the most widely recognized viral agents associated with food-borne and waterborne outbreaks of nonbacterial gastroenteritis and probably the most common cause of food-borne disease worldwide. These viruses cause epidemic viral gastroenteritis resulting in large outbreaks. The prototype norovirus, Norwalk virus, was first discovered by Kapikian et al. (1972) after an outbreak of gastroenteritis in a school in Norwalk, Ohio. Immune electron microscopy was used to examine feces from volunteers who consumed fecal filtrates from infected cases (Dolin et al., 1971; Kapikian et al., 1972). At that time, most cases of gastroenteritis that could not be attributed to a bacterial agent were termed *acute nonbacterial gastroenteritis*. The discovery of Norwalk virus provided the first evidence of a viral etiology for human diarrheal disease. Despite this discovery, noroviruses remained largely unrecognized until about 15 years ago because their detection was technically difficult and because the illness is generally mild and short-lived and is not reportable to public health authorities.

Noroviruses are primarily transmitted through the fecal-oral route, by consumption of fecally contaminated food or water, or by direct person-to-person spread. Secondary spread may also occur by airborne transmission. Outbreaks commonly occur in closed community situations such as rest homes, schools, camps, hospitals, resorts, and cruise ships and where the food and water sources are shared. Because norovirus infections are not notifiable, the total burden of disease is not known and is generally grossly underreported (Mead et al., 1999; Wheeler et al., 1999). However, some of the disease burden is recorded through the notification of gastroenteritis outbreaks to the public health disease surveillance systems in many developed countries. It has been estimated that noroviruses are responsible for approximately 60% of food-borne disease in the United States, including more than 9 million cases, 33% of hospitalizations, and 7% of deaths related to food-borne disease each year (Mead et al., 1999). Fankhauser et al. (2002) found that 93% of 284 nonbacterial gastroenteritis outbreaks in the United States were due to norovirus, and in 57% of these, contaminated food was the vehicle of infection. The majority of viral gastroenteritis outbreaks in Europe have been attributed to noroviruses, where they were reported to be responsible for more than 85% of nonbacterial gastroenteritis outbreaks between 1995 and 2000 (Lopman et al., 2003).

The sapoviruses, formerly described as the "Sapporo-like viruses," or SLVs, also belong to the Caliciviridae family and cause gastroenteritis among both children and adults, although association with food-borne transmission is rare. The sapoviruses are most commonly associated with pediatric disease in infants, and transmission is more likely to be from person to person.

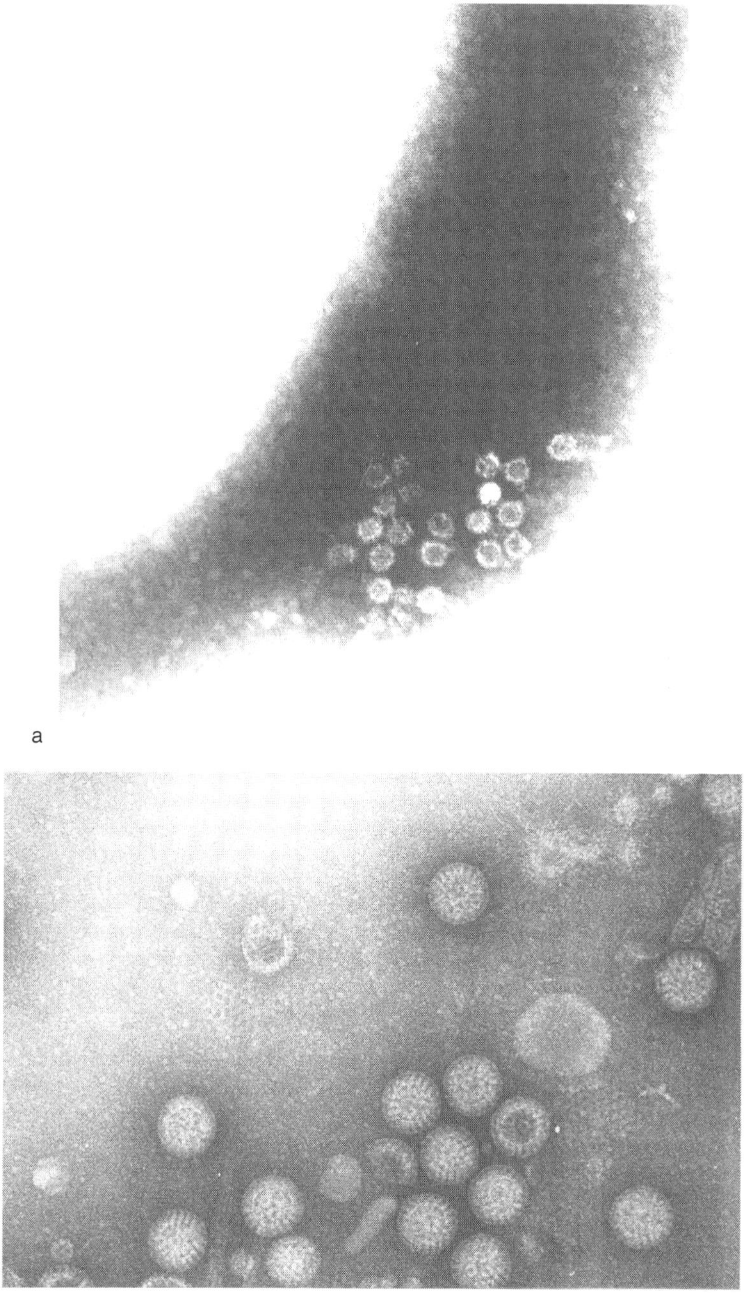

a

b

Figure 2.1. Electron micrographs of human enteric viruses. Negative staining. (a) Baculovirus-expressed recombinant Norwalk virus-like particles (VLPs); (b) rotavirus; (c) adenovirus.

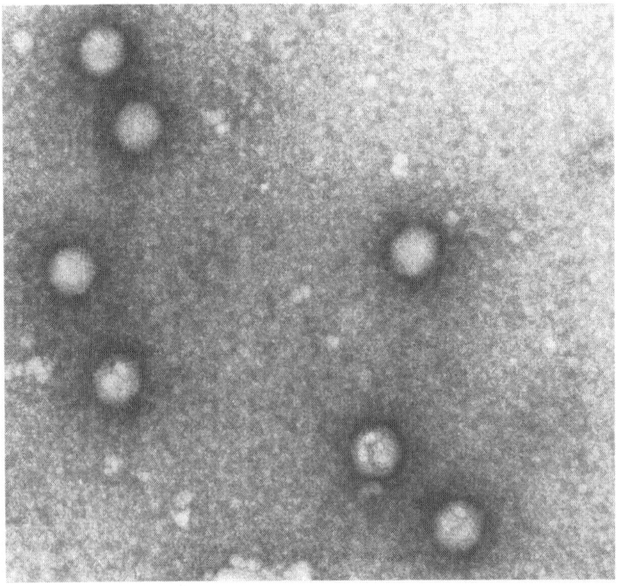

c

Figure 2.1. *Continued*

4.2. Taxonomy and Morphology

There are four genera in the Caliciviridae family: Norovirus and Sapovirus, which are both human pathogens, and Lagovirus and Vesivirus, which infect animals and are not known to be pathogenic for humans. The Norwalk-like viruses and the Sapporo-like viruses were renamed as Norovirus and Sapovirus in August 2002 by the ICTV (van Regenmortel et al., 2000). The noroviruses do not show the characteristic cup-shaped morphology of caliciviruses but instead show a "fuzzy" or ragged appearance by direct electron microscopy, which is why they were classified as a distinct group until 1995 (Fig. 2.1a). Sapoviruses have a morphological appearance more typical of the caliciviruses, with distinct cup-shaped indentations on the surface of the virions.

The noroviruses are 28- to 35-nm, nonenveloped, linear, positive-sense, single-stranded RNA viruses with a genome of approximately 7.6 kb and icosahedral capsid symmetry (Table 2.1). The buoyant density in cesium chloride gradient is 1.36–1.41 g/ml. The genome is composed of three open reading frames (ORFs), which code for the nonstructural proteins including the RNA polymerase (ORF1), the capsid protein (ORF2), and a minor structural protein (ORF3).

There is a single species, norovirus, which has seven designated strains: Norwalk virus, Snow Mountain virus, Hawaii virus, Southampton virus, Lordsdale virus, Mexico virus, Desert Shield virus, and one tentative species, swine calicivirus, listed in the ICTV database. Noroviruses have a defined nomenclature whereby strains are named after the geographic location of the outbreak from which they were first identified. A number of distinct genogroups and genotypes have been characterized based on DNA sequencing of PCR products from the RNA polymerase region in ORF1 (Ando et al., 1995). Sequencing of the genetically variable capsid gene (ORF2) has produced further strain discrimination and recognition of additional genotypes (Fankhauser et al., 2002; Green et al., 2000; Vinje et al., 2004). Currently, four norovirus genogroups (GI, GII, GIII, GIV, and GV) have been identified, and these are subdivided into at least 15 genetic clusters (Ando et al., 2000). Genogroup III includes the bovine enteric caliciviruses, including the Jena and Newbury agents, which are genetically closer to noroviruses than other known caliciviruses (Fig. 2.2). Genogroup V includes the recently identified murine norovirus, MNV-1.

The sapoviruses are 28- to 35-nm, nonenveloped, positive-sense, single-stranded RNA viruses with a genome of approximately 7.6 kb and icosahedral capsid symmetry (Table 2.1) and exhibit the properties of the Caliciviridae. They are small round viruses with a morphology similar to that of noroviruses by EM. The ICTV (van Regenmortel et al., 2000) lists six species of Sapovirus all named according to their first identification: Houston/86, Houston/90, London 29845, Manchester virus, Parkville virus, and Sapporo virus. The sapoviruses are genetically more similar to the members of the Lagovirus genus (rabbit calicivirus) than to those in the

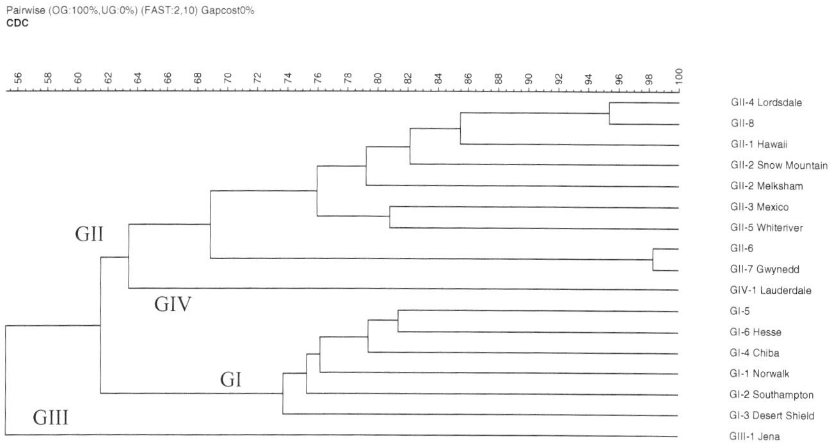

Figure 2.2. Dendrogram showing genetic relationships of norovirus sequences in a 172-bp region of the polymerase gene (region B) of the genome. References sequences from Genbank and CDC Calicinet database. Norovirus strains represented are Norwalk virus M87661, Chiba virus AB022679, Southampton virus L07418, Desert Shield virus U04469, Hesse virus AF093797, Jena virus AJ011099, Hawaii virus U07611, Snow Mountain virus L23831, Melksham virus X818879, Mexico virus U22498, Lordsdale virus X86557, White river virus AF414423, Gwynedd virus AF414409, Ft. Lauderdale virus AF414426, GI/5 AF414406, GII/6 AF414407, and GII/8 AY054299. Dendrogram created by unweighted pair-group method using arithmetic averages (UPGMA).

Norovirus genus. Three genogroups (I, II, and III) have been identified based on sequence analysis (van Regenmortel et al., 2000, Schuffenecker et al., 2001). These viruses are reported to be genetically more similar to animal caliciviruses than the noroviruses (Matson et al., 1995).

4.3. Growth and Biological Properties

Most information on the biology and properties of noroviruses has been obtained through human volunteer studies in the 1970s (Dolin et al., 1971, 1972; Green et al., 2001). Human noroviruses are nonculturable, and until recently no animal model had been identified. Inoculation of chimpanzees with Norwalk virus elicited immune responses but no symptoms developed and no virus was shed in feces (Wyatt et al., 1978). Sustained attempts have been made to culture human noroviruses over the past 10–15 years but without success. More than 26 different cell lines combined with many varied cell culture supplements and growth conditions have been evaluated, but no norovirus-induced CPE or replicating norovirus was obtained (Duizer et al., 2004). Noroviruses have now been identified that infect animals, including pigs, cattle, and mice, and progress in this field is now being made with the growth of a mouse norovirus in artificial culture (Wobus et al., 2004). This culture system will help to discover more about human noroviruses and their mechanisms of pathogenicity. The infected mice develop gastroenteritis, and so this discovery holds potential as a future model for human norovirus disease.

Prior to the development of molecular methods, there was limited knowledge about these viruses because their identification was difficult. The inability to culture noroviruses coupled with the problems associated with identification of the virus by EM restricted their detection for many years. Noroviruses are difficult to identify by direct EM in fecal samples and foods because of their small size and the nature of the background matrices. Immune electron microscopy (IEM) is frequently used to improve the sensitivity of detection, but the antibody coating can mask the appearance of the virus. The development of assays such as reverse transcription-polymerase chain reaction (RT-PCR) has facilitated the detection and identification of these viruses, and consequently the role of noroviruses in gastroenteritis outbreaks has been clarified. Noroviruses show great genetic diversity, which has complicated their identification by molecular assays. To date, none of the numerous norovirus primer sets designed have been able to detect 100% of known norovirus strains, but some sets have been found to be more sensitive and have a broader detection range than others (Vinje et al., 2003).

The lack of a culture system for noroviruses has hindered the development of traditional immunological and serological detection assays because it is not possible to cultivate sufficient noroviruses *in vitro* to generate antigen for antibody production. Recently, advances in routine detection of noroviruses have been made with the development of commercially available ELISA assays. These assays use monoclonal antibodies prepared against recombinant norovirus capsid proteins generated in a baculovirus expression system. The norovirus capsid proteins are also known as virus-like particles, or VLPs, and are essentially empty viral protein coats without the nucleic acid. The noninfectious VLPs are highly immunogenic, making them suitable for antiserum and vaccine production (Estes et al., 2000). However, commercially available ELISA assays are not able to detect all norovirus strains and are reported to have limited sensitivity and specificity (Richards et al., 2003; Burton-Mcleod et al., 2004).

Because norovirus is nonculturable, its infectivity can only be assessed in human dose-response experiments, hence there is little information on its survival characteristics. Studies using human volunteers showed that norovirus retains infectivity when heated to 60°C for 30 min and therefore is not inactivated by pasteurization treatment. The virus also retains infectivity after exposure to pH 2.7 for 3 hr at room temperature (Dolin et al., 1972; Green et al., 2001). Further evidence of its resistance to low pH was shown when norovirus was exposed to heat treatment and subsequent marination at pH 3.75 in mussels for 1 month. No decrease in norovirus titer was observed by real-time RT-PCR (Hewitt and Greening, 2004). There are anecdotal reports of people developing gastroenteritis after eating pickled shellfish. Norovirus, like other enteric viruses, remains infectious under refrigeration and freezing conditions, appears to survive well in the environment, and is resistant to drying. This was demonstrated when carpet layers became ill after lifting carpet that had become contaminated 12 days earlier in a rest-home outbreak (Cheesbrough et al., 2000).

Fecal pollution from sewage discharges, septic tank leachates, and boat discharges has caused contamination of shellfish beds, recreational water, irrigation water, and drinking water. It is probable that noroviruses persist in these environments for extended periods of time. In live oysters, noroviruses were still detectable after 4–6 weeks in natural growing conditions (Greening et al., 2003). These viruses are also resistant to treatment with 3.75 to 6.25 mg chlorine/L, which is equivalent to free residual chlorine of 0.5 to 1.0 mg/ml, a level of free chlorine consistent with that generally present in a chlorinated drinking water supply. However, the viruses were inactivated after treatment with 10 mg/L of chlorine, which is the concentration applied to water supplies after a contamination event (Green et al., 2001). Norovirus also retained infectivity after exposure to 20% ether at 4°C for 18 hr (Dolin et al., 1972).

As with noroviruses, sapoviruses have also not been cultured *in vitro* yet. In addition, sapoviruses have not been studied as intensively as noroviruses, hence little information is available on the biological and physical properties of these viruses. Detection and identification is generally by molecular methods (Jiang et al., 1999; Green et al., 2000; Schuffenecker et al., 2001).

4.4. Infection and Disease

Noroviruses are extremely infectious and cause epidemic gastroenteritis. The infectious dose is believed to be as low as 10–100 virus particles (Caul, 1996).

Recent dose-response studies show that both the infective dose and host susceptibility may vary according to the infecting norovirus strain (Moe et al., 2004; Lindesmith et al., 2005). The mechanism of pathogenicity of noroviruses is still not clearly understood because of the inability to propagate these viruses, but information is being obtained from the *in vitro* culture of a mouse norovirus (Wobus et al., 2004).

It is known that the mature enterocyte cells in the small intestine become infected and that malabsorption of fats, D-xylose, and lactose occurs for up to 2 weeks. Unusually, gastric emptying is also delayed, and this may explain the nausea and characteristic projectile vomiting associated with norovirus infection. Large numbers of noroviruses are excreted in feces from the onset of symptoms and continue to be shed in decreasing numbers for up to 2 weeks after infection. Animals infected with the Newbury agent, bovine caliciviruses assigned to Norovirus genogroup III, show similar symptoms, pathological changes and processes as seen in humans (Appleton, 2001).

In the absence of reliable laboratory tests for norovirus, Kaplan et al. (1982) developed epidemiological and clinical criteria for the diagnosis of noroviral gastroenteritis outbreaks. These criteria were stools negative for bacterial pathogens, a mean or median duration of illness of 12–60 hr, vomiting in ≥50% of cases, and a mean or median incubation period of 24–48 hr. These criteria are still widely used. The symptoms of acute-onset projectile vomiting, watery nonbloody diarrhea with abdominal cramps, and nausea may develop within 12 hr of exposure, and low-grade fever also occurs occasionally. Dehydration is a common complication that can particularly affect the young and elderly, necessitating rehydration therapy. There is no evidence of any long-term sequelae after norovirus infection. The symptoms associated with Sapovirus infection are similar to those of noroviral gastroenteritis, but the sapoviruses do not cause epidemic gastroenteritis.

The mechanism of immunity to norovirus infection is not clear. Infection normally stimulates production of both gut and serum antibody, and although immunity to the infecting norovirus strain may develop, it is generally short-lived, strain-specific, and does not confer protection against future infection. Reinfection with a different strain can occur soon after the initial infection. Thus, given the genetic variability of noroviruses, people are likely to be reinfected many times during their lifetimes. Recent research has suggested that there may be a genetic determinant involved in susceptibility to norovirus infection, with people belonging to histo-blood group O being at greater risk for severe infection (Hutson et al., 2002, 2004).

Projectile vomiting is a characteristic symptom that can contribute to secondary spread through droplet infection, where droplets containing virus may contaminate surfaces or be swallowed. Evidence that norovirus transmission occurs through aerosolization of vomit was clearly demonstrated at a U.K. hotel. During a meal, a guest vomited at the table, and norovirus infection spread in a radial pattern through the restaurant, progressively decreasing from 91% attack rate among those seated at the same table to an attack rate of 25% in those patrons who were seated the farthest distance away from the guest who vomited (Marks et al., 2000). Norovirus infection characteristically has an attack rate of 50–70% or even higher in some situations. This high attack rate combined with a low infectious dose, prolonged virus excretion, short-term immunity, and the environmental stability of the noroviruses contributes to the epidemic nature of noroviral gastroenteritis.

Norovirus infection was termed *winter vomiting disease* because outbreaks occurred most frequently in the winter months, especially in rest homes and institutions. This seasonality is no longer apparent as norovirus outbreaks are now reported to occur throughout the year.

4.5. Food-borne Disease

Noroviruses are the main cause of food-borne viral gastroenteritis worldwide with food-borne transmission accounting for a large proportion of norovirus outbreaks in many countries. Food-borne norovirus outbreaks resulting from preharvest contamination of foods such as shellfish and postharvest contamination through food handling have been reported worldwide. Among these are several outbreaks resulting from consumption of norovirus-contaminated shellfish (Dowell et al., 1995; Christensen et al., 1998; Berg et al., 2000; Simmons et al., 2001), bakery products (Kuritsky et al., 1984), delicatessen meats (Schwab et al., 2000), sandwiches (Parashar et al., 1998; Daniels et al., 2000), raspberries (Ponka et al., 1999), water and ice (Beller et al., 1997; Brugha et al., 1999; Beuret et al., 2002). Presymptomatic infection in food handlers has also been shown to cause outbreaks of food-borne norovirus infection (Lo et al., 1994; Gaulin et al., 1999).

Among the 284 outbreaks of norovirus illness reported to CDC from July 1997 to June 2000, the cause of transmission was not determined in 42, or 24% of outbreaks (Fankhauser et al., 2002). Determination of the original source of the virus is often problematic because several modes of transmission frequently operate during norovirus gastroenteritis outbreaks. Although the initial transmission route may be through consumption of contaminated foods, secondary transmission via direct contamination of the environment or person-to-person contact also often occurs. This results in wide dissemination where infection quickly spreads through institutions, schools, camps, resorts, and cruise ships and causes large-scale epidemics with more than 50% attack rates.

The use of DNA sequencing techniques for genotyping of noroviruses has greatly assisted the epidemiologic investigation of gastroenteritis outbreaks. The comparison of noroviral sequences from fecal specimens and contaminated foods, such as oysters, can clearly indicate if it is a common source outbreak or if individual cases are somehow related. In 1993, 23 gastroenteritis outbreaks across 6 states in the United States were shown to be related to consumption of oysters harvested from a single area and contaminated with the same norovirus strain (Dowell et al., 1995).

There are few reports of Sapovirus infection directly resulting from consumption of food. An outbreak of viral gastroenteritis among adults at a school in Parkville, Maryland, in 1997 was determined to be food-related. The causal agent was a Sapovirus later designated as the Parkville virus (Noel et al., 1997).

4.6. Zoonotic Transmission

Research in Japan, The Netherlands, and the United Kingdom has demonstrated calicivirus-like particles and calicivirus RNA sequences in the cecum of pigs and in fecal samples from calves (Sugieda et al., 1998; Dastjerdi et al., 1999; van der Poel et al., 2000). Molecular analysis of these enteric caliciviruses, now termed the *Jena* and *Newbury agents*, shows that they are genetically more closely associated with human noroviruses than with other known caliciviruses and they are now assigned to Norovirus genogroup III. Although the discovery of these noroviruses prompted concerns that calves and pigs may be a reservoir of infection for human noroviral disease, there is no documented evidence of transmission to humans (Oliver et al., 2003). Similarly, there are no reports of zoonotic transmission of sapoviruses.

5.0. ROTAVIRUS

5.1. Distribution and Transmission

Rotaviruses are the major cause of severe diarrhea and gastroenteritis in infants and young children. It is estimated that rotaviruses cause more than 130 million cases of diarrhea in children under 5 years of age annually worldwide (Glass and Kilgore, 1997). Rotaviral infection is a particularly serious problem in developing countries where up to 600,000 deaths occur annually among children. In the United States, rotaviruses are estimated to cause about 4 million infections per year resulting in almost 70,000 hospitalizations and more than 100 deaths annually (Kapikian et al., 2001; Sattar et al., 2001). Although the disease occurs in all age groups, it is generally considered to be a mild infection in adults, hence the true extent of adult infections is not known.

Rotaviruses are transmitted by the fecal-oral route and cause disease in both humans and animals, especially domestic animals, with subsequent serious economic loss. Although the animal and human strains are usually distinct, some human strains are closely related to animal strains, and cross-species infections do occur (Sattar et al., 2001). Infection is not generally recognized as food-borne, but outbreaks associated with food and water have been reported in a number of countries (Sattar et al., 2001).

5.2. Taxonomy and Morphology

Rotaviruses are classified in the genus Rotavirus in the family Reoviridae, a large family composed of nine genera. Electron micrographs of rotaviruses show a characteristic wheel-like appearance, hence the name rotavirus, derived from the Latin meaning "wheel" (Fig. 2.1b). These viruses are distinct in that they have a complex segmented genome that undergoes reassortment during replication. There are five species of Rotavirus, designated

Rotavirus A (simian rotavirus) through Rotavirus E (porcine rotavirus). Two possible species, Rotavirus F (avian) and Rotavirus G (avian) are also listed but they differ in their ability to reassort the genome segments. Most human infections are caused by Rotavirus A, B, and C, but all rotaviral species can infect a range of vertebrates, including primates, ruminants, rodents and birds (van Regenmortel et al., 2000; Sattar et al., 2001).

Rotaviruses are 60- to 80-nm, nonenveloped, linear segmented double-stranded RNA viruses with icosahedral capsid symmetry (Table 2.1). The 16- to 27-kb, genome is enclosed by a triple-layered capsid composed of a double protein shell and an inner core. Eleven segments of DNA code for six structural and five nonstructural proteins. Two of the structural proteins, VP7 (glycoprotein) and VP4 (protease or P protein), comprise the outer shell of the capsid and are important in virus infectivity. These two proteins are used to define the rotavirus serotype; there are 14 VP7 serotypes and 11 VP4 serotypes within the Rotavirus A species. The VP6 protein located on the inner capsid layer is designated the group-specific antigen and is the major target of rotavirus diagnostic assays. This protein is believed to play a role in the development of protective immunity. Genomic reassortment of the rotaviral RNA segments may occur during replication, particularly when there is coinfection with more than one strain. In the replication phase, the immature virus particles acquire a transient lipid envelope as they develop in the endoplasmic reticulum of the host cell.

5.3. Growth and Biological Properties

Although many rotaviruses can be grown in cell cultures, they have proved difficult to cultivate *in vitro*, and growth is restricted to a few cell lines derived mainly from monkey kidneys. Addition of trypsin to the culture medium is required to enhance viral growth in cell cultures. Rotaviruses do not show the same tolerance to extreme conditions as other enteric viruses, although they are stable in the environment and can be stored for several months at 4°C or even 20°C. They are resistant to drying and may survive on fomites and surfaces. Heating at 50°C for 30 min reduces their infectivity by 99%, and infectivity is rapidly lost at pH <3.0 and >10.0. Repeated cycles of freeze-thaw can also destroy infectivity. The viruses are resistant to solvents such as ether and chloroform and to non-ionic detergents such as deoxycholate. Chelating agents such as EDTA disrupt the outer shell and inactivate rotaviruses. Treatment with disinfectants such as chlorine, phenol, formalin, and 95% ethanol is also effective against rotavirus (Kapikian et al., 2001). Normal cooking temperatures are usually sufficient to inactivate rotaviruses. The viruses are found in water and sewage, are resistant to chlorine levels present in drinking water, and are persistent in the environment. Human rotavirus can survive for several weeks in river water at 4°C and 20°C.

5.4. Infection and Disease

The incubation period for rotavirus infection is 1 to 2 days. The characteristic symptoms of vomiting and watery diarrhea develop quickly and persist for 3 to 8 days, frequently accompanied by fever and abdominal pain. Dehydration is a key factor that contributes to the high infant death rate from rotavirus disease, especially in developing countries where rehydration therapy is often not readily available. Virus is shed in feces for 5 to 7 days. The main transmission route is fecal-oral. Because rotaviruses most often infect young children, the major route of transmission is believed to be person-to-person through care-givers and the general adult population. Rotaviruses can also infect adults and have also been occasionally associated with food and water borne outbreaks. In particular, Rotavirus B strains have caused large epidemics in human adults in China. Group C rotavirus causes sporadic outbreaks in children (Glass and Kilgore, 1997; Sattar et al., 2001). Rotavirus disease is more common during the winter months in countries with a temperate climate. In tropical regions, outbreaks can occur both in the cooler and drier months and throughout the year especially where transmission is related to contaminated water supplies and where no sewage treatment systems exist (Cook et al., 1990; Ansari et al., 1991).

Some immunity develops after infection although it does not give complete protection from future infections. However, repeat infections are often less severe than the original infection. An oral rotavirus vaccine was developed in the late 1980s, but distribution was delayed after lengthy investigations into possible complications associated with the vaccine. This vaccine has recently been approved for commercial global distribution.

5.5. Food-borne Disease

The virus is stable in the environment, hence infection can occur through consumption of contaminated water or food and contact with contaminated surfaces. Eleven food-borne outbreaks consisting of 460 cases of rotaviral gastroenteritis were reported in New York between 1985 and 1990. Seven outbreaks were associated with food-service premises, and the implicated foods included salad, cold foods, shepherd's pie, and water or ice (Sattar et al., 2001). In a recent study in the Netherlands, lack of food handling hygiene was identified as one of the main risk factors for rotavirus infection (de Wit et al., 2003).

Large-scale outbreaks of rotaviral gastroenteritis have been reported in Japanese primary schools with more than 3,000 cases recorded for one outbreak (Hara et al., 1978; Matsumoto et al., 1989). School lunches prepared at a central facility were suspected as the vehicle of infection, but no rotavirus was isolated from food or water. In Costa Rica, market lettuce was found to be contaminated with rotavirus and HAV at a time when there was a high incidence of rotaviral diarrhea in the community (Hernandez et al., 1997). Waterborne rotaviral outbreaks have been reported in many countries, including China, Germany, Israel, Sweden, Russia, and the United States (Ansari et al., 1991; Sattar et al., 2001). Large numbers of rotaviral particles are excreted in feces after infection, and calves infected with rotavirus are known to shed 10^{10} particles per gram of feces. Contamination of water supplies by animals could therefore be a source of waterborne disease. Links have been reported between human and animal rotaviral disease, and it is possible that zoonotic transmission of rotavirus may also occur.

6.0. ASTROVIRUS

6.1. Distribution and Transmission

Astroviruses are distributed worldwide and have been isolated from birds, cats, dogs, pigs, sheep, cows, and man. The main feature of astrovirus infection in both humans and animals is a self-limiting gastroenteritis. The astroviruses are a common cause of human gastroenteritis, with most cases of infection detected in young children under 1 year of age (Bresee and Glass, 1999; Appleton, 2001). A surveillance study in the United Kingdom reported that astroviruses were the most common viral cause of infectious gastrointestinal disease (Roderick et al., 1995). Although astroviruses cause a mild infection in adults, they have been associated with gastroenteritis in immunocompromised adults. Transmission is through the fecal-oral route via food, water, and person-to-person contact. Asymptomatic excretion occurs in 5–20% of neonates and young children and is a significant source of infection, especially in nurseries, childcare centers, and hospitals (Caul, 1996; Bresee and Glass, 1999; Appleton, 2001).

6.2. Taxonomy and Morphology

The astroviruses were first recognized in 1975 (Madeley and Cosgrove, 1975) and were named according to their star-like appearance under the electron microscope. They belong to the family Astroviridae, and human astrovirus is the single type species in the genus Mamastrovirus. Astroviruses are 28- to 30-nm, spherical, nonenveloped, positive-sense, single-stranded RNA viruses with a genome of about 6.8–8 kb and a buoyant density of 1.32 g/ml in potassium tartrate–glycerol gradient (Table 2.1). Because only 10% of astroviruses exhibit the typical 5- or 6-pointed star-like morphology by direct EM, the efficiency of detection was restricted until the introduction of molecular detection methods and improved culture techniques. At least eight human serotypes, two bovine serotypes, and one serotype of each of feline, ovine, and porcine astrovirus are recognized. The human strains are all antigenically distinct from the bovine and ovine strains. A second genus, Avastrovirus, contains the type species turkey astrovirus, of which there are two serotypes. This virus infects birds, including turkeys and ducks (van Regenmortel et al., 2000).

6.3. Growth and Biological Properties

Astroviruses have been isolated in cell cultures but are fastidious viruses to grow *in vitro*. Although human, bovine, feline, and porcine astroviruses have been isolated in primary embryonic kidney cell lines such as human embryonic kidney cells (HEK), only human and porcine astroviruses have been adapted to grow in established cell lines, and trypsin is required in the growth medium to boost infectivity. Although CaCo-2 continuous cell line has proved to be useful for the propagation of astroviruses (Willcocks et al., 1990), virus detection is carried out mainly by EM of stool specimens, molecular assays, or by combined culture-PCR methods. Astroviruses are resistant to extreme environmental conditions. Their heat tolerance allows them to survive 50°C for 1 hr. At 60°C, the virus titer falls by 3 $\log_{10}$ and 6 $\log_{10}$ after 5 and 15 min, respectively. The virus is also stable at pH 3.0 and is resistant to chemicals, including chloroform, lipid solvents, and alcohols and to non-ionic, anionic, and zwitterionic detergents (Appleton, 2001).

6.4. Infection and Disease

Clinically, astroviruses cause symptoms similar to those of caliciviruses after an incubation period of 3–4 days. Symptoms include diarrhea, fever, nausea, and general malaise with occasional vomiting. Normally, diarrhea persists for only 2–3 days but can be prolonged for up to 14 days with virus excretion in feces. Outbreaks commonly occur in institutional settings, especially pediatric wards. In temperate climates, a seasonal peak in winter and spring occurs, but infections may occur throughout the year.

6.5. Food-borne Disease

Epidemiological evidence of transmission by foods is limited, but infections via contaminated shellfish and water have been reported (Oishi et al., 1994; Appleton, 2001). In 1991, a large outbreak of acute gastroenteritis occurred in Japan involving thousands of children and adults from 14 different schools (Oishi et al., 1994). The outbreak was traced to food prepared by a common supplier for school lunches. Astrovirus type 6 was identified by immune electron microscopy and confirmed by molecular and culture methods. There are several Japanese reports of astrovirus genomes identified in shellfish, and there is evidence that astroviruses appear to contribute to food borne outbreaks of gastroenteritis mainly through the consumption of contaminated oysters (Kitahashi et al., 1999).

7.0. ADENOVIRUS

7.1. Distribution and Transmission

The adenoviruses are widespread in nature infecting birds and mammals including man. They commonly cause respiratory disease but may also be involved in other illnesses such as gastroenteritis and conjunctivitis. In particular, the enteric adenoviruses cause gastroenteritis and are the second most important cause, after rotaviruses, of acute gastroenteritis in children under 4 years of age (Allard et al., 1990, Bresee and Glass, 1999). Adenoviruses can be transmitted from person-to-person by direct contact or via fecal-oral, respiratory or environmental routes.

7.2. Taxonomy and Morphology

Adenoviruses belong to the Adenoviridae family and are classified into two genera: the Mastadenovirus, which infects mammals, and the Aviadenovirus, which infects birds. More than 100 members of the Adenoviridae have been isolated from humans and animals, including birds and amphibians. Adenoviruses are 80- to 110-nm, nonenveloped, linear double-stranded DNA viruses with icosahedral symmetry and a genome of 28–45 kb (Table 2.1; Fig. 2.1c). The buoyant density in cesium chloride is 1.32–1.35 g/ml. Six species of human adenoviruses (HAdV-A to HAdV-F) have been identified according to DNA homology (van Regenmortel et al., 2000). Between 50% and 90% DNA homology exists within these species, but only 5–20% homology exists between the species. To date, 51 human adenovirus serotypes have been recognized, including serotypes 40 and 41, the *enteric adenoviruses*, which comprise the HAdV-F species.

7.3. Growth and Biological Properties

Serotypes 40 and 41 of enteric adenoviruses are difficult to grow in cell cultures, whereas most of the nonfecal types are culturable. Adenoviruses are slow-growing compared with a majority of enteroviruses and can be quickly overgrown in some cell lines. The A549 and 293 cell lines have been successfully used for the isolation of adenoviruses from food and environmental samples. Adenoviruses are resistant to various chemical and physical agents including lipid solvents and to adverse pH conditions (Enriquez et al., 1995; Thurston-Enriquez et al., 2003a, 2003b). They can withstand freeze-thawing several times without a significant decrease in titer but are inactivated after heating at 56°C for more than 10 min. The adenoviruses are capable of prolonged survival in the environment and are considered to be more stable than enteroviruses in many environmental situations.

7.4. Infection and Disease

Most human adenovirus infections in normally healthy individuals are mild or subclinical but can be associated with respiratory, ocular, and gastrointestinal disease. In most cases of clinical infection, the symptoms are relatively mild. Of the many types of adenoviruses, only HAdV serotypes 40 and 41 are generally associated with fecal-oral spread and cause gastroenteritis, although all serotypes are shed enterically in feces. HAdV types 40 and 41 can be detected in large numbers in the feces of young children with acute gastroenteritis. In immunocompromised individuals, infection with types 40 and 41 may cause chronic diarrhea. Although rare, some deaths have been reported in immunocompromised children (Bresee and Glass, 1999). Adenoviruses can cause persistent asymptomatic infections and may become established in tonsils, adenoids, and intestines of infected hosts. It is not known whether they are capable of reactivation causing overt disease.

The virus is shed in large numbers in feces and respiratory secretions, often for months or years after infection. The main transmission routes are the fecal-oral route for the enteric adenoviruses and aerosols or direct contact for the nonenteric serotypes. Waterborne transmission of adenovirus has been associated with conjunctivitis in children. Enteric adenovirus infections are common all year round, whereas outbreaks of adenovirus-associated respiratory disease normally occur from late winter to early summer.

7.5. Food-borne Disease

Adenoviruses have been identified in a variety of environmental samples, including wastewater, sludge, shellfish, and in marine, surface and drinking waters. No food-borne or waterborne outbreaks associated with the enteric adenoviruses have been reported, but, as these viruses are common in the environment, it is possible that disease has occurred but the source of infection has not been recognized. There is no documented evidence for food-borne transmission or disease resulting from consumption of adenovirus-contaminated shellfish.

8.0. ENTEROVIRUSES

8.1. Distribution and Transmission

Enteroviruses include polioviruses, coxsackie A and B viruses, and echoviruses, many of which are culturable. They are transmitted by the fecal-oral route and are excreted in feces but do not generally cause gastroenteritis. Polioviruses were the first viruses to be shown to be food-borne, but because of the mass immunization campaigns, virulent wild-type strains are now rarely seen. Outbreaks of food-borne illness associated with coxsackieviruses and echoviruses have been reported (Cliver, 1997; Sattar and Tetro, 2001).

8.2. Taxonomy and Morphology

The enteroviruses are 28- to 30-nm, nonenveloped, positive-sense, single-stranded RNA viruses with icosahedral symmetry and a genome of 7.2–8.4 kb (Table 2.1). They are classified in the large Picornaviridae family, and seven species have been designated within the Enterovirus genus, namely bovine enterovirus, human enterovirus A, human enterovirus B, human enterovirus C, human enterovirus D, poliovirus, porcine enterovirus A, and porcine enterovirus B. Within these different species, numerous serotypes have been reported. The enteroviruses belong to the human enterovirus A (Enterovirus 71) and human enterovirus D (Enterovirus 68 and 70) species. Coxsackie A viruses belong to human enterovirus A, human enterovirus B, and human enterovirus C species. All of the coxsackie B viruses and echoviruses are members of the human enterovirus B species. The Poliovirus species is composed of three distinct serotypes. There are five unassigned tentative species and 22 serotypes within the genus Enterovirus, including two coxsackie A viruses (types CV-A4 and CV-A60) (van Regenmortel et al., 2000).

8.3. Growth and Biological Properties

Many of the enteroviruses are culturable, including all serotypes of poliovirus, echoviruses, and coxsackie B viruses. The enteroviruses are resistant to environmental stressors including heat, adverse pH, and chemicals. Because they are easily cultured *in vitro* and are stable in the environment, live attenuated vaccine strains of poliovirus have been used as indicator

Table 2.2 Classification of Enteroviruses

Family Picornaviridae; Genus Enterovirus
Species
Bovine enterovirus, BEV (2 serotypes)
Human enterovirus A, HEV-A (10 serotypes)
Human enterovirus (1 serotype)
Coxsackie A virus (9 serotypes)
Human enterovirus B, HEV-B (36 serotypes)
Coxsackie A virus (1 serotype)
Coxsackie B virus (6 serotypes)
Echovirus (28 serotypes)
Human enterovirus (1 serotype)
Human enterovirus C, HEV-C (11 serotypes)
Coxsackie A virus (11 serotypes)
Human enterovirus D, HEV-D (2 serotypes)
Human enterovirus (2 serotypes)
Poliovirus, PV (3 serotypes)
Porcine enterovirus, PEV-A (1 serotype)
Porcine enterovirus B, PEV-B (2 serotypes)

viruses for the presence of other virulent enteric viruses in food and water. They have also been used extensively in environmental and food virology research for methods development and to gather information on virus recovery, persistence, and behavior in these settings.

8.4. Infection and Disease

Enteroviruses cause a range of diseases, including viral meningitis and poliomyelitis. They are mainly spread by either the fecal-oral route or direct contact with respiratory secretions of an infected person. The virus is spread through the fecal-oral route mainly among small children who are not yet toilet trained, by adults changing the diapers of an infected infant, and through consumption of fecally contaminated food or water. The enteroviruses multiply mainly in the gastrointestinal tract but can also multiply in other tissues such as nerve and muscle, as does the poliovirus. The incubation period is usually between 3 and 7 days with virus transmission to others occurring from 3 to 10 days after symptoms develop. Enteroviral infection is most common in summer and early autumn, and many infections are asymptomatic. Only a few people (approximately 0.001%) develop aseptic or viral meningitis, and no long-term complications normally follow the mild illnesses or aseptic meningitis. On rare occasions, a person may develop myocarditis or encephalitis.

8.5. Food-borne Disease

The first recorded outbreak associated with food-borne viruses was an outbreak of poliomyelitis linked to consumption of raw milk in 1914 (Jubb, 1915). A further 10 outbreaks associated with raw milk consumption were reported in the United States and United Kingdom over the following 35 years (Sattar and Tetro, 2001). The widespread introduction of pasteurized milk in the 1950s decreased transmission by this route. There have been very few recorded food-borne outbreaks associated with enterovirus infection despite the regular occurrence of enteroviruses in the environment. Enteroviruses, including echoviruses and coxsackie A and B viruses, have been isolated from sewage, raw and digested sludge, marine and fresh waters, and shellfish. In two reported food-borne outbreaks associated with echoviruses in the United States, the source of the virus was not identified (Cliver, 1997). No outbreaks associated with the consumption of shellfish have been reported.

9.0. OTHER VIRUSES WITH POTENTIAL FOR FOOD-BORNE TRANSMISSION

Other viruses transmitted by the fecal-oral route and found in feces of humans and animals include the parvoviruses, coronaviruses, toroviruses, picobirnaviruses, and the tick-borne encephalitis virus (Table 2.1). The ability of many of these viruses to cause gastroenteritis in humans and or animals is still unproven, and there is little evidence to link them with food-borne disease (Glass, 1995; Cliver, 1997; Bresee and Glass, 1999).

9.1. Parvovirus

Parvoviruses have been proposed as causal agents of human gastroenteritis. Their role in viral gastroenteritis of some animal species has been well documented. Parvoviruses are single-stranded DNA viruses and are among the smallest known viruses at 18–26 nm in diameter. They have a smooth surface with no discernable features and were included with "small round viruses" before definitive classification of these viruses was completed. Three possible serotypes known as the Parramatta agent, the cockle agent, and the Wollan/Ditchling group have been identified by IEM. There is limited evidence of parvovirus association with food-borne disease, but it has been linked with consumption of contaminated shellfish (Appleton and Pereira, 1977; Appleton, 2001). The "cockle agent" parvovirus was implicated in a large U.K. outbreak related to consumption of contaminated cockles (Appleton, 2001). More than 800 confirmed cases of gastroenteritis occurred, and parvovirus was identified in all stools examined from this large gastroenteritis outbreak.

9.2. Coronavirus

Coronaviruses are large (80–220 nm) pleomorphic, enveloped, positive-strand RNA viruses belonging to the Coronaviridae family. They generally cause respiratory infections but can also cause gastroenteritis in animals and are excreted in feces. Their role in human gastroenteritis is unclear, although "coronavirus-like particles" can be identified in human feces (Glass, 1995). There are no reports of food-borne outbreaks, but the 2003 SARS (severe acute respiratory syndrome) outbreak in Hong Kong was linked to sewage or sewage-contaminated water.

9.3. Torovirus

Toroviruses are 100- to 150-nm, enveloped, positive-sense, single-stranded RNA viruses belonging to the genus Torovirus in the Coronaviridae family. They were first discovered in 1979 and were named Breda viruses (Woode et al., 1982; Glass, 1995). When observed by EM, they have a distinctive pleomorphic appearance with club-shaped projections extending from the capsid. Toroviruses are known to cause gastroenteritis in animals, especially dairy cattle, in which they cause a marked decrease in milk production. Although they have been isolated from feces of children and adults with diarrhea (Koopmans et al., 1991), their exact role in human gastroenteritis and food-borne disease is still unknown.

9.4. Picobirnavirus

The picobirnaviruses are small, 35-nm, positive-sense, single-stranded RNA viruses with a bisegmented genome and are classified in the family Birnaviridae (Glass, 1995). These viruses are known to cause gastroenteritis in a range of domestic animals. They have also been detected in humans with and without diarrhea but have only been associated with gastroenteritis in immunocompromised HIV patients. Little is known about these viruses, although they have been identified in humans from several countries, including Australia, Brazil, England, and the United States. Their role as true human pathogens is unproven, and there is no documented evidence of food-borne transmission.

9.5. Tick-borne Encephalitis Virus

Tick-borne encephalitis viruses (TBE) may also be food-borne (Cliver, 1997). Tick-borne encephalitis is a zoonotic arbovirus infection endemic to Eastern and Central Europe and Russia. However, the distribution of these viruses can extend to Northern Europe, China, Japan, and Korea. The TBE viruses and the other closely related arboviruses causing yellow fever, Japanese encephalitis, and dengue are all members of the Flaviviridae. Three subtypes of TBE virus cause tick-borne encephalitis: the Eastern European subtype, the Western European subtype, and the Siberian subtype. Most cases occur in spring and summer after bites of different species of *Ixodes* tick. Food-borne transmission is less common but can occur after consumption of unpasteurized dairy products from infected cattle and goats. The disease is serious and can result in long-term neurological sequelae or death. Increased tourism to the endemic areas has extended the risk of travellers acquiring TBE. Vaccines are now available in some countries.

9.6. Other Food-borne Routes of Virus Transmission

Transmission of viral disease can occur through human breastfeeding. Human breast milk is a transmission route for some blood-borne viruses.

There are reports of human immunodeficiency virus (HIV), human lymphotrophic virus-1 (HTLV-1), and cytomegalovirus (CMV) being transmitted to infants in milk from infected mothers during breastfeeding (Sattar and Tetro, 2001). This can present serious problems in less developed countries with a high incidence of HIV and few alternative options available to infected mothers.

10.0. SUMMARY AND CONCLUSIONS

Food-borne disease is increasing worldwide and has become a major public health problem. The majority of food-borne viral disease is caused by noroviruses and HAV. Food-borne transmission of other enteric viruses is less common. HEV is the only virus that appears to be a likely candidate for direct zoonotic transmission from animals to man, and to date there is little evidence to support its transmission by this route.

The overall contribution of viruses to global food-borne disease burden is unknown because accurate data on the prevalence of food-borne viral disease are not available for all countries. National epidemiological surveillance systems vary from country to country, and a large proportion of viral infections are not notifiable and therefore not reported. However, from epidemiological data that have been collected in Europe, the United States, and other countries, it is apparent that viruses play a significant role and that the economic burden of food-borne viral gastrointestinal disease can be substantial in terms of staff illness, time away from employment, and disruption to services.

Over the past 15 years, the introduction of molecular methods for detection and identification of enteric viruses, many of which are not culturable or are difficult to grow in culture, has greatly increased our understanding of their role in food-borne disease. The development of molecular techniques, such as PCR and real-time quantitative PCR, is rapidly increasing the knowledge base by facilitating studies on the behavior and persistence of these viruses in food matrices. However, it is important to recognize the limitations of these techniques. PCR-based methods currently detect both infectious and noninfectious viruses and are not able to determine viral infectivity, which is the key factor when assessing human health risks from food-borne pathogens. It is important that data generated solely from molecular-based assays is judiciously interpreted when studying these viruses. Use of cell culture combined with PCR methods (culture-PCR) can overcome some of these problems for those viruses that are difficult to grow. Unfortunately, the infectivity status of the main food-borne viral pathogen, norovirus, still cannot be determined by *in vitro* methods. This has limited our knowledge of the natural history and biological properties of this pathogen and has also slowed progress in the development of effective control and intervention strategies.

11.0. REFERENCES

Allard, A., Girones, R., Juto, P., and Wadell, G., 1990, Polymerase chain reaction for detection of adenoviruses in stool samples, *J. Clin. Microbiol.* 28:2659–2667.

Ando, T., Monroe, S. S., Gentsch, J. R., Jin, Q., Lewis, D. C., and Glass, R. I., 1995, Detection and differentiation of antigenically distinct small round-structured viruses (Norwalk-like viruses) by reverse transcription-PCR and southern hybridization. *J. Clin. Microbiol.* 33:64–71.

Ando, T., Noel, J. S., and Fankhauser, R. L., 2000, Genetic classification of "Norwalk-like viruses." *J. Infect. Dis.* 181(Suppl 2):S336–S348.

Ansari, S. A., Springthorpe, V. S., and Sattar, S. A., 1991, Survival and vehicular spread of human rotaviruses: possible relation to seasonality of outbreaks. *Rev. Infect. Dis.* 13:448–461.

Appleton, H., 2001, Norwalk virus and the small round viruses causing foodborne gastroenteritis, in: *Foodborne Disease Handbook: Viruses, Parasites, Pathogens and HACCP*, 2nd ed., vol. 2 (Y. H. Hui, S. A. Sattar, K. D. Murrell, W.-K. Nip, and P. S. Stanfield, eds.), Marcel Dekker, New York, pp. 77–97.

Appleton, H., and Pereira, M. S., 1977, A possible virus aetiology in outbreaks of food-poisoning from cockles. *Lancet* 1:780–781.

Balayan, M. S., Andjaparidze, A. G., Savinskaya, S. S., Ketiladze, E. S., Braginsky, D. M., Savinov, A. P., and Poleschuk, V. F., 1983, Evidence for a virus in non-A, non-B hepatitis transmitted via the fecal-oral route. *Intervirology* 20:23–31.

Beller, M., Ellis, A., Lee, S. H., Drebot, M. A., Jenkerson, S. A., Funk, E., Sobsey, M. D., Simmons, O. D., III, Monroe, S. S., Ando, T., Noel, J., Petric, M., Hockin, J., Middaugh, J. P., and Spika, J. S., 1997, Outbreak of viral gastroenteritis due to a contaminated well—International consequences. *J. Am. Med. Assoc.* 278:563–568.

Berg, D. E., Kohn, M. A., Farley, T. A., and McFarland, L. M., 2000, Multi-state outbreaks of acute gastroenteritis traced to fecal-contaminated oysters harvested in Louisiana. *J. Infect. Dis.* 181:S381–S386.

Beuret, C., Kohler, D., Baumgartner, A., and Luthi, T. M., 2002, Norwalk-like virus sequences in mineral waters: one-year monitoring of three brands. *Appl. Environ. Microbiol.* 68:1925–1931.

Bosch, A., Sanchez, G., Le Guyader, F., Vanaclocha, H., Haugarreau, L., and Pinto, R. M., 2001, Human enteric viruses in Coquina clams associated with a large hepatitis A outbreak. *Water Sci. Technol.* 43:61–65.

Bresee, J., and Glass, R., 1999, Astrovirus, enteric adenovirus and other enteric viral infections, in: *Tropical Infectious Diseases: Principles, Pathogens, and Practice* (R. L. Guerrant, D. H. Walker, and P. F. Weller, eds.), Churchill Livingstone, New York, pp. 1145–1153.

Brugha, R., Vipond, I. B., Evans, M. R., Sandifer, Q. D., Roberts, R. J., Salmon, R. L., Caul, E. O., and Mukerjee, A. K., 1999, A community outbreak of food-borne small round-structured virus gastroenteritis caused by a contaminated water supply. *Epidemiol. Infect.* 122:145–154.

Burton-Mcleod, J., Kane, E., Beard, R., Hadley, L., Glass, R., and Ando, T., 2004, Evaluation and comparison of two commercial enzyme-linked immunosorbent assay kits for detection of antigenically diverse human noroviruses in stool samples. *J. Clin. Microbiol.* 42:2587–2595.

Calder, L., Simmons, G., Thornley, C., Taylor, P., Pritchard, K., Greening, G., and Bishop, J., 2003, An outbreak of hepatitis A associated with consumption of raw blueberries. *Epidemiol. Infect.* 131:745–751.

Caul, E. O., 1996, Viral gastroenteritis: small round structured viruses, caliciviruses and astroviruses. Part II: the epidemiological perspective. *J. Clin. Pathol.* 49:959–964.

CDC, 2003, Hepatitis A outbreak associated with green onions at a restaurant—Monaca, Pennsylvania, 2003. *MMWR Morbid. Mortal. Weekly Rep.* 52:1155–1157.

Cheesbrough, J. S., Green, J., Gallimore, C. I., Wright, P. A., and Brown, D. W., 2000, Widespread environmental contamination with Norwalk-like viruses (NLV) detected in a prolonged hotel outbreak of gastroenteritis. *Epidemiol. Infect.* 125:93–98.

Christensen, B. F., Lees, D., Henshilwood, K., Bjergskov, T., and Green, J., 1998, Human enteric viruses in oysters causing a large outbreak of human food borne infection in 1996/97. *J. Shellfish Res.* 17:1633–1635.

Clemente-Casares, P., Pina, S., Buti, M., Jardi, R., Martin, M., Bofill-Mas, S., and Girones, R., 2003, Hepatitis E epidemiology in industrialized countries. *Emerg. Infect. Dis.* 9:448–454.

Cliver, D. O., 1997, Virus transmission via food. *World Health Stats. Q.* 50:90–101.

Coelho, C., Heinert, A. P., Simoes, C. M., and Barardi, C. R., 2003, Hepatitis A virus detection in oysters (*Crassostrea gigas*) in Santa Catarina State, Brazil, by reverse transcription-polymerase chain reaction. *J. Food Prot.* 66:507–511.

Conaty, S., Bird, P., Bell, G., Kraa, E., Grohmann, G., and McAnulty, J., 2000, Hepatitis A in New South Wales, Australia from consumption of oysters: the first reported outbreak. *Epidemiol. Infect.* 124:121–130.

Cook, S. M., Glass, R. I., LeBaron, C. W., and Ho, M. S., 1990, Global seasonality of rotavirus infections. *Bull. World Health Org.* 68:171–177.

Croci, L., De Medici, D., Scalfaro, C., Fiore, A., Divizia, M., Donia, D., Cosentino, A. M., Moretti, P., and Costantini, G., 2000, Determination of enteroviruses, hepatitis A virus, bacteriophages and *Escherichia coli* in Adriatic Sea mussels. *J. Appl. Microbiol.* 88:293–298.

Cromeans, T., Favorov, M. O., Nainan, O. V., and Margolis, H. S., 2001, Hepatitis A and E viruses, in: *Foodborne Disease Handbook: Viruses, Parasites, Pathogens and HACCP*, 2nd ed., vol. 2 (Y. H. Hui, S. A. Sattar, K. D. Murrell, W.-K. Nip, and P. S. Stanfield, eds.), Marcel Dekker, New York, pp. 23–76.

Daniels, N. A., Bergmire-Sweat, D. A., Schwab, K. J., Hendricks, K. A., Reddy, S., Rowe, S. M., Fankhauser, R. L., Monroe, S. S., Atmar, R. L., Glass, R. I., and Mead, P., 2000, A foodborne outbreak of gastroenteritis associated with Norwalk-like viruses: first molecular traceback to deli sandwiches contaminated during preparation. *J. Infect. Dis.* 181:1467–1470.

Dastjerdi, A. M., Green, J., Gallimore, C. I., Brown, D. W. G., and Bridger, J. C., 1999, The bovine Newbury agent-2 is genetically more closely related to human SRSVs than to animal caliciviruses. *Virology* 254:1–5.

de Wit, M., Koopmans, M. P. G., and Van Duynhoven, Y., 2003, Risk factors for Norovirus, Sapporo-virus, and Group A rotavirus gastroenteritis. *Emerg. Infect. Dis.* 9:1563–1570.

Dolin, R., Blacklow, N. R., DuPont, H., Formal, S., Buscho, R. F., Kasel, J. A., Chames, R. P., Hornick, R., and Chanock, R. M., 1971, Transmission of acute infectious nonbacterial gastroenteritis to volunteers by oral administration of stool filtrates. *J. Infect. Dis.* 123:307–312.

Dolin, R., Blacklow, N. R., DuPont, H., Buscho, R. F., Wyatt, R. G., Kasel, J. A., Hornick, R., and Chanock, R. M., 1972, Biological properties of Norwalk agent of acute infectious nonbacterial gastroenteritis. *Proc. Soc. Exp. Biol. Med.* 140: 578–583.

Dowell, S. F., Groves, C., Kirkland, K. B., Cicirello, H. G., Ando, T., Jin, Q., Gentsch, J. R., Monroe, S. S., Humphrey, C. D., and Slemp, C., 1995, A multistate outbreak of oyster-associated gastroenteritis: implications for interstate tracing of contaminated shellfish. *J. Infect. Dis.* 171:1497–1503.

Duizer, E., Schwab, K., Neill, F., Atmar, R., Koopmans, M., and Estes, M., 2004, Laboratory efforts to cultivate noroviruses. *J. Gen. Virol.* 85:79–87.

Emerson, S., and Purcell, R., 2003, Hepatitis E virus, *Rev. Med. Virol.* 13:145–154.

Enriquez, C. E., Hurst, C. J., and Gerba, C. P., 1995, Survival of the enteric adenoviruses 40 and 41 in tap, sea, and waste water. *Water Res.* 29:2548–2553.

Estes, M. K., Ball, J. M., Guerrero, R. A., Opekun, A. R., Gilger, M. A., Pacheco, S. S., and Graham, D. Y., 2000, Norwalk virus vaccines: challenges and progress. *J. Infect. Dis.* 181(Suppl 2):S367–373.

Fankhauser, R. L., Monroe, S. S., Noel, J. S., Humphrey, C. D., Bresee, J. S., Parashar, U. D., Ando, T., and Glass, R. I., 2002, Epidemiologic and molecular trends of "Norwalk-like viruses" associated with outbreaks of gastroenteritis in the United States. *J. Infect. Dis.* 186:1–7.

Fiore, A., 2004, Hepatitis A transmitted by food. *Clin. Infect. Dis.* 38:705–715.

Gaulin, C., Frigon, M., Poirier, D., and Fournier, C., 1999, Transmission of calicivirus by a foodhandler in the pre-symptomatic phase of illness. *Epidemiol. Infect.* 123:475–478.

Glass, R. I., 1995, Other viral agents of gastroenteritis, in: *Infections of the Gastrointestinal Tract* (M. J. Blaser, P. D. Smith, J. I. Ravdin, H. B. Greenberg, and R. L. Guerrant, eds.), Raven Press, New York, pp. 1055–1063.

Glass, R. I., and Kilgore, P. E., 1997, Etiology of acute viral gastroenteritis, in: *Diarrheal Disease*, vol. 38 (M. Gracey and J. A. Walker-Smith, eds.), Vevey/Lippincott-Raven, Philadelphia, pp. 39–53.

Green, J., Vinje, J., Gallimore, C. I., Koopmans, M., Hale, A., Brown, D. W., Clegg, J. C., and Chamberlain, J., 2000, Capsid protein diversity among Norwalk-like viruses. *Virus Genes* 20:227–236.

Green, K., Kapikian, A. Z., and Chanock, R. M., 2001, Human caliciviruses, in: *Fields Virology*, 4th ed., vol. 1 (D. M. Knipe and P. M. Howley, eds.), Lippincott Williams & Wilkins, Philadelphia, pp. 841–874.

Greening, G. E., Hewitt, J., Hay, B. E., and Grant, C. M. 2003. Persistence of Norwalk-like viruses over time in Pacific oysters grown in the natural environment, in: *Molluscan Shellfish Safety*, Proceedings of the 4th International Conference on Molluscan Shellfish Safety (A. Villalba, B. Reguera, J. L. Romalde, and R. Beiras, eds.), Consellería de Pesca e Asuntos Marítimos da Xunta de Galicia and Intergovernmental Oceanographic Commission of UNESCO, Santiago de Compostela, Spain, pp. 367–377.

Halbur, P. G., Kasorndorkbua, C., Gilbert, C., Guenette, D., Potters, M. B., Purcell, R. H., Emerson, S. U., Toth, T. E., and Meng, X. J., 2001, Comparative pathogenesis of infection of pigs with hepatitis E viruses recovered from a pig and a human. *J. Clin. Microbiol.* 39:918–923.

Halliday, L. M., Kang, L. Y., Zhou, T. K., Hu, M. D., Pan, Q. C., Fu, T. Y., Huang, Y. S., and Hu, S. L., 1991, An epidemic of hepatitis A attributable to the ingestion of raw clams in Shanghai, China. *J. Infect. Dis.* 164:852–859.

Hara, M., Mukoyama, J., Tsuruhara, T., Ashiwara, Y., Saito, Y., and Tagaya, I., 1978, Acute gastroenteritis among schoolchildren associated with reovirus-like agent. *Am. J. Epidemiol.* 107:161–169.

Hernandez, F., Monge, R., Jimenez, C., and Taylor, L., 1997, Rotavirus and hepatitis A virus in market lettuce (*Latuca sativa*) in Costa Rica. *Int. J. Food Microbiol.* 37:221–223.

Hewitt, J., and Greening, G. E., 2004, Survival and persistence of norovirus, hepatitis A virus and feline calicivirus in marinated mussels. *J. Food Prot.* 67:1743–1750.

Hollinger, F. B., and Emerson, S. U., 2001, Hepatitis A virus, in: *Fields Virology*, 4th ed., vol. 1 (D. M. Knipe and P. M. Howley, eds.), Lippincott Williams & Wilkins, Philadelphia, pp. 799–840.

Hutson, A., Atmar, R., Graham, D., and Estes, M., 2002, Norwalk virus infection and disease is associated with ABO histo-blood group type. *J. Infect. Dis.* 185:1335–1337.

Hutson, A. M., Atmar, R. L., and Estes, M. K., 2004, Norovirus disease: changing epidemiology and host susceptibility factors. *Trends Microbiol.* 12:279–287.

Jaykus, L. A., Hemard, M. T., and Sobsey, M. D., 1994, Human enteric pathogenic viruses, in: *Environmental Indicators and Shellfish Safety* (M. D. Pierson and C. R. Hackney, eds.), Pierson Associates, Newport, CT, pp. 92–153.

Jiang, X., Huang, P. W., Zhong, W. M., Farkas, T., Cubitt, D. W., and Matson, D. O., 1999, Design and evaluation of a primer pair that detects both Norwalk- and Sapporo-like caliciviruses by RT-PCR. *J. Virol. Methods* 83:145–154.

Jothikumar, N., Aparna, K., Kamatchiammal, S., Paulmurugan, R., Saravanadevi, S., and Khanna, P., 1993, Detection of hepatitis E virus in raw and treated wastewater with the polymerase chain reaction. *Appl. Environ. Microbiol.* 59:2558–2562.

Jubb, G., 1915, The third outbreak of epidemic poliomyelitis at West Kirby. *Lancet* 1:67.

Kapikian, A. Z., Wyatt, R. G., Dolin, R., Thornhill, T. S., Kalica, A. R., and Chanock, R. M., 1972, Visualization by immune electron microscopy of a 27-nm particle associated with acute infectious nonbacterial gastroenteritis. *J. Virol.* 10:1075–1081.

Kapikian, A. Z., Hoshino, Y., and Chanock, R. M., 2001, Rotaviruses, in: *Fields Virology*, 4th ed., vol. 2 (D. M. Knipe and P. M. Howley, eds.), Lippincott Williams & Wilkins, Philadelphia, pp. 1787–1833.

Kaplan, J. E., Feldman, R., Campbell, D. S., Lookabaugh, C., and Gary, G. W., 1982, The frequency of a Norwalk-like pattern of illness in outbreaks of acute gastroenteritis. *Am. J. Public Health* 72:1329–1332.

Kingsley, D. H., Hoover, D. G., Papafragkou, E., and Richards, G. P., 2002, Inactivation of hepatitis A virus and a calicivirus by high hydrostatic pressure. *J. Food Prot.* 65:1605–1609.

Kitahashi, T., Tanaka, T., and Utagawa, E., 1999, Detection of HAV, SRSV and astrovirus genomes from native oysters in Chiba City, Japan. *Kansenshogaku Zasshi* 73:559–564.

Koopmans, M., van Wujckhuise-Sjouke, L., Schukken, Y., Cremers, H., and Horzinek, M., 1991, Association of diarrhea in cattle with torovirus infections on farms. *Am. J. Vet. Res.* 52:1769–1773.

Kruse, H., Brown, D., Lees, D., Le Guyader, S., Lindgren, S., Koopmans, M. P. G., Macri, A., and Von Bonsdorff, C. H, 2002, Scientific Committee on Veterinary Measures relating to Public Health, 2002, Scientific opinion on Norwalk-like viruses, European Commission—Health and Consumer Protection Directorate—General. Available at: http://europa.eu.int/comm/food/fs/sc/scv/out49_en.pdf.

Kuritsky, J. N., Osterholm, M. T., Greenberg, H. B., Korlath, J. A., Godes, J. R., Hedberg, C. W., Forfang, J. C., Kapikian, A. Z., McCullough, J. C., and White, K.

E., 1984, Norwalk gastroenteritis: a community outbreak associated with bakery product consumption. *Ann. Intern. Med.* 100:519–521.

Lees, D., 2000, Viruses and bivalve shellfish, *Int. J. Food Microbiol.* 59:81–116.

Lindesmith, L., Moe, C., LePendu, J., Frelinger, J. A., Treanor, J., and Baric, R. S., 2005, Cellular and humoral immunity following Snow Mountain virus challenge. *J. Virol.* 79:2900–2909.

Lo, S. V., Connolly, A. M., Palmer, S. R., Wright, D., Thomas, P. D., and Joynson, D., 1994, The role of the pre-symptomatic food handler in a common source outbreak of food-borne SRSV gastroenteritis in a group of hospitals. *Epidemiol. Infect.* 113:513–521.

Lopman, B. A., Reacher, M. H., Van Duijnhoven, Y., Hanon, F. X., Brown, D., and Koopmans, M., 2003, Viral gastroenteritis outbreaks in Europe, 1995–2000. *Emerg. Infect. Dis.* 9:90–96.

Madeley, C. R., and Cosgrove, B. P., 1975, Letter: 28 nm particles in faeces in infantile gastroenteritis. *Lancet* 2(7932):451–452.

Marks, P. J., Vipond, I. B., Carlisle, D., Deakin, D., Fey, R. E., and Caul, E. O., 2000, Evidence for airborne transmission of Norwalk-like virus (NLV) in a hotel restaurant. *Epidemiol. Infect.* 124:481–487.

Matson, D. O., Zhong, W. M., Nakata, S., Numata, K., Jiang, X., Pickering, L. K., Chiba, S., and Estes, M. K., 1995, Molecular characterization of a human calicivirus with sequence relationships closer to animal caliciviruses than other known human caliciviruses. *J. Med. Virol.* 45:215–222.

Matsumoto, K., Hatano, M., Kobayashi, K., Hasegawa, A., Yamazaki, S., Nakata, S., Ciba, S., and Kimura, Y., 1989, An outbreak of gastroenteritis associated with acute rotaviral infection in schoolchildren. *J. Infect. Dis.* 160:611–615.

Mead, P. S., Slutsker, L., Dietz, V., McCaig, L. F., Bresee, J. S., Shapiro, C., Griffin, P. M., and Tauxe, R. V., 1999, Food-related illness and death in the United States. *Emerg. Infect. Dis.* 5:607–625.

Meng, X. J., Purcell, R. H., Halbur, P. G., Lehman, J. R., Webb, D. M., Tsareva, T. S., Haynes, J. S., Thacker, B. J., and Emerson, S. U., 1997, A novel virus in swine is closely related to the human hepatitis E virus. *Proc. Natl. Acad. Sci. U. S. A.* 94:9860–9865.

Meng, X.-J., Halbur, P. G., Shapiro, M. S., Govindarajan, S., Bruna, J. D., Mushahwar, I. K., Purcell, R. H., and Emerson, S. U., 1998, Genetic and experimental evidence for cross-species infection by swine hepatitis E virus. *Am. Soc. Microbiol.* 72:9714–9721.

Meng, X. J., Wiseman, B., Elvinger, F., Guenette, D. K., Toth, T. E., Engle, R. E., Emerson, S. U., and Purcell, R. H., 2002, Prevalence of antibodies to hepatitis E virus in veterinarians working with swine and in normal blood donors in the United States and other countries. *J. Clin. Microbiol.* 40:117–122.

Moe, C., Teunis, P., Lindesmith, L., McNeal, E.-J. C., LePendu, J., Treanor, J., Herrmann, J., Blacklow, N. R., and Baric, R. S., 2004, Norovirus dose response. Second International Calicivirus Conference, Dijon, France.

Niu, M. T., Polish, L. B., Robertson, B. H., Khanna, B. K., Woodruff, B. A., Shapiro, C. N., Miller, M. A., Smith, J. D., Gedrose, J. K., Alter, M. J., and Margolis, H. S., 1992, Multistate outbreak of hepatitis A associated with frozen strawberries. *J. Infect. Dis.* 166:518–524.

Noel, J. S., Liu, B. J., Humphrey, C. D., Rodriguez, E. M., Lambden, P. R., Clarke, I. N., Dwyer, D. M., Ando, T., Glass, R. I., and Monroe, S. S., 1997, Parkville virus: a novel genetic variant of human calicivirus in the Sapporo virus clade, associated with an outbreak of gastroenteritis in adults. *J. Med. Virol.* 52:173–178.

Oishi, I., Yamazaki, K., Kimoto, T., Minekawa, Y., Utagawa, E., Yamazaki, S., Inouye, S., Grohmann, G. S., Monroe, S. S., Stine, S. E., Carcamo, C., Ando, T., and Glass, R. I., 1994, A large outbreak of acute gastroenteritis associated with astrovirus among students and teachers in Osaka, Japan. *J. Infect. Dis.* 170:439–443.

Oliver, S. L., Dastjerdi, A. M., Wong, S., El-Attar, L., Gallimore, C., Brown, D. W., Green, J., and Bridger, J. C., 2003, Molecular characterization of bovine enteric caliciviruses: a distinct third genogroup of noroviruses (Norwalk-like viruses) unlikely to be of risk to humans. *J. Virol.* 77:2789–2798.

Parashar, U. D., Dow, L., Fankhauser, R. L., Humphrey, C. D., Miller, J., Ando, T., Williams, K. S., Eddy, C. R., Noel, J. S., Ingram, T., Bresee, J. S., Monroe, S. S., and Glass, R. I., 1998, An outbreak of viral gastroenteritis associated with consumption of sandwiches: implications for the control of transmission by food handlers. *Epidemiol. Infect.* 121:615–621.

Pebody, R. G., Leino, T., Ruutu, P., Kinnunen, L., Davidkin, I., Nohynek, H., and Leinikki, P., 1998, Foodborne outbreaks of hepatitis A in a low endemic country: an emerging problem? *Epidemiol. Infect.* 120:55–59.

Pina, S., Jofre, J., Emerson, S. U., Purcell, R. H., and Girones, R., 1998, Characterization of a strain of infectious hepatitis E virus isolated from sewage in an area where hepatitis E is not endemic. *Appl. Environ. Microbiol.* 64:4485–4488.

Ponka, A., Maunula, L., von Bonsdorff, C., and Lyytikainen, O., 1999, An outbreak of calicivirus associated with consumption of frozen raspberries. *Epidemiol. Infect.* 123:469–474.

Ramsay, C. N., and Upton, P. A., 1989, Hepatitis A and frozen raspberries, *Lancet* 1:43–44.

Reid, T. M., and Robinson, H. G., 1987, Frozen raspberries and hepatitis A. *Epidemiol. Infect.* 98:109–112.

Richards, A. F., Lopman, B., Gunn, A., Curry, A., Ellis, D., Cotterill, H., Ratcliffe, S., Jenkins, M., Appleton, H., Gallimore, C. I., Gray, J. J., and Brown, D. W., 2003, Evaluation of a commercial ELISA for detecting Norwalk-like virus antigen in faeces. *J. Clin. Virol.* 26:109–115.

Robertson, B. H., Khanna, B. K., Nainan, O. V., and Margolis, H. S., 1991, Epidemiologic patterns of wild-type hepatitis A virus strains determined by genetic variation. *J. Infect. Dis.* 163:286–292.

Robertson, B. H., Jansen, R. W., Khanna, B., Totsuka, A., Nainan, O. V., Siegl, G., Widell, A., Margolis, H. S., Isomura, S., Ito, K., Ishizu, T., Moritsugu, Y., and Lemon, S. M., 1992, Genetic relatedness of hepatitis A virus strains recovered from different geographic regions. *J. Gen. Virol.* 73:1365–1377.

Roderick, P., Wheeler, J., Cowden, J., Sockett, P., Skinner, R., Mortimer, P., Rowe, B., and Rodriques, L., 1995, A pilot study of infectious intestinal disease in England. *Epidemiol. Infect.* 114:277–288.

Sattar, S. A., Springthorpe, V. S., and Tetro, J. A., 2001, Rotavirus, in: *Foodborne Disease Handbook, Viruses, Parasites, Pathogens, and HACCP*, 2nd ed., vol. 2 (Y. H. Hui, S. A. Sattar, K. D. Murrell, W.-K. Nip, and P. S. Stanfield, eds.), Marcel Dekker, New York, pp. 99–126.

Sattar, S. A., and Tetro, J. A., 2001, Other foodborne viruses, in: *Foodborne Disease Handbook: Viruses, Parasites, Pathogens, and HACCP*, 2nd ed., vol. 2 (Y. H. Hui, S. A. Sattar, K. D. Murrell, W.-K. Nip, and P. S. Stanfield, eds.), Marcel Dekker, New York, pp. 127–136.

Schlauder, G., Dawson, G., Erker, J., Kwo, P., Knigge, M., Smalley, D., Desai, S., and Mushahwar, I., 1998, The sequence and phylogenetic analysis of a novel hepatitis

E virus isolated from a patient with acute hepatitis reported in the United States. *J. Gen. Virol.* 79:447–456.

Schuffenecker, I., Ando, T., Thouvenot, D., Lina, B., and Aymard, M., 2001, Genetic classification of "Sapporo-like viruses." *Arch. Virol.* 146:2115–2132.

Schwab, K. J., Neill, F. H., Fankhauser, R. L., Daniels, N. A., Monroe, S. S., Bergmire-Sweat, D. A., Estes, M. K., and Atmar, R. L., 2000, Development of methods to detect "Norwalk-like viruses" (NLVs) and hepatitis A virus in delicatessen foods: application to a food-borne NLV outbreak. *Appl. Environ. Microbiol.* 66:213–218.

Scientific Committee on Veterinary Measures Relating to Public Health, 2002, Scientific opinion on Norwalk-like viruses, European Commission—Health and Consumer Protection Directorate—General.

Simmons, G., Greening, G., Gao, W., and Campbell, D., 2001, Raw oyster consumption and outbreaks of viral gastroenteritis in New Zealand: evidence for risk to the public's health. *Aust. N. Z. J. Public Health* 25:234–240.

Sobsey, M., Shields, P., Hauchman, F., Davis, A., Rullman, V., and Bosch, A., 1988, Survival and persistence of hepatitis A virus in environmental samples, in: *Viral Hepatitis and Liver Disease* (A. Zuckermann, ed.), Alan Liss, New York, pp. 121–124.

Sugieda, M., Nagaoka, H., Kakishima, Y., Ohshita, T., Nakamura, S., and Nakajima, S., 1998, Detection of Norwalk-like virus genes in the caecum contents of pigs. *Arch. Virol.* 143:1215–1221.

Tei, S., Kitajima, N., Takahashi, K., and Mishiro, S., 2003, Zoonotic transmission of hepatitis E virus from deer to human beings. *Lancet* 362:371–373.

Tei, S., Kitajima, N., Ohara, S., Inoue, Y., Miki, M., Yamatani, T., Yamabe, H., Mishiro, S., and Kinoshita, Y., 2004, Consumption of uncooked deer meat as a risk factor for hepatitis E virus infection: an age and sex-matched case-control study. *J. Med. Virol.* 74:67–70.

Thurston-Enriquez, J. A., Haas, C. N., Jacangelo, J., and Gerba, C. P., 2003a, Chlorine inactivation of adenovirus type 40 and feline calicivirus. *Appl. Environ. Microbiol.* 69:3979–3985.

Thurston-Enriquez, J. A., Haas, C. N., Jacangelo, J., Riley, K., and Gerba, C. P., 2003b, Inactivation of feline calicivirus and adenovirus type 40 by UV radiation. *Appl. Environ. Microbiol.* 69:577–582.

van der Poel, W. H., Vinje, J., van der Heide, R., Herrera, M. I., Vivo, A., and Koopmans, M. P., 2000, Norwalk-like calicivirus genes in farm animals. *Emerg. Infect. Dis.* 6:36–41.

van Regenmortel, M. H. V., Fauquet, C. M., Bishop, D. H. L., Carstens, E. B., Estes, M. K., Lemon, S. M., Maniloff, J., Mayo, M. A., McGeoch, D. J., Pringle, C. R., and Wickner, R. B., 2000, *Virus Taxonomy: Classification and Nomenclature of Viruses. Seventh Report of the International Committee on Taxonomy of Viruses*, Academic Press, San Diego.

Vinje, J., Hamidjaja, R. A., and Sobsey, M. D., 2004, Development and application of a capsid VP1 (region D) based reverse transcription PCR assay for genotyping of genogroup I and II noroviruses. *J. Virol. Methods* 116:109–117.

Vinje, J., Vennema, H., Maunula, L., von Bonsdorff, C. H., Hoehne, M., Schreier, E., Richards, A., Green, J., Brown, D., Beard, S. S., Monroe, S. S., de Bruin, E., Svensson, L., and Koopmans, M. P., 2003, International collaborative study to compare reverse transcriptase PCR assays for detection and genotyping of noroviruses. *J. Clin. Microbiol.* 41:1423–1433.

Wheeler, J., Sethi, D., Cowden, J., Wall, P., Rodrigues, L., Tompkins, D., Hudson, M., and Roderick, P., 1999, Study of intestinal disease in England: rates in the community, presenting to general practice, and reported to national surveillance. *Br. Med. J.* 318:1046–1050.

Wilkinson, N., Kurdziel, A. S., Langton, S., Needs, E., and Cook, N., 2001, Resistance of poliovirus to inactivation by hydrostatic pressure. *Innov. Food Sci. Emerg. Technol.* 2:95–98.

Willcocks, M. M., Carter, M. J., Laidler, F. R., and Madeley, C. R., 1990, Growth and characterisation of human faecal astrovirus in a continuous cell line. *Arch. Virol.* 113:73–81.

Wobus, C. E., Karst, S. M., Thackray, L. B., Chang, K. O., Sosnovtsev, S. V., Belliot, G., Krug, A., Mackenzie, J. M., Green, K. Y., and Virgin, H. W., 2004, Replication of Norovirus in cell culture reveals a tropism for dendritic cells and macrophages. *PLoS Biol.* 2:e432. Available at www.plosbiology.org.

Woode, G. N., Reed, D. E., Runnels, P. L., Herig, M. A., and Hill, H. T., 1982, Studies with an unclassified virus from diarrheal calves. *Vet. Microbiol.* 7:221–240.

Wyatt, R. G., Greenberg, H. B., Dalgard, D. W., Allen, W. P., Sly, D. L., Thornhill, T. S., Chanock, R. M., and Kapikian, A. Z., 1978, Experimental infection of chimpanzees with the Norwalk agent of epidemic viral gastroenteritis. *J. Med. Virol.* 2:89–96.

Yazaki, Y., Mizuo, H., Takahashi, M., Nishizawa, K., Sasaki, N., and Gotanda, Y., 2003, Sporadic acute or fulminant hepatitis E in Hokkaido, Japan, maybe foodborne, as suggested by the presence of hepatitis E virus in pig liver as food. *J. Gen. Virol.* 84:2351–2357.

CHAPTER 3

Molecular Virology of Enteric Viruses (with Emphasis on Caliciviruses)

Javier Buesa and Jesús Rodríguez-Díaz

1.0. CALICIVIRUSES

Human caliciviruses are members of the family Caliciviridae and are responsible for a majority of the outbreaks of acute nonbacterial gastroenteritis. In fact, they now are considered a common cause of sporadic cases of diarrhea in the community (Glass et al., 2000b; Koopmans et al., 2002; Lopman et al., 2002; Hutson et al., 2004). These viruses were implicated in as many as 95% of the reported viral gastroenteritis outbreaks examined over a 4.5-year period in the United States (Fankhauser et al., 2002), and similar high incidence rates have been found in other studies (Maguire et al., 1999; Glass et al., 2000b; Koopmans et al., 2000; Lopman et al., 2003). Common features of the Caliciviridae include the presence of a single major structural protein from which the capsid is constructed and 32 cup-shaped depressions on the surface of the virion arranged in an icosahedral symmetry. The name of the family was derived from the Latin word *calix*, which means cup or goblet, and refers to the surface hollows (Madeley, 1979). Another major feature of Caliciviridae is the absence of a methylated cap at the 5′ end of the viral RNA. Instead, a small protein (VPg) of ~10×10^3 to 12×10^3 kDa is covalently linked to the viral RNA and is considered essential for the infectivity of the RNA (Black et al., 1978; Ando et al., 2000).

Norwalk virus was the first human enteric calicivirus to be discovered, this after a gastrointestinal outbreak affecting both children and adults in an elementary school in Norwalk, Ohio (Kapikian et al., 1972). Over the next several years, other agents of epidemic viral gastroenteritis were described including Hawaii virus, Montgomery County agent, Snow Mountain virus, Southampton virus, Toronto virus, and so forth. An interim scheme to classify these viruses on a morphological and physicochemical basis was proposed by Caul and Appleton (1982). Some of these viruses were antigenically related to Norwalk virus by IEM (immune electron microscopy) and cross-challenge studies, whereas others, like the Sapporo virus (Chiba et al., 1979), were found to be antigenically distinct (Nakata et al., 1996).

Human caliciviruses have been classified into two distinct genera namely, Norovirus (previously called "Norwalk-like viruses" (NLVs) or "small round structured viruses") and Sapovirus (formerly "Sapporo-like viruses," or SLVs) (Green et al., 2000b; Mayo, 2002). The noroviruses and sapoviruses form distinct phylogenetic clades within Caliciviridae (Berke et al., 1997), and certain features of their viral genome organization distinguish them from

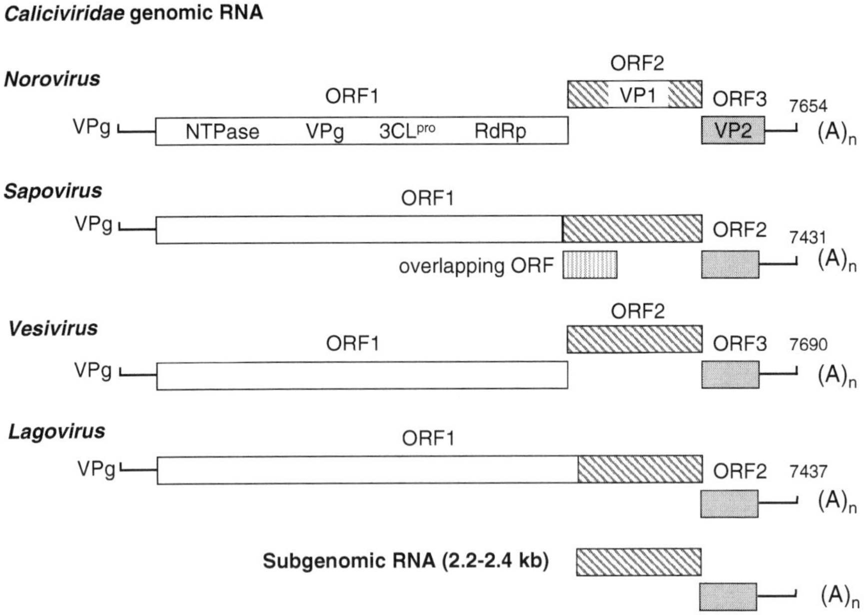

Figure. 3.1 Genome organization of the four different genera of Caliciviridae. The genome of Norovirus and Vesivirus has three open reading frames (ORFs) that encode the nonstructural proteins, the major capsid protein (VP1), and a minor structural basic protein (VP2). The genera Sapovirus and Lagovirus encode the capsid protein contiguous with the large nonstructural polyprotein (ORF1). An additional small overlapping ORF in a +1 frameshift has been described in certain strains of sapoviruses. A subgenomic RNA that covers the entire 3′ end of the genome, from the capsid gene to the 3′ end, has been detected in calicivirus-infected cells.

each other and from other genera of the family (Fig. 3.1). These viruses also differ in their epidemiology and host range. For example, noroviruses infect all age groups and are usually responsible for outbreaks of acute gastroenteritis frequently associated with contaminated food or water. Sapoviruses, on the other hand, are associated with sporadic cases of acute gastroenteritis and mainly infect infants and young children (Green et al., 2001), although outbreaks in adults have also been described (Noel et al., 1997).

Enteric caliciviruses of animals, associated with gastroenteritis in pigs, calves, chickens, mink, dogs, and cats, have also been described (Guo et al., 1999b; Liu et al., 1999a; van der Poel et al., 2000; Guo and Saif, 2003). The Caliciviridae has two additional genera, Lagovirus and Vesivirus, each of which includes caliciviruses that infect animals such as the rabbit hemorrhagic disease virus (RHDV) and the vesicular exanthema of swine virus (VESV), respectively. The proposed nomenclature for calicivirus strains is host species from which the virus was isolated/virus genus/virus name/strain designation/year of isolation/country of origin. The type strain for Norovirus and Vesivirus are Norwalk virus (Hu/NV/Norwalk/8fIIa/1968/US) and VESV (Sw/VV/VESV/A48/1948/US), respectively (Atmar and Estes, 2001).

On examination by negative contrast electron microscopy (EM), caliciviruses show characteristic cup-shaped structures on their surfaces. These cuplike depressions are more prominent in sapoviruses, leading to the characteristic six-pointed "star of David" appearance, especially when viewed along the major two-, three-, and fivefold axes of symmetry. The noroviruses have a polyadenylated, positive-sense, single-stranded RNA genome with three major open reading frames (ORFs) (Jiang et al., 1993b; Lambden et al., 1993). The virions have a buoyant density of 1.33 to 1.41 g/cm^3 in cesium chloride ($CsCl_2$) (Caul and Appleton, 1982; Madore et al., 1986) and usually lack the distinctive calicivirus cuplike morphology when viewed by EM. The sapoviruses have a polyadenylated, positive-sense, single-stranded RNA genome with two (or three) major ORFs (Liu et al., 1995), have a buoyant density of 1.37 to 1.38 g/cm^3 in CsCl (Terashima et al., 1983), and often possess distinctive calicivirus cup-like morphology (Madeley, 1979).

The molecular era of norovirus research started with the successful cloning of the genomes of Norwalk and Southampton viruses obtained from stool samples (Jiang et al., 1990; Lambden et al., 1993). Despite numerous attempts, human calicivirus infections have not been induced in experimental animals nor have these viruses been propagated successfully in cultured cells, thus hampering many aspects of their research (Duizer et al., 2004). Unlike human caliciviruses, some animal caliciviruses, including primate calicivirus (Smith et al., 1983), feline calicivirus (FCV) (Love and Sabine, 1975), and San Miguel sea lion virus (SMSV) (Smith et al., 1973), have been successfully propagated in cell cultures. These viruses have provided a direct approach for the study of virus infections, genome transcription, viral protein translation, and virus replication (Green et al., 2002). In addition, information gained by the study of caliciviruses that grow efficiently in cell cultures, such as FCV and VESV, or that have an animal model and a limited cell culture system (such as RHDV), has been important in the identification of features that are likely to be shared among members of the Caliciviridae (Marin et al., 2000; Morales et al., 2004).

1.1. Structure, Composition, and Molecular Biology

The Norwalk virus capsid is composed of a single major structural protein, known as VP1, and a few copies of a second small basic structural protein named VP2 (Prasad et al., 1999; Glass et al., 2000a; Green et al., 2001). Cloning and expression of norovirus proteins VP1 and VP2 in insect cells using the baculovirus expression system resulted in the self-assembly of the viral capsid and the production of recombinant virus-like particles (rVLPs) that were antigenically and structurally similar to the native virions (Jiang et al., 1992; Green et al., 1997; Hale et al., 1999; Kobayashi et al., 2000).

The three-dimensional structure of Norwalk rVLPs was first determined by cryo-electron microscopy and computer-image processing to a resolution

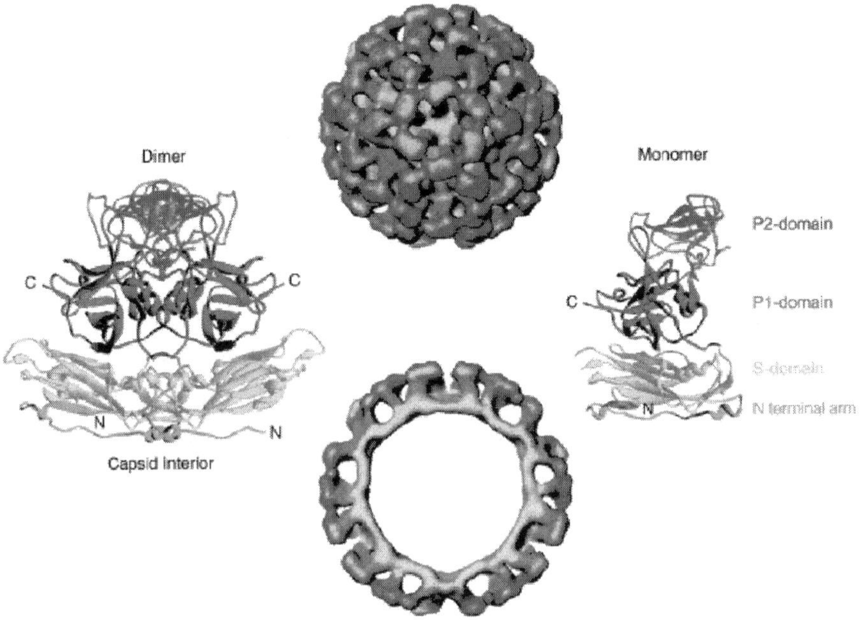

Figure. 3.2 The structure of Norwalk virus-like particles (NV VLPs) has been solved by cryo-electron microscopic reconstruction to 22 Å (top, surface representation; bottom, cross section) and by x-ray crystallography to 3.4 Å. The NV VLPs have 90 dimers of capsid protein (left, ribbon diagram) assembled in T = 3 icosahedral symmetry. Each monomeric capsid protein (right, ribbon diagram) is divided into an N-terminal arm region (green) facing the interior of the VLP, a shell domain (S domain, yellow) that forms the continuous surface of the VLP, and a protruding domain (P domain) that emanates from the S domain surface. The P domain is further divided into subdomains, P1 and P2 (red and blue, respectively) with the P2 subdomain at the most distal surface of the VLPs. (Reproduced with permission from Prasad et al., 1999, and Bertolotti-Ciarlet et al., 2002; see color insert.)

of 22 Å. This analysis showed that the particles (38 nm in diameter by this technique) have a distinct architecture and exhibit T = 3 icosahedral symmetry (Fig. 3.2). The capsid contains 180 copies of the capsid protein assembled into 90 dimers with an arch-like structure. The arches are arranged in such a way that there are large hollows at the icosahedral five- and threefold positions, and these hollows are seen as cuplike structures on the viral surface (Prasad et al., 1994, 1996a, 1999). To form a T = 3 icosahedral structure, the capsid protein has to adapt to three quasi-equivalent positions, and the subunits located at these positions are conventionally referred to as A, B, and C. The only high-resolution (3.4 Å) structure is that of the Norwalk virus capsid determined by x-ray crystallography (Prasad et al., 1999).

Each subunit or monomeric capsid protein folds into an N-terminal region facing the inside of the capsid, a shell (S) domain that forms the

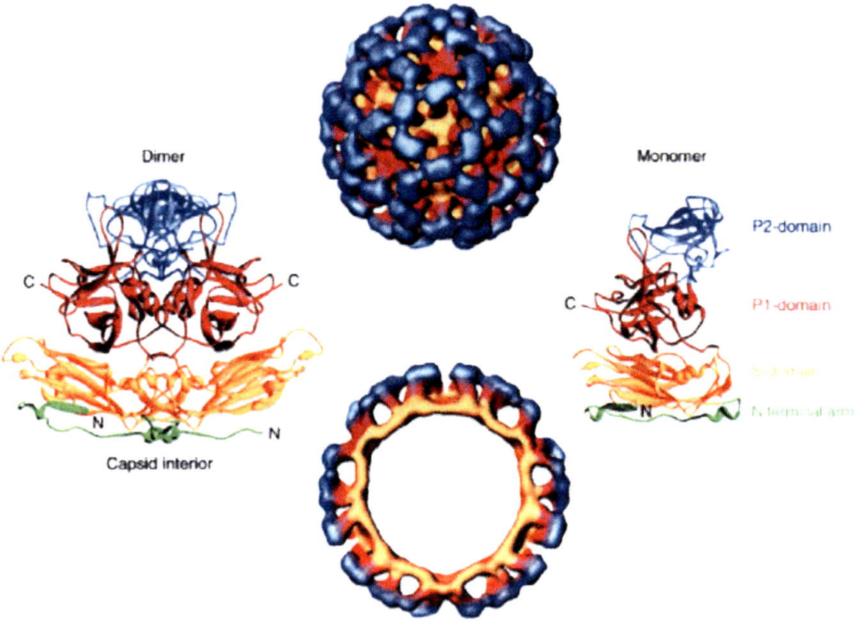

Figure. 3.2 The structure of Norwalk virus-like particle (NV VLP) has been resolved by cryo-electron microscopic reconstruction to 22 Å (top, surface representation; bottom, cross-section) and by X-ray crystallography to 3.4 Å. The NV VLPs have 90 dimers of capsid protein (left, ribbon diagram) assembled in T = 3 icosahedral symmetry. Each monomeric capsid protein (right, ribbon diagram) is divided into an N-terminal arm region (green) facing the interior of the VLP, a shell domain (S-domain, yellow) that forms the continuous surface of the VLP, and a protruding domain (P-domain) that emanates from the S-domain surface. The P-domain is further divided into subdomains, P1 and P2 (red and blue, respectively) with the P2-subdomain at the most distal surface of the VLPs. Reproduced, with permission, from Prasad et al., 1999 and Bertolotti-Ciarlet et al., 2002.

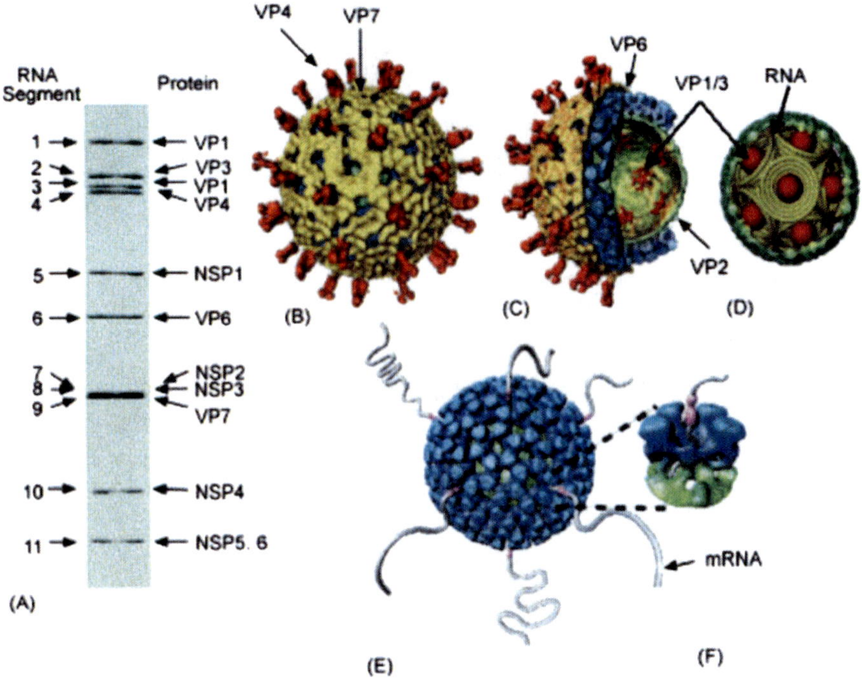

Figure. 3.3 Architectural features of rotavirus. (A) PAGE gel showing, 11 dsRNA segments comprising the rotavirus genome. The gene segments are numbered on the left and the proteins they encode are indicated on the right. (B) Cryo-EM reconstruction of the rotavirus triple-layered particle. The spike protein VP4 is colored in orange and the outermost VP7 layer in yellow. (C) A cutaway view of the rotavirus TLP showing the inner VP6 (blue) and VP2 (green) layers and the transcriptional enzymes (shown in red) anchored to the VP2 layer at the five-fold axes. (D) Schematic depiction of genome organization in rotavirus. The genome segments are represented as inverted conical spirals surrounding the transcription enzymes (shown as red balls) inside the VP2 layer in green. (E and F) Model from cryo-EM reconstruction of transcribing DLPs. The endogenous transcription results in the simultaneous release of the transcribed mRNA from channels located at the five-fold vertex of the icosahedral DLP. Reproduced, with permission, from Jayaram et al., 2004.

continuous surface of the VLP, and a protruding (P) domain that forms the protrusions (Fig. 3.2). A flexible hinge of eight amino acids links the S and P domains. The P domain is located at the exterior of the capsid and is likely to contain determinants of genotype specificity. The NH_2-terminal (N) arm, located within the S domain, consists of residues 10 to 49 and faces the interior of the capsid. The part of the S domain that forms a β-barrel consists of amino acids 50 to 225. The entire S domain (amino acids 1 to 225) corresponds to the N-terminal region of the capsid protein that is relatively conserved among noroviruses in sequence comparisons.

Amino acid residues 226 to 530 form the P domain, which corresponds to the C-terminal half of the capsid protein and forms the arch-like structures extending from the shell. The S domain is required for the assembly of the capsid and participates in multiple intermolecular interactions of dimers, trimers, and pentamers, whereas the P domain is mainly involved in dimeric interactions (Prasad et al., 1999). The P domain is considered to have subdomains, P1 formed by amino acids 226 to 278 and 406 to 530, and P2 encompassing amino acids 279 to 405. The P2 subdomain is the most variable region of the capsid protein among noroviruses (Hardy et al., 1996) and is believed to play an important role in immune recognition and receptor interaction. The isolated P domain forms dimers and binds to histo-blood group antigen receptors, not requiring the formation of VLPs (Tan et al., 2004). In addition, a binding pocket in the P domain is responsible for viral receptor binding, and formation of this pocket involves only intramolecular interactions (Tan et al., 2003). Prasad et al. (1999) have demonstrated that, although the S domain has a canonical eight-stranded β-barrel structure, the P2 subdomain in the P domain has a fold similar to that observed in domain 2 of the elongation factor Tu (EF-Tu), a structure never before seen in a viral capsid protein. Moreover, the fold of the P1 subdomain is unlike any other polypeptide observed so far (Bertolotti-Ciarlet et al., 2003).

A comparison of the capsid proteins from various caliciviruses reveals significant variations both in their sequences and in their sizes. In general, the capsid proteins of human caliciviruses are smaller than those of animal caliciviruses (Chen et al., 2004). The Norwalk virus recombinant capsid protein can also self-assemble into smaller VLPs (23 nm) with suspected $T = 1$ symmetry that is thought to be composed of 60 copies of the capsid protein (White et al., 1997).

It has been postulated that the N-terminal region of the capsid protein or the genomic RNA acts as a switching region that controls variations in the conformation of the coat protein of $T = 3$ viruses (Rossmann and Johnson, 1989). Norovirus particles are different from other $T = 3$ viruses because recombinant capsid protein readily forms rVLPs without RNA (Prasad et al., 1999). It has been suggested that the determinants for the $T = 3$ capsid assembly for Norwalk virus may lie outside of the N-terminus and that the interaction between subunits B and C is not mandatory for the formation of the capsid (Bertolotti-Ciarlet et al., 2002).

Caliciviruses have a linear, single-stranded, positive-sense RNA genome of 7.5–7.7 kb (Green et al., 2001) (see Fig. 3.1). The RNA genome of Norwalk virus, the prototype strain for the genus Norovirus, is 7,654 nucleotides in length and is polyadenylated at the 3′ end (Jiang et al., 1993b). The lack of a cap structure typical of eukaryotic mRNA and the absence of an internal ribosomal entry site suggest that the VPg protein may play a role in the initiation of protein synthesis on calicivirus RNA through unique protein-protein interactions with the cellular translation machinery (Daughenbaugh et al., 2003). Removal of VPg from calicivirus RNA by proteinase K digestion results in loss of infectivity and dramatically reduces translation of FCV RNA *in vitro* (Herbert et al., 1997). The genome of calicivirus is organized into two or three major ORFs. The nonstructural proteins encoded in the calicivirus ORF1 were predicted on the basis of their sequence similarities with picornavirus nonstructural proteins (Neill, 1990). Amino acid sequence motifs in common with the poliovirus 2C NTPase, 3C protease, and 3D RNA-dependent RNA polymerase (RdRp) were readily identified and provided templates for further characterization of the calicivirus nonstructural proteins. Proteolytic mapping and enzymatic studies of *in vitro*–translated polyprotein or recombinant protein expression have confirmed the presence of an NTPase (p41), a 3C-like protease ($3CL^{pro}$), an RdRp, and the location in the polyprotein of the genome-linked protein VPg (Liu et al., 1996; Dunham et al., 1998; Pfister and Wimmer, 2001). The proposed six nonstructural proteins encoded in the norovirus ORF1 defined so far proceed N to C terminus, p48-NTPase-p22-VPg-$3CL^{pro}$-RdRp (Ettayebi and Hardy, 2003). It has been recently reported that the 3C-like proteinase ($3CL^{pro}$) inhibits host cell translation by cleavage of poly(A)-binding protein (PABP), a key protein involved in the translation of polyadenylated mRNAs (Kuyumcu-Martinez et al., 2004).

In the genera Norovirus and Vesivirus, the capsid structural protein VP1 is encoded in a separate ORF (ORF2), whereas viruses in the genera Sapovirus and Lagovirus encode the capsid protein contiguous with the large nonstructural polyprotein (ORF1) (see Fig. 3.1). Viruses in the latter two genera have two major ORFs (ORF1 and ORF2). In the genus Sapovirus, the ORF1/2 junction consists of a one- or four-nucleotide overlap between the stop codon of ORF1 and the first AUG codon of ORF2. A third ORF has been described in certain strains that overlaps the capsid protein gene in a +1 frameshift, which is not found in the Norovirus genome (Liu et al., 1995; Guo et al., 1999b; Clarke and Lambden, 2000). The presence of a conserved translation initiation motif GCAAUGG at the 5′ end of this overlapping ORF suggests that a biologically active protein may be encoded in this ORF (Schuffenecker et al., 2001). Viruses in the genera Norovirus and Vesivirus have three major ORFs (ORF1, ORF2, and ORF3). In noroviruses, the first and third ORFs are in the same reading frame and Norwalk virus ORF3 encodes a 212-amino-acid minor structural protein of the virion (Glass et al., 2000a). All calicivirus genomes begin with a 5′-end terminal GU. This 5′ end sequence is repeated internally in the genome and most likely corresponds to the beginning of a subgenomic-sized RNA transcript (2.2–2.4 kb) that is coterminal with the 3′ end of the genome and that has been observed in FCV-

and RHDV-infected cells as well as packaged into virions (Herbert et al., 1996). A comparison of the 5′-end sequences of representative viruses within each of the four genera and the corresponding repeated internal sequences suggests that this is a characteristic feature of caliciviruses. The synthesis of a subgenomic RNA in calicivirus-infected cells is a major difference between the replication strategy of caliciviruses and picornaviruses, although several of the replicative enzymes share distant homology (Green et al., 2001). It is not clear whether caliciviruses have a picornavirus-like internal ribosomal entry site (IRES) for the initiation of translation but is considered unlikely because translation of the FCV capsid precursor protein begins at the first AUG of the subgenomic RNA (Sosnovtsev and Green, 2003).

1.2. Molecular Diversity of Caliciviruses

Early studies demostrating the great variability of noroviruses soon led to the belief that it was important to distinguish between strains to better understand their epidemiology. Because no antigenic analysis of norovirus strains was available due to the lack of immune reagents, genome characterization by sequence analysis was used to provide an interim system of genotyping (Koopmans et al., 2003). As the genotypes ideally would correlate with serotypes, the sequence of the major structural protein gene was used as the basis for phylogenetic analyses (Ando et al., 2000; Koopmans et al., 2001). To determine which areas of the genome should be analyzed, sequenced regions in the RNA-dependent RNA polymerase, the capsid gene, and part of the 3′ ORF were compiled and analyzed (Green et al., 2000a; Vinje et al., 2000). By a similar approach, Ando et al. (2000) proposed a numerical system for genotypes based on phylogenetic grouping according to genetic relatedness in the major capsid protein.

Noroviruses were classified into two genetic groups, GI and GII, on the basis of sequence homologies across highly conserved regions of the genome, such as the RNA-dependent RNA polymerase and the capsid gene, although subdivision into five genetic groups or genogroups has been proposed recently (Karst et al., 2003) (Table 3.1). Molecular characterization of bovine enteric caliciviruses suggests that they be included into a proposed GIII group, which so far contains only viruses found in cattle (Ando et al., 2000; Oliver et al., 2003; Smiley et al., 2003). The porcine noroviruses cluster within GII (Sugieda and Nakajima, 2002). Phylogenetic analysis places at least two human noroviruses within a proposed genogroup IV: strains Alphatron (GenBank accession no. AF195847) and Ft. Lauderdale (GenBank accession no. AF414426) (Fankhauser et al., 2002).

The recently described murine norovirus has been included into a proposed GV group, the members of which are closer to GII than those of GI by sequence alignment (Karst et al., 2003). In the major capsid protein VP1, human norovirus strains within the same genogroup share at least 60% amino acid sequence identity, whereas most GI strains share less than 50% amino acid identity with GII strains (Green et al., 2001; Koopmans et al., 2003). Within genogroups, noroviruses can be further divided into at least 22

Table 3.1 Current genogroups and genotypes of noroviruses [adapted from Green et al. (2001) and Koopmans et al. (2003)]

Genogroup	*Genotype*	*Prototype strain*	*GenBank accession number (capsid gene)*	*Other strains*
I	1	Norwalk/1968/UK	M87661	KY/89/JP
	2	Southampton/1991/UK	L07418	White Rose, Crawley
	3	Desert Shield/395/1990/SA	U04469	Birmingham 291
	4	Chiba 407/1987/JP	D38547	Thistle Hall; Valetta, Malta
	5	Musgrove/1989/UK	AJ277614	Butlins
	6	Hesse 3/1997/GE	AF093797	Sindlesham, Mikkeli, Lord Harris
	7	Winchester/1994/UK	AJ277609	Lwymontley
II	1	Hawaii/1971/US	U07611	Wortley, Girlington
	2	Melksham/1994/UK	X81879	Snow Mountain
	3	Toronto 24/1991/CA	U02030	Mexico, Auckland, Rotterdam
	4	Bristol/1993/UK	X76716	Lordsdale, Camberwell, Grimsby
	5	Hillingdon/1990/UK	AJ277607	White river
	6	Seacroft/1990/UK	AJ277620	
	7	Leeds/1990/UK	AJ277608	Gwynedd, Venlo, Creche
	8	Amsterdam/1998/NL	AF195848	
III		Jena	AJ011099	Bovine strains
IV	1	Alphatron/1998/NL	AF195847	Ft. Lauderdale
V		MNV-1	AY228235	Murine norovirus 1

genetic clusters of genotypes based on genetic homology in the complete ORF2 sequence (Koopmans et al., 2000; Green et al., 2001; Vinje et al., 2004). Within these broad groupings there are many different genetic variants of noroviruses circulating in the community (Jiang et al., 1996; Green et al., 2000a; Buesa et al., 2002; Lopman et al., 2003, 2004). Picornavirus serotypes generally have >85% amino acid identity across the VP1 gene, which is in the range of the cutoff for calicivirus genotypes (>80% amino acid identity) (Oberste et al., 1999).

Serologically, the norovirus genotypes have been classified into different antigenic groups on the basis of solid-phase immuneelectron microscopy (SPIEM) studies (Vinje et al., 2000) or by enzyme immunoassays using type-specific antibodies generated against recombinant capsid proteins (Atmar and Estes, 2001). The molecular diagnosis of noroviruses relies mainly on a relatively small sequence in the RNA polymerase region (Ando et al., 1995; Vinje and Koopmans, 1996; Jiang et al., 1999b) or the 3′ end of ORF1 (Fankhauser et al., 2002). Although robust for generic norovirus detection, RT-PCR (reverse transcription–polymerase chain reaction) assays targeting small regions of the genome that do not code for structural viral proteins will neither reflect the epitopes of the viruses important for attachment to the host cell nor correlate with their antigenic properties. Therefore, neither RNA polymerase nor 3′ end of ORF1 is suitable for genotyping, although, in combination with capsid typing, they may become important for identifying norovirus recombinants (Vinje et al., 2000; Vinje et al., 2004).

The Sapovirus capsid gene is fused to and is in frame with the polyprotein gene (Lambden et al., 1994; Numata et al., 1997). All Sapovirus strains except a human strain, London/92, and a porcine strain, PEC Cowden, contain an additional ORF predicted in +1 frame, overlapping the N terminus of the capsid gene (Guo et al., 1999b; Jiang et al., 1999a; Clarke and Lambden, 2001). By analogy with noroviruses, the sapoviruses were previously divided into four genotypes, belonging to two genogroups (Liu et al., 1995; Noel et al., 1997; Clarke and Lambden, 2000). More recently, human sapoviruses have been classified into four or five genogroups (Schuffenecker et al., 2001; Farkas et al., 2004) (Table 3.2). In addition, the porcine enteric calicivirus (PEC) strain Cowden has been shown to be related to the Sapovirus genus and to belong to a differentiated cluster (Guo et al., 1999b). Based on capsid sequences, it has recently been proposed to classify the currently known sapoviruses into nine genetic clusters within five genogroups, including one genogroup represented by the PEC strain Cowden (Farkas et al., 2004).

The occurrence of recombination among human caliciviruses in nature was suspected when it was found that the capsid nucleotide sequences of Snow Mountain virus and Melksham virus were almost identical (94%) but their RNA polymerase sequences were significantly different (79%). When Melksham virus was compared with Mexico virus and the Japanese strain

Table 3.2 Current genogroups and genotypes of sapoviruses [adapted from Schuffenecker et al. (2001) and Farkas et al. (2004)]

Genogroup/cluster	*Prototype strain*	*GenBank accession number (capsid gene)*	*Other strains*
GI/1	Sapporo/82	U65427	Houston/86, Plymouth/92, Manchester/93, Lyon30388/98
GI/2	Parkville/94	U73124	Houston/90
GI/3	Stockholm/97	AF194182	Mexico14917/00
GII/1	London/92	U95645	Lyon/598/97
GII/2	Mexico340/90	AF435812	
GII/3	Cruise ship/00	AY289804	
GIII	PEC/Cowden	AF182760	
GIV	Houston7-1181/90	AF435814	
GV	Argentina39	AY289803	

Oth-25, no significant difference was found in sequence identities of the capsid and RNA polymerase regions and hence Snow Mountain virus was considered a recombinant virus (Hardy et al., 1997). Caliciviruses contain subgenomic RNA that covers the entire 3′ end of the genome from the capsid gene to the 3′ end (Fig. 3.1). It has been hypothesized that the subgenomic RNA could act as an independent unit participating in the recombination events. It remains unknown how subgenomic RNA is involved in virus replication, but it is clear that both genomic and subgenomic RNAs share highly conserved 5′ end sequences and that both RNAs are assembled into virions. It is possible that interaction between genomic and subgenomic RNAs occurs by the same mechanism as that of genomic-genomic interaction, thereby significantly increasing the chance of recombination events (Jiang et al., 1999a). If RNA recombination is a common phenomenon among caliciviruses, a high diversity might be expected in Caliciviridae, which would facilitate the emergence of new variants and make genotyping more difficult. In addition, recombination may permit caliciviruses to escape host immunity, analogous to antigenic shifts in influenza viruses, but by a different molecular mechanism. It has also been reported that accumulation of mutations in the protruding P2 domain of the capsid protein may result in predicted structural changes, including disappearance of helix structure of the protein, and thus a possible emergence of new phenotypes (Nilsson et al., 2003).

1.3. Virus Replication

Studies on the replication strategy of human caliciviruses have been hindered by the lack of an efficient cell culture system. Nevertheless, expression of recombinant proteins from cDNA clones has allowed the generation of proteolytic processing maps for the nonstructural proteins of several caliciviruses, like Southampton virus (a norovirus) and RHDV (a lagovirus) (Liu et al., 1996; Wirblich et al., 1996). Analysis of individual recombinant proteins from these noncultivable caliciviruses allowed the identification of NTPase and 3C-like cysteine protease activities for RHDV and noroviruses (Liu et al., 1999b) and a 3D-like RNA-dependent RNA polymerase for RHDV (López-Vazquez et al., 1998). Studies on the replication mechanisms of cultivable caliciviruses, such as feline calicivirus (FCV), have contributed to a better understanding of the basic features of calicivirus replication (Sosnovtsev and Green, 2003). FCV replicates by producing two major types of polyadenylated RNAs: a positive-sense genomic RNA of approximately 7.7 kb and a subgenomic RNA of 2.4 kb (Herbert et al., 1996). The genomic RNA serves as a template for the synthesis of nonstructural protein encoded by ORF1, and the subgenomic RNA serves as a template for the translation of structural proteins (Carter, 1990).

1.4. Virus-Cell Interactions

Human and animal enteric caliciviruses are assumed to replicate in the upper intestinal tract, causing cytolytic infection in the villous enterocytes but not in the crypt enterocytes of the proximal small intestine. Biopsies of the jejunum taken from experimentally infected volunteers who developed gastrointestinal disease after oral administration of Norwalk, Montgomery County, or Hawaii viruses showed histopathologic lesions, consisting of blunting of the villi, crypt cell hyperplasia, infiltration with mononuclear cells, and cytoplasmic vacuolization (Blacklow et al., 1972; Dolin et al., 1975). Experiments with recombinant Norwalk VLPs and human gastrointestinal biopsies showed binding of the rVLPs to epithelial cells of the pyloric region of the stomach and to enterocytes on duodenal villi. Attachment of these rVLPs occurred only to cells as well as to saliva from histo-blood group antigen-secreting individuals (Marionneau et al., 2002). It was previously known that RHDV attaches to H type 2 histo-blood group oligosaccharide present on rabbit epithelial cells (Ruvöen-Clouet et al., 2000). Significant attachment and entry of Norwalk rVLPs to differentiated Caco-2 cells has also been demonstrated (White et al., 1996). Differentiated Caco-2 cells resemble mature enterocytes, express the H antigen, and were derived from a group O individual (Amano and Oshima, 1999).

To date, the Cowden strain of porcine enteric calicivirus (PEC) is the only cultivable enteric calicivirus (Flynn and Saif, 1988). However, in order to replicate, it requires the incorporation of an intestinal content preparation (ICP) from uninfected gnotobiotic pigs as a medium supplement. Different porcine intestinal enzymes such as trypsin, pancreatin, alkaline phosphatase, enterokinase, elastase, protease, and lipase were tested as medium supplements, but none of them alone promoted the growth of PEC in cell cultures (Parwani et al., 1991). It was speculated that some enzymes or factors in porcine ICP could activate the viral receptor, promote signaling of host cells, or may cleave viral capsid for successful uncoating (Guo and Saif, 2003).

Although noroviruses are highly infectious (it has been estimated that less than 10 virions may be enough to infect an adult), studies with volunteers have shown that some subjects remain uninfected despite having been challenged with high infectious doses (Matsui and Greenberg, 2000). These individuals may remain disease-free because of innate resistance or because of preexisting immunity to the virus (Lindesmith et al., 2003). An increased risk of Norwalk virus infection has been associated with blood group O; Norwalk VLPs bind to gastroduodenal cells from individuals who are secretors (Se+) but not to those from nonsecretors (Se−) (Hutson et al., 2002; Marionneau et al., 2002). The gene responsible for the secretor phenotype, *FUT2*, encodes an $\alpha(1,2)$ fucosyltransferase that produces the carbohydrate H type 1 found on the surface of epithelial cells and in mucosal secretions (Lindesmith et al., 2003). The form of H type 1 secreted depends on additional glycosyltransferases, including the Lewis, A and B enzymes present in epithelial and red blood cells (Marionneau et al., 2001). The recent discovery that noroviruses attach to cells in the gut only if the individuals express specific, genetically determined carbohydrates is a breakthrough in understanding norovirus-host interactions and susceptibility to norovirus disease (Lindesmith et al., 2003; Hutson et al., 2004).

Recent studies also suggest that some animal caliviviruses may cross the species barrier and potentially infect humans. The hypothetical existence of animal reservoirs and the possibility of interspecies transmission have been suggested by phylogenetic links of bovine and porcine viruses within the genera Norovirus and Sapovirus, respectively (Clarke and Lambden, 1997; Dastjerdi et al., 1999; Liu et al., 1999a; van Der Poel et al., 2000). However, information concerning the frequency of interspecies transmission among caliciviruses is limited.

It has recently been demonstrated that gnotobiotic pigs inoculated by the intravenous route with wild-type PEC Cowden developed diarrhea and also presented histological lesions in the duodenum and jejunum similar to those observed in gnotobiotic pigs inoculated orally with wild-type PEC (Guo et al., 2001). Moreover, PEC RNA and low titers of virus antigen were detected in sera from pigs inoculated with wild-type PEC both orally and intravenously (Guo and Saif, 2003). Thus, viremia may occur after natural PEC infection, and acute sera may contain infectious virus. It has been hypothesized that human calicivirus infections might also induce viremia, interacting with erythrocytes in the blood (Guo and Saif, 2003). If this is true, how the enteric caliciviruses reach the bloodstream from the gut and vice versa remains unexplained.

2.0. ROTAVIRUSES

Rotaviruses are the leading etiologic agents of viral gastroenteritis in infants and young children worldwide and in the young of a large variety of animal species (Kapikian, 2001). Rotavirus infections in humans continue to occur throughout their lives, but the resulting disease is mild and often asymptomatic (Bishop, 1996). In addition to sporadic cases of acute gastroenteritis, outbreaks of rotavirus diarrhea in school-aged children and adults have increasingly been reported (Griffin et al., 2002; Mikami et al., 2004).

Rotaviruses are responsible for an estimated 500,000 deaths each year in developing countries (Parashar et al., 2003). There is an urgent need to develop an effective vaccine and to implement therapeutic strategies to prevent and treat these infections. A better understanding of the molecular mechanisms of rotavirus replication and interaction with the host cell and of antigenic variability of these viruses are fundamental to achieving these goals.

Rotaviruses are classified into at least five groups (A to E), and there are possibly two more groups (F and G) based on the reactivities of the VP6 middle layer protein (Estes, 2001). Group A rotaviruses are most commonly associated with human infections (Kapikian, 2001). Within group A, four subgroups (I, II, I + II, and non-I/non-II) are recognized based on the reactivity of VP6 with two monoclonal antibodies. The two outer capsid proteins, VP7 and VP4, form the basis of the current dual classification system of group A rotaviruses into G and P types (Estes, 2001). At least 14 different G-serotypes and 20 P-types have been identified among human and animal rotaviruses, depending on VP7 and VP4, respectively. G serotypes correlate fully with G genotypes as determined by sequence analysis of the VP7 gene. However, to date only 12 of the 20 P genotypes have been correlated with P serotypes (Estes, 2001). Because VP4 and VP7 are coded for by different RNA segments (segment 4 and segments 7–9, respectively), various combinations of G- and P-types can be observed both in humans and in animals. Viruses carrying G1P[8], G2[P4], G3[P8], and G4[P8] represent more than 90% of human rotavirus strains cocirculating in most countries, although other G and P combinations are being isolated in increasing numbers (Cunliffe et al., 1999; Buesa, 2000; Iturriza-Gomara et al., 2000; Adah et al., 2001; Armah et al., 2003; Zhou et al., 2003).

Group B rotaviruses are common animal pathogens infecting pigs, cows, sheep, and rats but have also been found to infect humans (Sanekata et al., 2003). Group C rotaviruses are commonly found in animals including pigs and dogs and can cause outbreaks in humans (Castello et al., 2000).

2.1. Structure of the Virion

Rotavirus particles are of icosahedral symmetry, have three concentric layers of protein, and measure ~1,000 Å in diameter including the spikes (Estes, 2001). The core layer is formed by a protein known as VP2 (which surrounds the entire viral genome) and proteins VP1 and VP3, which are transcription enzymes and are attached as a heterodimeric complex to the inside surface of VP2 at the fivefold symmetry positions (Prasad et al., 1996b) (Fig. 3.3). VP1, the RNA-dependent RNA polymerase, interacts with VP3, guanylytransferase, and methylase (Liu et al., 1992). This innermost layer is composed of 120 copies of VP2, a RNA-binding protein (Labbé et al., 1991). The addition of VP6 to the VP2 layer produces double-layered particles (DLP). VP6 forms 260 trimers interrupted by 132 aqueous channels of three different kinds in relation to the capsid's symmetry. The outer capsid of the triple-layered particles (TLP) is composed of two proteins, VP7 and VP4.

The smooth surface of the virus is made up of 260 trimers of VP7, and 60 spikes emerging from the viral surface consist of dimers of VP4 (Prasad et al., 1988; Yeager et al., 1990). Rotavirus DLP and TLP contain 132 porous channels that allow the exchange of compounds to the inside of the particle. There are 12 type I channels, each located at the icosahedral fivefold vertices of the TLP and DLP. Each of the type I channels is surrounded by five type II channels, and finally 60 type III channels are placed at the hexavalent positions immediately neighboring the icosahedral threefold axis (Prasad et al., 1988; Yeager et al., 1990).

During cell entry, the TLP loses the VP7 and VP4 proteins, and the resulting DLP becomes transcriptionally active inside the cytoplasm (Estes, 2001). VP4 is a nonglycosylated protein of 776 amino acids and has essential functions in the virus cell cycle, including receptor binding, cell penetration, hemagglutinin activity, and permeabilization of cellular membranes (Fiore et al., 1991; Burke and Desselberger, 1996; Gilbert and Greenberg, 1998;

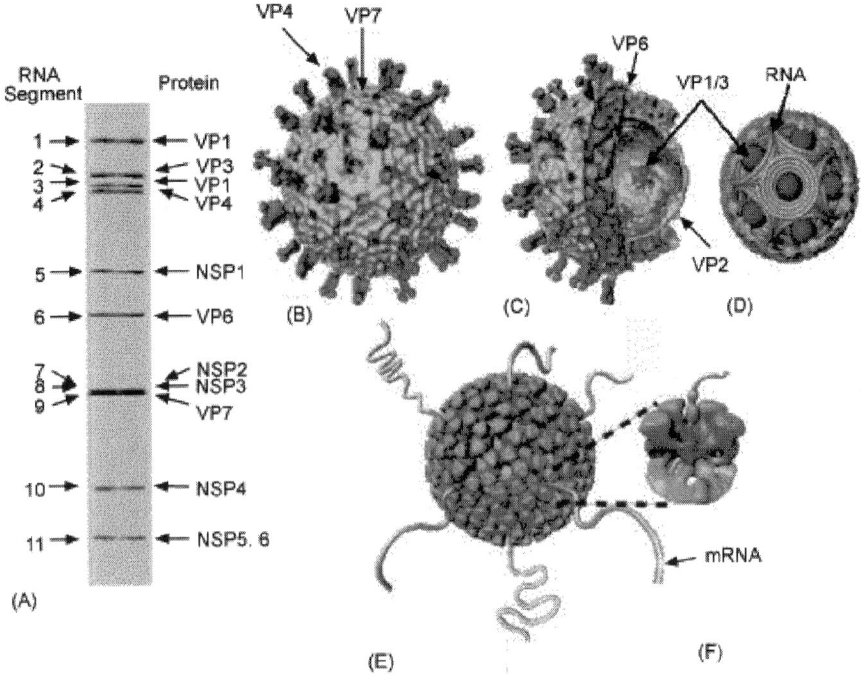

Figure. 3.3 Architectural features of rotavirus. (A) PAGE gel showing 11 dsRNA segments composing the rotavirus genome. The gene segments are numbered on the left and the proteins they encode are indicated on the right. (B) Cryo-EM reconstruction of the rotavirus triple-layered particle. The spike proteins VP4 are colored in orange and the outermost VP7 layer in yellow. (C) A cutaway view of the rotavirus TLP showing the inner VP6 (blue) and VP2 (green) layers and the transcriptional enzymes (shown in red) anchored to the VP2 layer at the fivefold axes. (D) Schematic depiction of genome organization in rotavirus. The genome segments are represented as inverted conical spirals surrounding the transcription enzymes (shown as red balls) inside the VP2 layer in green. (E and F) Model from cryo-EM reconstruction of transcribing DLPs. The endogenous transcription results in the simultaneous release of the transcribed mRNA from channels located at the fivefold vertex of the icosahedral DLP. (Reproduced with permission from Jayaram et al., 2004; see color insert.)

Denisova et al., 1999; Zárate et al., 2000b). VP4 is post-transcriptionally cleaved into the larger VP5* and the smaller VP8* subunits, and cleavage of VP4 enhances viral infectivity by several-fold. It has been shown that trypsin cleavage confers icosahedral ordering on the VP4 spikes, which is essential for the virus to enter the cell (Crawford et al., 2001). Moreover, biochemical studies of recombinant VP4 indicate that proteolysis of monomeric VP4 yields dimeric VP5* (Dormitzer et al., 2001).

VP7 is a calcium-binding glycoprotein of 326 amino acids, with nine variable regions contributing to type-specificity (Nishikawa et al., 1989; Hoshino et al., 1994). In addition, VP7 interacts with integrins αxβ2 and α4β1 (Coulson et al., 1997; Hewish et al., 2000) and induces polyclonal intestinal B-cell activation during rotavirus infection (Blutt et al., 2004).

When the DLP is located intracellularly, it becomes transcriptionally competent, and new mRNA transcripts are translocated from the particle through type I channels at the fivefold axes (Lawton et al., 1997). Prasad et al. (1988) were the first to propose that these channels in the VP6 layer could be used by mRNA to exit. Cryo-EM studies have confirmed that DLPs maintain their structural integrity during the process of transcription. In a pseudo-atomic model of the T = 13 VP6 layer (Mathieu et al., 2001), a β-hairpin motif of VP6 with a highly conserved sequence that protrudes into the mRNA exit channel may play a functional role in the translocation of the nascent transcripts (Lawton et al., 2000). A detailed mutational analysis of the VP6 layer has helped to elucidate the determinants of VP6 required for its assembly on VP2 and how VP6 may affect endogenous transcription (Charpilienne et al., 2002).

2.2. The Genome

The genome of rotavirus (simian rotavirus strain SA11) consists of 11 segments of double-stranded RNA (dsRNA) with conserved 5′ and 3′ ends, ranging from 667 bp (segment 11) to 3,302 bp (segment 1) in size, totaling 6,120 kDa. Six structural proteins (VP1, VP2, VP3, VP4, VP6, VP7) and six nonstructural proteins (NSP1 to NSP6) are encoded. The coding assignments, functions, and many properties of the proteins encoded by each of the 11 genome segments are now well established (Table 3.3). Protein assignments have been determined by *in vitro* translation using mRNA or denatured dsRNA and by analyses of reassortant viruses (Estes and Cohen, 1989; Estes, 2001). Cryo-EM analysis of rotavirus has provided first visualization of the structural organization of the viral genome (Prasad et al., 1996b). The dsRNA forms a dodecahedral structure in which the RNA double helices, interacting closely with the VP2 inner layer, are packed around the transcription complexes located at the icosahedral vertices (Fig. 3.3D). VP2 has RNA binding properties and may be responsible for the icosahedral ordering and for the closely interacting portions of the RNA. A study by Pesavento et al. (2001) demonstrated reversible condensation and expansion of the rotavirus genome within the capsid interior under various chemical conditions. This condensation is concentric with respect to the particle center, and a dark mass of density is consistently seen in the center of each of the particles. At high pH in the presence of ammonium ions, the genome condenses to a radius of 180 Å from the original radius of 220 Å, and when brought back to physiological pH the genome expands to its original radius (Pesavento et al., 2001). This study suggests that VP2, through its RNA-binding properties, plays an important role in maintaining appropriate spacing between the RNA strands in the native expanded state. A plausible and elegant model for the structural organization of the genome that emerges from these studies is that each dsRNA segment is spooled around a transcription enzyme complex at the fivefold axes, inside the innermost capsid layer (Prasad et al., 1996b; Pesavento et al., 2001, 2003).

2.3. Mechanisms of Evolution and Strain Diversity

Rotaviruses, like many other RNA viruses, display a great degree of diversity. As well as showing different G and P types and a variety of combinations, there is also intratypic variation. Three mechanisms, singly or in combination, are important for the evolution and diversity of rotaviruses, although it is yet unclear what their relative contributions are to the burden of the disease.

2.3.1. Antigenic Drift

The rate of mutation within rotavirus genes is relatively high because RNA replication is error-prone. The mutation rate has been calculated to be $<5 \times 10^{-5}$ per nucleotide per replication, which implies that on average a rotavirus genome differs from its parental genome by at least one mutation (Blackhall et al., 1996). Point mutations can accumulate and give rise to intratypic variation as identified by the existence of lineages within the VP7 and VP4 genes of particular G and P types. Point mutations can lead to antigenic changes, which may result in the emergence of escape mutants (Palombo et al., 1993; Cunliffe et al., 1997; Maunula and von Bonsdorf, 1998; Iturriza-Gómara et al., 2002).

2.3.2. Antigenic Shift

Shuffling of gene segments through reassortment can occur during dual infection of one cell. Reassortment can therefore contribute to the diversity of rotaviruses, and there is increasing evidence that reassortment takes place *in vivo* (Ramig, 2000). There is evidence that reassortment, through alteration in protein-protein interactions, possibly leads to changes in conformational epitopes and may contribute to the evolution of antigenic types (Chen et al., 1992; Lazdins et al., 1995). Interspecies transmission and subsequent reassortment have enormous potential to increase the diversity of cocirculating rotaviruses. In addition, human rotavirus genes encoding proteins to which the human population is immunologically naïve may allow a rapid spread of the reassortant strain (Iturriza-Gomara et al., 2000).

2.3.3. Gene Rearrangement

The concatemerization or truncation of genome segments and their ORFs has the potential to contribute to the evolution of rotaviruses through the production of new proteins with altered functions (Desselberger, 1996).

2.4. Genome Replication

The RNA polymerase activity of DLPs catalyzes synthesis of 11 mRNAs of rotaviruses, which range in size from ~0.7 to 3.3 kb. With the exception of gene 11, they are monocistronic (Estes, 2001). The nascent transcripts are extruded through channels present at the fivefold axes of the DLPs (Lawton et al., 1997). The 5′- and 3′-untranslated regions (UTRs) are 9 to 49 and 17 to 182 nucleotides in length, respectively. The viral mRNAs serve as templates for the synthesis of minus-strand RNA to form dsRNA molecules (Chen et al., 1994). The synthesis of viral dsRNA and the assembly of cores and DLPs occur in viroplasms in the cytoplasm (Estes, 2001). There are three species of RNA-containing replication intermediates (RIs) in the infected cells: pre-core RIs, which contain the structural proteins VP1 and VP3; core-RIs, with VP1, VP2, and VP3; and double-layered RIs, which contain VP1, VP2, VP3, and VP6. The 11 genomic segments are produced and packaged in equimolar amounts in rotavirus-infected cells, demonstrating that RNA packaging and replication are coordinated processes (Patton and Gallegos, 1990). The absence of naked dsRNA in infected cells suggests that packaging takes place before replication (Patton et al., 2003).

The only primary sequences that are conserved among the rotavirus mRNAs are located within the UTRs. Because all 11 mRNAs are replicated by the same VP1-VP2-VP3 polymerase complex, the viral mRNAs almost certainly must share common *cis*-acting signals recognized by the polymerase (Patton et al., 2003). The most remarkable feature of the 3′ ends of rotavirus mRNAs and of other members of the Reoviridae is the absence of a poly(A) tail. Instead, all rotavirus genes and mRNAs end with the same short sequence UGACC, which is conserved among all the segments in group A rotaviruses. Site-specific mutagenesis has revealed that it is the 3′-CC of the 3′ consensus sequence that is most critical for minus-strand synthesis (Chen et al., 2001).

In addition, it has been demostrated that the promoter for minus-strand synthesis is formed by base-pairing in *cis* of complementary sequences proximal to the 5′- and 3′-ends of the viral mRNAs (Chen and Patton, 1998). The 3′-consensus sequence also contains a *cis*-acting signal that acts as a translation enhancer, whose activity is mediated by NSP3 that specifically recognizes the last four to five nucleotides of the 3′-consensus sequence (Poncet et al., 1994). NSP3 also interacts with the initiation factor eIF4GI and facilitates the circularization of viral mRNAs in polysomes, thus increasing the efficiency of viral gene expression (Piron et al., 1998).

The development of a cell-free system that supports the synthesis of dsRNA from exogenous mRNA represents an important milestone in the study of rotavirus replication, providing a mechanism by which the elements in viral mRNAs that promote minus-strand synthesis can be analyzed. This system is based in virion-derived cores that have been disrupted or "opened" by incubation in hypotonic buffer (Chen et al., 1994). Although several investigators have tried to develop a reverse genetics system for rotaviruses, this goal has not been achieved so far (Patton et al., 2003). Our understanding of even the most basic mechanisms that occur during rotavirus replication would be greatly enhanced by the development of the reverse genetics approach.

2.5. Cell Infection

Rotaviruses have a specific tropism *in vivo* infecting primarily the mature enterocytes of the villi of the small intestine. Recent reports suggest that extraintestinal spread of the virus takes place during infection, indicating a wider tropism than previously considered (Blutt et al., 2003; Mossel and Ramig, 2003). Rotaviruses can bind to a wide variety of cell lines, although only a subset of them (including cells of renal and intestinal origin, and transformed cell lines from breast, stomach, lung, and bone) is efficiently infected (Ciarlet et al., 2002a; López and Arias, 2004). Most studies on rotavirus replication have been carried out using the simian kidney epithelial MA104 cells, which are routinely used to produce progeny virus. However, new investigations into the pathophysiological mechanisms of rotavirus infection are now being performed using *in vitro* polarized cells such as the human intestinal HT-29 and Caco-2 cells (Servin, 2003).

Rotaviruses enter the cell by a complex multistep process in which different domains of the viral surface interact with different cell surface molecules that act as receptors for the attachment of the viruses (Guerrero et al., 2000; Ciarlet and Estes, 2001; López and Arias, 2004). Some animal rotavirus strains interact with sialic acid (SA) residues to attach to the cell surface, and the infectivity of these strains is diminished by the treatment of cells with neuraminidase (NA). By contrast, many animal strains and some human strains are NA-resistant (Ciarlet and Estes, 1999). The interaction of rotavirus with SA has been shown to depend on the VP4 genotype of the virus and not the species of origin (Ciarlet et al., 2002b). Ganglioside GM3 has been suggested to act as the SA-containing receptor for the porcine rotavirus strain OSU (Rolsma et al., 1998), and ganglioside GM1 (NA-resistant) has been identified as the receptor for the NA-resistant human rotavirus strains KUN and MO (Guo et al., 1999a). The VP8* domain of VP4 is involved in interactions with SA, whereas VP5* is implicated in interactions with integrins. The interaction of rotavirus with integrin $\alpha 2\beta 1$ has been shown to be mediated by the DGE integrin-recognition motif, located at amino acids 308 to 310 of VP4, within VP5* (Zárate et al., 2000a). VP4 also contains the tripeptide IDA at amino acids 538 to 540, which is ligand motif for integrin $\alpha 4\beta 1$ (Coulson et al., 1997). However, the functionality of this site has not been demonstrated. Integrin $\alpha v\beta 3$ is also involved in the cell entry of several rotavirus strains at a postattachment step (Graham et al., 2003). Besides, cell-surface heat shock cognate protein hsc70 has also been implicated as a postattachment receptor for both NA-sensitive and NA-resistant rotavirus strains (Guerrero et al., 2002). Studies with polarized epithelial cell lines show that the viral entry of SA-dependent strains is restricted to the apical membrane, whereas SA-independent strains enter either apically or basolaterally (Ciarlet et al., 2001).

It has been suggested that lipid rafts might play an important role in the cell entry of rotavirus (Isa et al., 2004), probably serving as platforms to allow an efficient interaction of cell receptors with the viral particle (Manes et al., 2003; López and Arias, 2004). Rotavirus infection in polarized, fully differentiated Caco-2 cells is followed by a defect in brush-border hydrolase expression (Jourdan et al., 1998). Sucrase-isomaltase activity and apical expression are specifically decreased by rotavirus infection without any apparent cell destruction (Jourdan et al., 1998). In addition, viral infection induces an increase in intracellular calcium concentration, damages the microvillar cytoskeleton, and promotes structural and functional injuries at the tight junctions in cell-cell junctional complexes of cultured Caco-2 cells without damaging the integrity of the monolayers (Brunet et al., 2000; Obert et al., 2000).

2.6. The NSP4 Enterotoxin

The rotavirus nonstructural glycoprotein NSP4 was shown to function as an intracellular receptor that mediates the acquisition of a transient membrane envelope as subviral particles bud into the endoplasmic reticulum (ER). It has been demonstrated that NSP4 binds intracellularly newly made DLPs, interacting with VP6. This receptor role of NSP4 is confirmed by the observation that DLPs bind to ER membranes containing only NSP4 (Taylor et al., 1993; Estes, 2001). Many structural motifs or protein regions have been implicated in the NSP4 biological function. Amino acids 17 to 20 from the C-terminus extreme are necessary and sufficient for inner capsid particle binding (O'Brien et al., 2000) and the region involved in the retention of the NSP4 protein into the endoplasmic reticulum has been mapped between the amino acids 85 and 123 in the cytoplasmatic region of the protein (Mirazimi et al., 2003). Residues at positions 48 to 91, a region that includes a potential cationic amphipatic helix, have been shown to be involved in membrane destabilization (Tian et al., 1996; Browne et al., 2000). Purified NSP4 or a peptide corresponding to NSP4 residues 114–135 induce diarrhea in young mice after an increase in intracellular calcium levels, suggesting a role for NSP4 in rotavirus pathogenesis (Tian et al., 1994; Ball et al., 1996; Horie et al., 1999). *In vitro* studies have shown that after rotavirus replication in cells, a functional 7-kDa peptide of NSP4 (amino acids 112–175) is released into the medium from virus-infected cells by a nonclassic, Golgi-independent cellular secretory pathway (Zhang et al., 2000). This endogenously produced peptide binds to an as yet unidentified apical membrane receptor to mobilize intracellular calcium through phospholipase C signaling.

NSP4, or its active $NSP4_{114-135}$ peptide, induces age-dependent diarrhea and age-dependent chloride permeability changes in mice lacking the cystic fibrosis transductance regulator (CFTR) that exhibit no functional cAMP-dependent secretory pathway (Morris et al., 1999). These observations indicate that NSP4 is a novel secretory agonist, because the classical secretagogues carbachol and the diterpene forskolin that induce chloride changes by activating cyclic adenosine monophosphate, instead of by mobilizing $[Ca^{2+}]_i$ as a secondary mediator, fail to cause disease in CFTR knockout mice (Morris et al., 1999). NSP4 or its active peptide may induce diarrhea in neonatal mice through the activation of an age- and Ca^{2+}-dependent plasma membrane anion permeability distinct from CFTR. The molecular identity of the responsible channel remains to be determined (Morris and Estes, 2001).

It has been suggested that NSP4 may directly inhibit the functioning of the cellular Na^+-dependent glucose transporter SGLT-1 (Halaihel et al., 2000). In addition, extracellular and/or intracellular NSP4 may contribute to diarrheal pathogenesis by altering the dynamics of intracellular actin distribution and intercellular contacts (Ciarlet and Estes, 2001). NSP4 can affect the cytoskeleton in polarized epithelial cells, but how these pleiotropic properties of NSP4 influence the function of NSP4 in morphogenesis or pathogenesis remains unclear (Tafazoli et al., 2001). The significance of immune response to NSP4 in protection against rotavirus infection in humans is still unknown, although it has been shown that NSP4 elicits both humoral and cell-mediated immune responses (Johansen et al., 1999; Ray et al., 2003).

3.0. ASTROVIRUSES

Human astroviruses are nonenveloped viruses with a positive-sense, single-stranded, polyadenylated RNA genome about 6,800 nucleotides in length (Matsui, 1997). Astroviruses are members of the Astroviridae family and were originally described in 1975 in association with outbreaks of gastroenteritis in newborns (Appleton and Higgins, 1975). Astroviruses produce infections mainly in young children, although illness rates increase again in the elderly (Lewis et al., 1989). However, they also cause disease in adults and immunocompromised patients (Cubitt et al., 1999; Coppo et al., 2000; Liste et al., 2000). Persistent gastroenteritis in children with no background disease has also been reported, mainly in association with serotype 3 strains (Caballero et al., 2003). Astrovirus infections occur worldwide, and their incidence ranges from 2% to 9% in both developing and developed countries (Gaggero et al., 1998; Bon et al., 1999; Mustafa et al., 2000; Guix et al., 2002). Astroviruses are transmitted by the fecal-oral route; outbreaks have been associated with consumption of sewage-polluted shellfish and ingestion of water from contaminated sources (Pintó et al., 1996; Pintó et al., 2001). The virus is frequently shed in stools in significant numbers at the onset of the illness, which is in contrast with low numbers of calicivirus produced in the stools of infected patients. Unlike caliviruses, astroviruses replicate *in vitro* in cell lines, and hence detailed studies on their replication are available (Matsui, 1997; Willcocks et al., 1999; Matsui and Greenberg, 2001). Human astroviruses were originally isolated in HEK cells and were subsequently adapted to grow in LLCMK2 cells in trypsin-containing media, although without demonstrable cytopathic effect (CPE). The colonic carcinoma cell line Caco-2, in contrast with LLCMK2 cells, is directly susceptible to fecally derived astrovirus and displays CPE as early as 2 days postinfection (Willcocks et al., 1990; Pinto et al., 1994).

The average diameter of astrovirus particles is approximately 28 nm (Madeley, 1979), although it may vary depending on the source of the virus and the method of preparation for EM (Woode et al., 1984). More detailed ultrastructural analysis of human astrovirus serotype 2 grown in LLCMK2 cells in the presence of trypsin revealed particles with icosahedral symmetry and an array of spikes emanating from the surface (Risco et al., 1995). These particles had an external diameter of 41 nm (including spikes). They did not display the star-like surface characteristically found on fecally shed virus. However, the star-like morphology was inducible after alkaline treatment (pH 10). Intact virions generally band at densities of 1.35 to 1.37 g/ml in CsCl gradients (Caul and Appleton, 1982), although banding at densities of 1.39–1.40 has also been reported (Konno et al., 1982; Midthun et al., 1993). The astroviruses band at a density of 1.32 g/ml in potassium tartrate–glycerol gradient, which preserves better than CsCl the structural integrity of the virus (Ashley and Caul, 1982).

The RNA genomes of several cell culture–adapted human astrovirus strains have been cloned and sequenced (Jiang et al., 1993a; Lewis et al., 1994; Willcocks et al., 1994; Geigenmuller et al., 1997), providing new perspectives for studying the molecular biology of astroviruses. The organization of the genome includes three long open reading frames (ORFs), designated ORF1a, ORF1b, and ORF2. ORF1a (~2700 nt) is located at the 5′ end of the genome and contains transmembrane helices, a 3C-like serine protease motif, a putative protease-dependent cleavage site, and a nuclear localization signal (Jiang et al., 1993a; Willcocks et al., 1994). ORF1b (~1550 nt) contains a RNA-dependent RNA polymerase motif, whereas ORF2 encodes the structural proteins (Matsui and Greenberg, 2001). The nonstructural proteins of the virus are translated from the genomic viral RNA as two polyproteins (Jiang et al., 1993a). One of them contains only ORF1a and the other includes ORF1a/1b and is translated via −1 ribosomal frameshifting event between ORF1a and ORF1b (Jiang et al., 1993a). Both proteins are proteolitically processed, generating a variety of polypeptides, although it is unclear whether the viral protease is responsible for all the cleavages (Geigenmuller et al., 2002a, 2002b; Kiang and Matsui, 2002).

A subgenomic RNA (approximately 2,400 nucleotides) that contains ORF2 is found in abundance in the cytoplasm of astrovirus-infected cultured cells (Monroe et al., 1991). The subgenomic RNA is translated as a 87-kDa capsid precursor protein that is believed to give rise to three to five smaller mature capsid proteins in a process that involves trypsin and a putative cellular protease (Monroe et al., 1991; Bass and Qiu, 2000; Mendez et al., 2002). The 87-kDa capsid protein is rapidly converted intracellularly to a 79-kDa form, which is found in smaller amounts in the cell supernatants. Bass and Qiu (2000) identified a trypsin cleavage site in a highly conserved region of the ORF2 product. Trypsin-free particles were minimally infectious in cultured Caco-2 cells but became highly infectious after trypsin treatment but not chymotrypsin treatment. This trypsin-enhanced infectivity correlated with conversion of the 79-kDa capsid protein to three smaller peptides of approximately 26, 29, and 34 kDa. However, the apparent molecular weight reported for the smaller mature proteins has depended on the astrovirus serotype and whether the virus studied was derived from infected cultured cells or from stools (Matsui and Greenberg, 2001).

Table 3.3 Genome segments and proteins of simian rotavirus SA11: Coding assignments, functions and biological properties of the encoded proteins. Adapted from Estes (2001) and Ramig and Estes (2003)

Genome segment size (bp)	*ORFs*	*Gene product(s)*	*Protein size aa (Da)*	*Location in virus particle*	*Functions and properties*
1 (3,302)	18-3282	VP1	1,088 (125,000)	Inner capsid 5-fold axis	RNA-dependent RNA polymerase Part of minimal replication complex Virus specific 3′-mRNA binding Part of virion transcription complex with VP3
2 (2,690)	17-2659	VP2	881 (102,431)	Inner capsid	Core matrix protein Non-specific ss and dsRNA-binding activity Myristylated Assembly and RNA-binding activity Part of minimal replication complex Leucine zipper
3 (2,591)	50-2554	VP3	835 (98,120)	Inner capsid, 5-fold axis	Guanylyltransferase Methyltransferase Located at the vertices of the core Part of virion transcription complex with VP1 Non-specific ssRNA binding
4 (2,362)	10-2337	VP4	776 (86,782)	Outer capsid spike	VP4 dimers form outer capsid spike Interacts with VP6
	—	VP5*	529 (60,000)		Virus infectivity enhanced by trypsin cleavage of VP4 into VP5* and VP8*
	—	VP8*	247 (28,000)		Hemagglutinin and cell attachment protein P-type neutralization antigen VP5* permeabilizes membranes Protection
5 (1,611)	31-1515	NSP1	495 (58,654)	Nonstructural	Associates with cytoskeleton Extensive sequence diversity between strains Two conserved cysteine-rich zinc-finger motifs Virus specific 5′-mRNA binding Interacts with host IFN regulatory factor 3
6 (1,356)	24-1214	VP6	397 (44,816)	Middle capsid	Major virion protein Middle capsid structural protein Homotrimeric structure Subgroup antigen

					Myristoylated Protection (mechanism?)
7 (1,105)	26-970	NSP3	315 (34,600)	Nonstructural	Homodimer Virus-specific 3′-mRNA binding Binds eIF4G1 and circularizes mRNA on initiation complex Involved in translational regulation and host shut-off
8 (1,059)	47-997	NSP2	317 (36,700)	Nonstructural	NTPase activity Helix destabilization activity Non-specific ssRNA-binding Involved in viroplasm formation with NSP5 Functional octamer Binds NSP5 and VP1 Induces NSP5 hyperphosphorylation
9 (1,062)	49-1026	VP7	326 (7,368)	Outer capsid glycoprotein	Outer capsid structural glycoprotein G-type neutralization antigen N-linked high mannose glycosylation and trimming RER transmembrane protein, cleaved signal sequence Ca^{2+} binding Protection
10 (751)	41-569	NSP4	175 (20,290)	Nonstructural	Enterotoxin Receptor for budding of double-layer particle through ER membrane RER transmembrane glycoprotein Ca^{2+}/Sr^{2+} binding site N-linked high mannose glycosylation Protection Host cell $[Ca^{2+}]_i$ mobilization
11 (667)	22-615	NSP5	198 (21,725)	Nonstructural	Interacts with VP2, NSP2 and NSP6 Homomultimerizes O-linked glycosylation (Hyper-) phosphorylated Autocatalytic kinase activity enhanced by NSP2 interaction Non-specific ssRNA binding
	80-355	NSP6	92 (11,012)	Nonstructural	Product of second, out-of-frame ORF Interacts with NSP5 Localizes to viroplasm

Recently, astrovirus virus-like particles (VLPs) have been generated by cloning the cDNA corresponding to ORF2 from a human astrovirus serotype 2 into vaccinia virus (Dalton et al., 2003). Protein composition of these purified VLPs revealed no substantial difference from that of authentic astrovirus virions when analyzed by Western blotting. Trypsin cleavage seems to be necessary to process the capsid polyprotein into mature structural proteins.

Human astroviruses are classified into eight serotypes (HAstV-1 to HAstV-8) according to the antigenic reactivity of the capsid proteins (Lee and Kurtz, 1982, 1994; Taylor et al., 2001). Molecular analysis of a region at the 5′ end of ORF2 can be used to confirm antigenic typing or to characterize strains with ambiguous serotyping results (Noel et al., 1995; Belliot et al., 1997; Monroe et al., 2001). HAstV-1 is the most prevalent of the human astroviruses that cause actue gastroenteritis, whereas HAstV-6 and HAstV-7 have rarely been isolated (Lee and Kurtz, 1994; Noel and Cubitt, 1994; Gaggero et al., 1998; Unicomb et al., 1998).

4.0. ENTEROVIRUSES

4.1. Polioviruses

Enteroviruses replicate in the gastrointestinal tract, but the resulting infection is frequently asymptomatic. Symptoms, when they occur, range from paralysis to fever. Enteroviruses, named after the site of replication, rarely cause gastroenteritis. In addition, in many cases the enterovirus isolated might merely have been a passenger virus unrelated to the disease (Melnick, 1996).

The capsids of enteroviruses (family Picornaviridae) are composed of four structural proteins (VP1–VP4) arranged in 60 repeating protomeric units with icosahedral symmetry. Among the family members, the capsid proteins are arranged similarly, but the surface architecture varies. These differences account for not only the different serotypes but also the different modes of interaction with cell receptors. The basic building block of the picornavirus capsid is the protomer, which contains one copy of each structural protein. The capsid is formed by VP1 to VP3, and VP4 lies on its inner surface. VP1, VP2, and VP3 have no sequence homology, yet all three proteins have the same topology: they form an eight-stranded antiparallel "β-barrel" core structure (Racaniello, 2001). The external loops that connect the beta strands are responsible for differences in the antigenic diversity of enteroviruses. Neutralization sites are more densely clustered on VP1. The structures of many human picornaviruses have been resolved, indicating that they share a number of conserved structural motifs. For example, the capsids of polioviruses, rhinoviruses, and coxsackieviruses have a groove, or canyon, surrounding each fivefold axis of symmetry. In contrast, cardioviruses and aphthoviruses do not have canyons (Racaniello, 2001). Immediately beneath the canyon floor of each protomer is a hydrophobic pocket occupied by a lipid moiety. These molecules, termed *pocket factors*, have been shown to stabilize the capsid, and their removal from the pocket is a necessary prerequisite to uncoating. On the basis of the electron densities and uncoating studies, the pocket factors are thought to be short-chain fatty acids (Smyth et al., 2003).

The enterovirus positive-sense RNA genome is approximately 7.4 kb long and serves as a template for both viral protein translation and virus replication (Racaniello, 2001). The 5′ end is covalently linked to a VPg protein (for genome-linked virus protein). The genome is organized into a long (~740 nucleotides) nontranslated region (5′NTR), which contains the internal ribosome entry site (IRES) and precedes a single ORF. The ORF is subdivided into three regions, P1 to P3. The P1 region codes for structural proteins, whereas P2 and P3 regions encode for nonstructural proteins essential for virus replication (2A–C, 3A–D). Translation of the ORF gives rise to a single large polyprotein that is post-translationally modified by virus encoded proteases. Immediately downstream of the protein-coding region is the 3′ nontranslated region (3′-NTR), which plays a role in viral RNA replication, followed by a terminal poly(A) tail (Racaniello, 2001).

The three serotypes of human poliovirus are considered a species within the genus Enterovirus. A redefinition of the criteria for species demarcation within the genus Enterovirus has recently been issued by the International Committee on Taxonomy of Viruses (ICTV). Based on these molecular and biochemical characteristics, Enteroviruses are now classified as human enteroviruses A through D, bovine enterovirus, and porcine enterovirus A and B (King et al., 2000).

Poliovirus infections, once responsible for high morbidity and mortality throughout the world, are now under control, and their eradication is a priority for the World Health Organization (WHO). In the past 15 years, since the Global Polio Eradication Initiative was launched by WHO, the number of cases has fallen from an estimated 350,000 cases in 1988 to 1,919 in 2002. In the same time period, the number of polio-infected countries has been reduced from 125 to 7. There is a historic opportunity to forever stop the transmission of poliovirus. If the world seizes this opportunity and acts immediately, no child will ever again know the effects of this devastating disease (Dowdle et al., 2001).

4.2. Kobuviruses

Aichi virus, a cytopathic small round virus, was isolated for the first time in 1989 from fecal samples of patients involved in an oyster-associated gastroenteritis outbreak (Yamashita et al., 1991). Since then, several Aichi virus strains have been isolated in BS-C-1 cells from patients with gastroenteritis (Yamashita et al., 1995). The virus is commonly found in outbreaks of gastroenteritis in Japan and is often associated with the consumption of oysters (Yamashita et al., 1995). Genetic analysis of Aichi virus revealed that it belongs to the Picornaviridae family but that it is different from any other genus such as the Entero-, Rhino-, Cardio-, Aphtho-, Hepato-, Tescho-, or

Parechovirus (Yamashita et al., 1998). It has been recently proposed by the ICTV that this virus be assigned to a new genus named Kobuvirus (King et al., 1999). *Kobu* means "knob" in Japanese; the reference based on the characteristic morphoplogy of the virion.

Aichi virus has been cultivated in BS-C-1 and Vero cells, producing CPE characterized by detachment of cells. It does not replicate in other human cell lines such as HeLa, RD, or HEL cells. By EM, the virion shows a rough surface and measures around 30 nm in diameter. Three capsid proteins of 42, 30, and 22 kDa have been described (Yamashita et al., 1998). Aichi virus possesses a genome of single-stranded RNA of 8,280 nucleotides, excluding a poly(A) tract. It contains a large ORF with 7,299 nucleotides that encodes a polyprotein precursor of 2,432 amino acids, which is preceded by 744 nucleotides and followed by 237 nucleotides and a poly(A). The precise secondary structure of the 5′ nontranslated region (5′-NTR) has not been defined, although an internal ribosomal entry site (IRES) similar to that of other members of the Picornaviridae family has been reported (Sasaki et al., 2001). The complete nucleotide sequence of Aichi virus (GenBank accession no. AB010145) has been determined (Yamashita et al., 1998).

The organization of the deduced amino acid sequence of the polyprotein encoded by the Aichi virus genome is analogous to that of the other picornaviruses. Preceded by a leader (L) protein, there is the P1 region, which corresponds to the structural proteins (VP0, VP3, and VP1, with molecular weights of 42, 30, and 22 kDa, respectively) followed by the P2 and P3 regions, which contain sequences encoding the nonstructural proteins (2A–C, 3A–D). It has been reported that the 2A protein of Aichi virus contains conserved motifs that are characteristic of the H-rev107 family of cellular proteins involved in the control of cellular proliferation (Hughes and Stanway, 2000). Amino acid sequences of the 2C, 3C, and 3D regions are well aligned with the corresponding sequences of other picornaviruses. The 3B protein corresponds to the VPg protein. The 3C protein is the protease and contains conserved motifs characteristic of all picornaviruses (Yamashita and Sakae, 2003). The relationship of kobuviruses to other picornaviruses has been analyzed based on the 3D amino acid sequence (which corresponds to the RNA polymerase) and the polyprotein sequence (Hughes, 2004). Recently, bovine kobuvirus has been isolated and characterized (accession no. AB084788) (Yamashita et al., 2003). The genus Kobuvirus, including bovine kobuvirus and Aichi virus, cluster near the genera Teschovirus, Cardiovirus, Erbovirus, and Aphthovirus (Hughes, 2004).

5.0. HEPATITIS A VIRUS

Hepatitis A virus (HAV), the prototype of the genus Hepatovirus in the family Picornaviridae (Minor, 1991), is a hepatotropic virus that represents a significant problem for human health (Hollinger and Emerson, 2001). HAV infection is mainly propagated via the fecal-oral route, and although transmission remains primarily from person to person, waterborne and food-borne outbreaks of the disease have been reported (De Serres et al., 1999; Hutin et al., 1999; Fiore, 2004). HAV was originally classified as enterovirus type 72 because its biophysical characteristics are similar to those of enteroviruses. However, later studies demonstrated that HAV nucleotide and amino acid sequences are different from those of other picornaviruses, as are the predicted sizes of several HAV proteins (Cohen et al., 1987b). This virus is difficult to cultivate in cell cultures and usually replicates very slowly without producing CPE. It is resistant to temperatures and drugs that inactivate other picornaviruses and is stable at pH 1. There is only a single serotype of human HAV with one immunodominant neutralization site (Lemon and Binn, 1983). However, a significant degree of nucleic acid variability has been observed among different isolates from different regions of the world (Robertson et al., 1992; Costa-Mattioli et al., 2001). The molecular basis of this genetic variability may be the high error rate of the viral RNA-dependent RNA polymerase and the absence of proofreading mechanisms (Sánchez et al., 2003).

The virion is composed of a genome of linear, single-stranded RNA of messenger sense polarity, approximately 7.5 kb in length, and a capsid containing multiple copies of three or four proteins named VP1, VP2, VP3, and a putative VP4, encoded in the P1 region of the genome (Coulepis et al., 1982; Racaniello, 2001). The presence of a fourth protein VP4 has been described repeatedly, but the reported molecular weight (7–14 kDa) does not correspond to that predicted from nucleic acid sequence data (1.5 or 2.3 kDa) (Weitz and Siegl, 1998). The P2 and P3 regions encode for nonstructural proteins associated with replication.

HAV has a buoyant density of 1.32 to 1.34 g/cm^3 in CsCl and sediments with 156S during sucrose gradient centrifugation (Coulepis et al., 1982). Infectious viral particles with higher (1.44 g/cm^3) as well as lower (1.27 g/cm^3) buoyant densities have also been reported (Lemon et al., 1985). Phylogenetic analysis based on a 168-base segment encompassing the VP1/2A junction region of the HAV genome has established the classification of human and simian isolates into seven different genotypes (I–VII), with genotypes I, II, III, and VII including human isolates (Robertson et al., 1992). About 80% of the human isolates belong to genotype I, which has been subdivided into two subgenotypes, IA and IB (Fujiwara et al., 2003).

5.1. The Genome

Detailed analysis of the HAV genome has been accomplished with cloned or RT-PCR–amplified cDNA. Cultured cells can be infected with RNA transcribed from cloned HAV cDNA (Cohen et al., 1987a). HAV contains a genome that differs from that of Caliciviridae in that the genes encoding the nonstructural proteins are located at the 3′ portion of the genome, whereas those encoding the structural proteins are located at the 5′ end (Weitz and Siegl, 1998). The HAV genome is divided into a 5′ nontranslated region of 735 nucleotides, a long open reading frame of 6,681 nucleotides encoding a polyprotein of 2,227 amino acids, and a 3′ nontranslated region 63 nucleotides in length.

The 5′-NTR contains a *cis*-acting internal ribosome entry site (IRES) that directs initiation of cap-independent translation directed to a particular AUG triplet several hundreds of nucleotides downstream (Glass et al., 1993; Brown et al., 1994; Ali et al., 2001; Borman et al., 2001; Kang and Funkhouser, 2002). It has been demonstrated that the IRES is located between nucleotides 151 and 734, and that it is able to direct internal initiation of translation in a cap-independent manner (Brown et al., 1994). However, a cap-dependent message effectively competes with the IRES of HAV (Glass et al., 1993). Translational efficiency of this IRES may be dependent on the availability of intact cellular proteins such as the p220 subunit of the eukaryotic initiation factor eIF-4F and requires association between the cap-binding translation initiation factor (eIF4E) and eIF4G (Ali et al., 2001; Borman et al., 2001). The 5′-NTR is the most highly conserved region in the HAV genome among all strains sequenced to date, with a 95% nucleotide identity. By contrast, the 3′-NTR region shows the highest (up to 20%) degree of variability. The presence of a 23-nucleotide-long *cis*-acting element has been described that specifically interacts with proteins involved in the establishment and maintenance of the persistent type of infection characteristic for replication of most HAV isolates in cell cultures. Several unidentified cytoplasmic and ribosomal proteins of infected cells bind to the 3′ end of HAV RNA, indicating an intimate and dynamic interaction between host proteins and viral RNA (Kusov et al., 1996). The poly(A) tail is also involved in formation of RNA/protein complexes.

In analogy with other picornaviruses, the coding region can be subdivided into P1, P2, and P3 regions, which specify proteins 1A–D, 2A–C, and 3A–D, respectively. The polyprotein is further processed to the four structural and seven nonstructural proteins by proteinases encoded in and around the 3C region (Probst et al., 1998). Proteins 1A–D correspond to structural proteins VP1–VP4. As in other picornaviruses, the 5′ end of HAV genome is covalently linked to VPg protein, which is specified by 3B, instead of the classical m7G cap structure (Weitz et al., 1986). The 2C gene carries a guanidine resistance marker and is assumed to play a role in viral RNA replication (Cohen et al., 1987b). The region 3D is considered the RNA-dependent RNA polymerase. This region shows the highest degree of homology (29%) with the corresponding sequence in the poliovirus type 1 protein (Cohen et al., 1987b). A region of 18 amino acids considered to be essential for an active polymerase is present in the 3D region. This sequence contains a conserved motif of two asparatate residues flanked by hydrophobic amino acids that might function as a GTP-binding domain (Kamer and Argos, 1984). Replication efficiency seems to be controled by amino acid substitutions in the 2B and 2C regions (Yokosuka, 2000).

5.2. Proteins

The genomes of all picornaviruses encode a single polyprotein, which is co- and post-translationally cleaved by virus-encoded proteinase(s). In contrast with well-characterized proteolytic events in the polyprotein of Enterovirus and Rhinovirus, this processing has been difficult to characterize in Hepatovirus. It has been shown that the primary cleavage of HAV polyprotein occurs at the 2A/2B junction, which has been mapped by the N-terminal sequencing of 2B. This primary cleavage of the polyprotein is mediated by $3C^{pro}$ proteinase, which is the only proteinase known to be encoded by the virus (Martin et al., 1995; Gosert et al., 1996). The P1-2A capsid protein precursor is probably released from the nonstructural protein precursor (P3-2BC) as soon as $3C^{pro}$ is synthesized, as the full-length polyprotein has not been observed in these studies. A P1-2A precursor produced in a cell-free translation system has been shown to be cleaved *in vitro* by recombinant $3C^{pro}$ to generate VP0 (VP4–VP2), VP3, and VP1-2A (Malcolm et al., 1992). This VP1-2A polypeptide is unique to hepatoviruses; it associates with VP0 and VP3 to form pentamers, intermediates in the morphogenesis of HAV particles (Borovec and Andersen, 1993). The mature capsid protein VP1 is subsequently derived from the VP1-2A precursor later in the morphogenesis process. It has been hypothesized that the maturation of VP1 is dependent on $3C^{pro}$ processing of the VP1-2A precursor (Probst et al., 1998). However, using recombinant vaccinia viruses expressing relevant HAV substrates, it has also been shown that $3C^{pro}$ is incapable of directing the cleavage of VP1 from the VP1-2A precursor, indicating that maturation of VP1 could not depend on processing by $3C^{pro}$ proteinase (Martin et al., 1999).

5.3. Virus Replication

Hepatitis A virus typically has a protracted and noncytolytic replication cycle in cell cultures and fails to shut down host cell metabolism (Lemon and Robertson, 1993). Even after successful adaptation to grow in cell cultures, replication of HAV is a slow process that terminates in a state of persistent infection (de Chastonay and Siegl, 1987). Cytopathic HAV strains have been recovered only from persistently infected cell cultures. Maximal levels of viral RNA synthesis can be detected at 24h after infection, and exponential production of progeny virus continues for up to 4 days postinfection. Lysis of infected cells may become apparent within 3 to 9 days postinfection, and yield of progeny virus rarely exceeds 10^7 $TCID_{50}$/ml (Siegl et al., 1984).

6.0. HEPATITIS E VIRUS

Hepatitis E virus (HEV), the prototype of the proposed Hepevirus, is a nonenveloped RNA virus, previously classified as a calicivirus but provisionally considered as a separate group of HEV-like viruses (Berke and Matson, 2000). Sequence comparison and phylogenetic analysis suggest that these viruses are more closely related to the Togaviridae (Koonin et al., 1992).

Hepatitis E virus is a major cause of epidemics and sporadic cases of acute hepatitis in Southeast and Central Asia, the Middle East, and many areas of Africa and Mexico (Purcell and Emerson, 2001). Sporadic cases have also been identified in Europe, Japan, and the United States (Schlauder et al., 1998, 1999; Takahashi et al., 2001). Transmission of HEV infection during outbreaks primarily occurs through contaminated water. Provision of clean drinking water is considered the best preventive strategy. The illness is self-limited and no chronic HEV infection has been described, although high mortality rate has been observed in pregnant women, with death rate as high as 25% (Balayan, 1997). HEV has been isolated from humans and swine (Meng et al., 1997, 1998a; Hsieh et al., 1999; Garkavenko et al., 2001). Antibodies to the HEV capsid protein have been detected in many animal species, including rats and non-human primates (Kabrane-Lazizi et al., 1999a; Arankalle et al., 2001).

To date, four genotypes have been identified (Tam et al., 1991); a genotype 3 strain of HEV isolated from swine has been passed experimentally to monkeys, and a genotype 3 strain isolated from humans has been passed to swine (Meng et al., 1998b). Intriguingly, attempts to infect swine with other HEV human isolates have failed (Meng et al., 1998a). The question of whether HEV infection is a zoonosis is still being discussed (Meng, 2000). The viral genome is approximately 7.2 kb in length and has three partially overlapping ORFs: ORF1 encodes nonstructural proteins (1,693 amino acids), ORF2 encodes the capsid protein (660 amino acids), and ORF3 encodes a very small protein (123 amino acids) that has been shown to bind *in vitro* to a number of proteins involved in cellular signal transduction (Korkaya et al., 2001). The virus has not been propagated efficiently in cell cultures. Recently, various regions of the viral genome have been cloned and expressed *in vitro*. The ORF1 gene encodes the largest protein, which contains motifs characteristic of a methyl-transferase, a papain-like protease, a helicase, and an RNA-dependent RNA polymerase (Koonin et al., 1992). It has been demonstrated that the HEV genome contains a m^7G cap at its 5′ end (Kabrane-Lazizi et al., 1999b; Zhang et al., 2001). There is still controversy over whether the 5′ cap is critical for HEV replication. Although it has been shown that the cap is required for infectivity of recombinant genomes *in vivo*, uncapped viral genomes can replicate in transfected HepG2 cells (Panda et al., 2000; Emerson et al., 2001).

The full-length ORF2 gene encodes a 72-kDa capsid protein consisting of 660 amino acids. ORF2 is the major, if not the only protein in the virion, but its real size in infectious virions is not known. When synthesized in insect cells, the initial 72-kDa protein is processed to smaller proteins of approximately 63, 56, and 53 kDa (Robinson et al., 1998). It has been demostrated that antibodies against ORF2 antibodies can neutralize HEV (Mast et al., 1998; Meng et al., 2001). Persons with preexisting anti-ORF2 did not develop hepatitis E after exposure to HEV in an outbreak in Pakistan (Bryan et al., 1994). Similarly, anti-ORF2–positive monkeys did not develop hepatitis after challenge with infectious HEV (Tsarev et al., 1994; Arankalle et al., 1999). A number of antigenic determinants have been identified within ORF2 (Khudyakov et al., 1999; Li et al., 2000; Riddell et al., 2000). However, it appears that only antibodies directed against the C-terminus of ORF2 are neutralizing (Mast et al., 1998). The ORF2-encoded 56-kDa protein, truncated at its N- and C-termini, served as a highly reactive antigen in detecting anti-HEV antibodies (Zhang et al., 1997).

The ORF3 protein contains only 123 amino acids, and its biological role has yet to be elucidated. Recombinant ORF3 protein expressed in eukaryotic cells is an immunogenic phosphoprotein that accumulates in the cytoplasm and associates with the cytoskeleton (Zafrullah et al., 1997). This association is dependent on amino acids 1 to 32 of ORF3, which comprise a hydrophobic domain. In addition, ORF3 contains two conserved proline-rich domains. By the yeast two-hybrid assay, it has been demonstrated that the second of these proline-rich domains binds to certain Src homology 3 domains present in a subset of cellular proteins. It has been hypothesized that ORF3 may be a viral regulatory protein involved in cellular signal transduction (Korkaya et al., 2001). Also, recombinant ORF3 protein has been shown to interact in yeast two-hybrid assays and by co-immunoprecipitation assays with the nonglycosylated form of the major capsid protein, ORF2 protein (Tyagi et al., 2002). These observations lead to the hypothesis that ORF3 has a well-regulated role in HEV structural assembly (Emerson and Purcell, 2003). One HEV vaccine candidate, a baculovirus-expressed protein spanning 112–607 amino acids of ORF2, was demonstrated to be safe and immunogenic in volunteers and is currently being evaluated in clinical trials in Nepal (Emerson and Purcell, 2003).

7.0. ENTERIC ADENOVIRUSES

Adenoviruses are nonenveloped, icosahedral viruses measuring 70 to 90 nm in diameter. They have a buoyant density in cesium chloride of 1.33 to 1.34 g/cm^3. The capsid is composed of 252 capsomeres, of which 240 are hexons and 12 are pentons. Inside the capsid is a single molecule of linear, double-stranded DNA (Shenk, 2001). Two genera are recognized within the Adenoviridae family: Mastadenovirus includes viruses that infect mammals, and Aviadenovirus contains viruses that infect birds (Benkö et al., 2000). Adenoviruses are species-specific and generally replicate only in cells derived from their native host. Human adenoviruses are associated with a variety of infectious diseases affecting the respiratory, urinary, and the gastrointestinal tracts and the eyes (Horwitz, 2001). To date, 51 serotypes of human adenoviruses have been recognized, which are classified into six subgroups, A to F, based on immunological properties, oncogenicity in rodents, DNA homologies, and morphological properties (Benkö et al., 2000).

Enteric adenoviruses were originally identified in stool samples of infants with acute gastroenteritis (Flewett et al., 1973) and have been consistently associated with gastroenteritis in children through epidemiological and clinical studies (Uhnoo et al., 1983; Uhnoo et al., 1984). They are responsible for 5% to 20% cases of acute diarrhea in children (Uhnoo et al., 1984; Kotloff et al., 1989; Uhnoo et al., 1990; Bon et al., 1999) and are found in clinical samples throughout the year with little seasonal variation (de Jong et al., 1983).

The enteric serotypes 40 and 41 have been designated as subgroup F adenoviruses. They share the adenovirus group antigen and are distinguished from each other and from other nonenteric serotypes by serology and DNA restriction patterns (Wadell, 1984). These two serotypes are shed in large numbers from the gut of infected patients and were originally described as being noncultivable or "fastidious" adenoviruses, because they could not be cultivated in cell cultures that generally supported the propagation of other adenovirus types. Later, however, it was discovered that enteric adenoviruses could be propagated in Graham 293 cells, a cell line of human embrionic kidney (HEK) cells transformed with adenovirus 5 early (E) region 1 (Graham et al., 1977; Takiff et al., 1981), although at lower levels than other serotypes. This suggests that E1 functions are poorly expressed in cells infected with adenovirus types 40 and 41, and therefore it was postulated that the inability of these serotypes to grow in cell lines normally supportive for other adenovirus types was due to the relative inability of the adenovirus 41 E1A gene to transactivate other adenovirus 41 genes (Takiff et al., 1984; van Loon et al., 1985a). The Graham 293 cell retains the E1A and E1B regions of the adenovirus genome covalently linked to the host DNA. The mechanism of facilitation of the growth of the EAd40 in 293 cells seems to be a function of the E1B-55kd protein (Mautner et al., 1989). Efficient replication of adenovirus types 40 and 41 has also been achieved in other cell lines, including Hep-2 cells, Chang conjuntiva cells, CaCo-2 cells (Perron-Henry et al., 1988; Pintó et al., 1994), and PLC/PRF/5 cells (a primary liver carcinoma cell line) (Grabow et al., 1992).

The genome of adenovirus 40 has been sequenced (Davidson et al., 1993) (GenBank accession no. L19443) and described in detail (Mautner et al., 1995). The main difference between adenovirus 40 and the other human adenovirus serotypes is the presence of two distinct fiber genes, a single VA gene involved in late translation and a highly divergent E3 region. The growth restriction of adenovirus 41 in cell cultures seems to be less severe than that of serotype 40 because a number of cell lines have been found to support the propagation of serotype 41 but not of serotype 40 (de Jong et al., 1983; Uhnoo et al., 1983; van Loon et al., 1985b). The blockade in adenovirus 41 replication occurs within the early phase of the infectious cycle (Tiemessen et al., 1996). In a study on the ability of adenovirus 40 E1A encoded products (proteins 249R and 221R) for *trans*-activation (van Loon et al., 1987; Ishino et al., 1988), it was found that the adenovirus 40 E1A promoter does indeed contain transcription factor binding sites sufficient for *trans*-activation by the adenovirus 5 E1A 289R protein. It is possible that adenovirus 40 has evolved to use components of the RNA processing machinery that are unique to enterocytes (Stevenson and Mautner, 2003). Hence, a better understanding of the replication and pathogenicity of adenovirus 40 will require the development of intestinal cell cultures.

8.0. OTHER ENTERIC VIRUSES

Although rotaviruses, caliciviruses, astroviruses, and enteric adenoviruses are the most prominent viral pathogens of acute gastroenteritis, other candidates such as coronaviruses, toroviruses, and picobirnaviruses are also considered to have capability to cause diarrhea in humans.

8.1. Coronaviruses

Coronavirus is a genus of the Coronaviridae family in the order Nidovirales (Enjuanes et al., 2000). The first reports of coronavirus-like particles in stools of patients with diarrhea were documented by Caul and Clarke (1975), who reported cultivation of these viral particles in human embryo intestinal organ culture and human embryo kidney monolayers (Caul and Clarke, 1975). Since then, the existence of human enteric coronavirus (HEC) has been controversial (Mathan et al., 1975; Caul and Egglestone, 1977). Some investigators have considered these particles as viruses and as potential pathogens in human diarrhea, necrotizing enterocolitis, and malabsorption. Other authors cite their variable size, their presence in stools of normal subjects, and their failure to be cultivated *in vitro* as evidence that they might represent cell debris or even portions of other microorganisms (Macnaughton and Davies, 1981). However, isolation and propagation of enteric coronaviruses have been reported (Resta et al., 1985; Luby et al., 1999). The enveloped particles are 120–160 nm in diameter and possess a genome of linear, positive-sense, single-stranded RNA of approximately 30 kb in size. The genome is surrounded by the nucleocapsid protein (N) with helical symmetry, which is contained in an envelope, in which the spike (S), envelope (E), hemagglutinin esterase (HE), and membrane (M) proteins are embedded. Proteins are expressed from RNA molecules that are in most cases subgenomic, and all mRNAs carry a leader sequence derived from the 5′ end of the genome (Lai and Holmes, 2001).

8.2. Toroviruses

Toroviruses (family Coronaviridae, order Nidovirales) are enveloped, positive-sense RNA viruses that have been implicated in enteric disease in cattle and possibly in humans. Despite their potential veterinary and clinical relevance, little is known about the epidemiology and molecular genetics of toroviruses (Koopmans and Horzinek, 1994; Smits et al., 2003).

Toroviruses have not been propagated in cell cultures, with the sole exception so far of the equine Berne virus. Early seroepidemiological surveys based on Berne virus cross-neutralization assays and enzyme-linked immunosorbent assay (ELISA) indicated that toroviruses occur in a wide variety of ungulate hosts (cattle, sheep, goats, and swine) (Koopmans and Horzinek, 1994). Breda virus was detected in the stools of diarrheic calves but could not be isolated in cell cultures (Woode et al., 1982). Thereafter, two antigenically distinct serotypes of Breda virus have been identified, referred to as bovine toroviruses 1 and 2 (BoTV-1 and BoTV-2) (Woode et al., 1985).

Morphologically similar pleomorphic particles characterized by a well-defined fringe around the outer edges have also been found in human stools (Beards et al., 1984; Duckmanton et al., 1997). These viruses have a unique morphology with a helical nucleocapsid in a torus-shaped configuration within an envelope carrying large spikes. In samples from children and adults with diarrhea, torovirus-like particles cross-reactive with Berne virus–specific and Breda virus–specific antisera were detected by immunoelectron microscopy (Beards et al., 1984; Beards et al., 1986; Duckmanton et al., 1997), and torovirus antigens were detected by ELISA (Koopmans et al., 1997). Recently, the nucleotide sequence of the Berne virus genomic RNA has been completed (Smits et al., 2003).

With a genome length of 28 kb, toroviruses are among the largest RNA viruses, rivaled in size only by the coronaviruses. The 5′-most two-thirds of the genome is taken up by two huge overlapping open reading frames, 1a and 1b, encoding polyproteins from which various subunits of the viral replicase/transcriptase are derived (Snijder et al., 1990a). Downstream of ORF1b, there are four cistrons of 5 kb, 0.7 kb, 1.2 kb, and 0.5 kb (as ordered from 5′ to 3′). These encode the structural proteins, the spike (S), membrane (M), hemagglutinin-esterase (HE), and nucleocapsid (N) proteins, respectively (Snijder and Horzinek, 1993). The structural proteins are translated from a 3′-coterminal nested set of subgenomic mRNAs, produced by discontinuous and nondiscontinuous RNA synthesis (Snijder et al., 1990b; van Vliet et al., 2002). Four genotypes, displaying 30% to 40% sequence divergence, have been recently distinguished, exemplified by bovine torovirus (BToV) Breda, porcine torovirus (PToV), equine torovirus Berne, and the putative human torovirus (Smits et al., 2003). It remains unclear, however, whether the toroviruses are host specific and how the torovirus genotypes are geographically distributed.

8.3. Picobirnaviruses

Picobirnavirus is the tentative name for a new virus genus in the family Birnaviridae (Leong et al., 2000). Picobirnaviruses were first observed in 1988 in fecal samples of humans and rats (Pereira et al., 1988a, 1988b). The viruses have subsequently been found in many mammals and birds (Rosen, 2003) and have been frequently detected in cases of human gastroenteritis associated with human immunodeficiency virus infections (HIV) (Grohmann et al., 1993; Gonzalez et al., 1998; Liste et al., 2000). Association of picobirnavirus with nonbacterial gastroenteritis outbreaks has also been reported recently (Banyai et al., 2003).

Morphological, physicochemical, and genomic characteristics suggest that picobirnaviruses belong to a distinct group of viruses. The small size of the virion and the nature of the viral genome prompted the name picobirnaviruses, referring to *pico* (small) and *birna* (two RNA segments). The virion is 25–41 nm in diameter, nonenveloped, and has a buoyant density of 1.38 to 1.42 g/cm^3 (Pereira et al., 1988b; Ludert et al., 1995). However, virions with a diameter of around 35 nm have been found most often (Ludert et al.,

1991; Gallimore et al., 1995). The genome consists of two segments of double-stranded RNA with lengths estimated around 2.3–2.6 Kpb and 1.5–1.9 Kpb, respectively (Gallimore et al., 1995; Chandra, 1997). These viruses are non-cultivable. They are detected by EM and polyacrylamide gel electrophoresis (PAGE) (Ludert and Liprandi, 1993; Gallimore et al., 1995; Cascio et al., 1996). Picobirnavirus genomic fragments migrate in PAGE within the migration range of the genomic segments of group A rotaviruses (Pereira et al., 1988a; Ludert et al., 1991; Rosen et al., 2000). This fact led to the discovery of these viruses when fecal samples containing both rotaviruses and picobirnaviruses were subjected to PAGE (Pereira et al., 1988a, 1988b). Currently, RT-PCR is used to detect human picobirnaviruses in fecal samples (Rosen et al., 2000).

Limited information is available on the genetic sequence of picobirnaviruses. To date, genomic segment 1 of a rabbit and a human picobirnavirus and segment 2 of two human picobirnaviruses have been cloned (Green et al., 1999; Rosen et al., 2000). Human picobirnavirus sequences have been determined mostly for the smaller genomic segment that encodes the RNA-dependent RNA polymerase (Rosen et al., 2000). The genomic segment 1 has been postulated to encode for the capsid protein (Green et al., 1999; Rosen et al., 2000). Based on the sequence data of the RNA polymerase, there are two major genogroups within human picobirnaviruses, for which prototype strains are 1-CHN-97 and 4-GA-91 (Rosen et al., 2000).

9.0. SUMMARY

Food-borne and waterborne viruses mainly cause gastroenteritis (caliciviruses, rotaviruses, astroviruses, and enteric adenoviruses), hepatitis (hepatitis A virus, hepatitis E virus), and other diseases (enteroviruses). Other enteric viruses like kobuviruses, coronaviruses, toroviruses, and picornaviruses also cause diarrhea, although the causative role for some of these viruses in humans is still controversial.

Human caliciviruses have been recognized as the leading cause of acute gastroenteritis outbreaks and sporadic cases in children and adults worldwide. Enteric caliciviruses belonging to the Norovirus and Sapovirus genera remain refractory to cell culture propagation. This limitation has hampered our ability to investigate their biology, pathogenesis, and host immunity, although molecular approaches are providing new insights into these areas. The morphology, composition, and structure of several enteric viruses have been elucidated in recent years. Cryo-electron microscopy and x-ray crystallography have been crucial for this purpose. Biochemical and structural studies of virus-like particles produced by recombinant baculoviruses are contributing to better understand the structure-function relationships of the capsid proteins. Viral genome organization is being clarified for all these viruses, as well as their replication and gene expression strategies. Most of the proteins encoded by the viral genomes have been characterized and their functions identified. Sequence analysis of viral genes is currently being applied in molecular epidemiological studies. However, many questions still remain to be answered. The development of efficient reverse genetics systems would be extremely useful in the analysis of the mechanisms of viral replication and of gene expression of the different enteric viruses.

Biochemical characterization of viral interactions with cells and analysis of the functional properties of the viral proteins are providing a better understanding of the pathogenesis of enteric viruses. Rotavirus NSP4 is the first viral enterotoxin to be characterized. Several cell membrane molecules have been identified recently as being receptors for different enteric viruses (i.e., integrins and hsc70 for rotaviruses, ABH histo-blood group carbohydrates for noroviruses). Studies on human susceptibility to norovirus infections have characterized some resistant nonsecretor (Se-) individuals in the population, which is a breakthrough in our knowledge of norovirus-host interactions. Similarly, molecular analyses of orally transmitted viruses causing hepatitis are clarifying the phylogenetic relationships between these viruses and other viral genera, as well as their pathophysiological mechanisms.

10.0. REFERENCES

Adah, M. I., Wade, A., and Taniguchi, K., 2001, Molecular epidemiology of rotaviruses in Nigeria: detection of unusual strains with G2P[6] and G8P[1] specificities. *J. Virol.* 39:3969–3975.

Ali, I. K., McKendrick, L., Morley, S. J., and Jackson, R. J., 2001, Activity of the hepatitis A virus IRES requires association between the cap-binding translation initiation factor (eIF4E) and eIF4G. *J. Virol.* 75:7854–7863.

Amano, J., and Oshima, M., 1999, Expression of the H type 1 blood group antigen during enterocyte differentiation of Caco-2 cells. *J. Biol. Chem.* 274:21209–21216.

Ando, T., Monroe, S. S., Gentsch, J. R., Jin, Q., Lewis, D. C., and Glass, R. I., 1995, Detection and differentiation of antigenically distinct small round-structured viruses (Norwalk-like viruses) by reverse transcription-PCR and southern hybridization. *J. Clin. Microbiol.* 33:64–71.

Ando, T., Noel, J. S., and Fankhauser, R. L., 2000, Genetic classification of "Norwalk-like viruses." *J. Infect. Dis.* 181(Suppl 2):S336–S348.

Appleton, H., and Higgins, P. G., 1975, Viruses and gastroenteritis in infants. *Lancet* 1:1297.

Arankalle, V. A., Chadha, M. S., and Chobe, L. P., 1999, Long-term serological follow up and cross-challenge studies in rhesus monkeys experimentally infected with hepatitis E virus. *J. Hepatol.* 30:199–204.

Arankalle, V. A., Joshi, Y. K., Kulkarni, A. M., Gandhe, S. S., Chobe, L. P., Rautmare, S. S., Mishra, A. C., and Padbhidri, S. P., 2001, Prevalence of anti-hepatitis E virus antibodies in different Indian animal species. *J. Viral Hepatol.* 8:223–227.

Armah, G. E., Steele, A. D., Binka, F. N., Esona, M. D., Asmah, R. H., Anto, F., Brown, D., Green, J., Cutts, F., and Hall, A., 2003, Changing patterns of rotavirus genotypes in Ghana: emergence of human rotavirus G9 as a major cause of diarrhea in children. *J. Clin. Microbiol.* 41:2317–2322.

Ashley, C. R., and Caul, E. O., 1982, Potassium tartrate-glycerol as a density gradient substrate for separation of small, round viruses from human feces. *J. Clin. Microbiol.* 16:377–381.

Atmar, R. L., and Estes, M. K., 2001, Diagnosis of noncultivatable gastroenteritis viruses, the human caliciviruses. *Clin. Microbiol. Rev.* 14:15–37.

Balayan, M. S., 1997, Epidemiology of hepatitis E virus infection. *J. Viral Hepatol.* 4:155–165.

Ball, J. M., Tian, P., Zeng, C. Q.-Y., Morris, A. P., and Estes, M. K., 1996, Age-dependent diarrhea induced by a rotaviral nonstructural glycoprotein. *Science* 272:101–103.

Banyai, K., Jakab, F., Reuter, G., Bene, J., Uj, M., Melegh, B., and Szucs, G., 2003, Sequence heterogeneity among human picobirnaviruses detected in a gastroenteritis outbreak. *Arch. Virol.* 148:2281–2291.

Bass, D. M., and Qiu, S., 2000, Proteolytic processing of the astrovirus capsid. *J. Virol.* 74:1810–1814.

Beards, G. M., Hall, C., Green, J., Flewett, T. H., Lamouliatte, F., and Du Pasquier, P., 1984, An enveloped virus in stools of children and adults with gastroenteritis that resembles the Breda virus of calves. *Lancet* 1:1050–1052.

Beards, G. M., Brown, D. W., Green, J., and Flewett, T. H., 1986, Preliminary characterisation of torovirus-like particles of humans: comparison with Berne virus of horses and Breda virus of calves. *J. Med. Virol.* 20:67–78.

Belliot, G., Laveran, H., and Monroe, S. S., 1997, Detection and genetic differentiation of human astroviruses: phylogenetic grouping varies by coding region. *Arch. Virol.* 142:1323–1334.

Benkö, M., Harrach, B., and Russell, W. C., 2000, Family *Adenoviridae*, in: *Virus Taxonomy. Classification and Nomenclature of Viruses. Seventh Report of the International Committe on Taxonomy of Viruses* (M. H. V. van Regenmortel, C. M. Fauquet, D. H. L. Bishop, E. B. Carstens, M. K. Estes, S. M. Lemon, J. Maniloff, M. A. Mayo, D. J. McGeoch, C. R. Pringle, and R. B. Wickner, eds.), Academic Press, San Diego, pp. 227–238.

Berke, T., and Matson, D. O., 2000, Reclassification of the Caliciviridae into distinct genera and exclusion of hepatitis E virus from the family on the basis of comparative phylogenetic analysis. *Arch. Virol.* 145:1421–1436.

Berke, T., Golding, B., Jiang, X., Cubitt, D. W., Wolfaardt, M., Smith, A. W., and Matson, D. O., 1997, Phylogenetic analysis of the Caliciviruses. *J. Med. Virol.* 52:419–424.

Bertolotti-Ciarlet, A., White, L. J., Chen, R., Prasad, B. V., and Estes, M. K., 2002, Structural requirements for the assembly of Norwalk virus-like particles. *J. Virol.* 76:4044–4055.

Bertolotti-Ciarlet, A., Chen, R., Estes, M. K., and Prasad, B. V., 2003, Structure of Norwalk virus: the prototype human calicivirus, in: *Viral Gastroenteritis* (U. Desselberger and J. Gray, eds.), Elsevier, Amsterdam, pp. 455–466.

Bishop, R. F., 1996, Natural history of human rotavirus infections. *Arch. Virol.* Suppl. 12:119–128.

Black, D. N., Burroughs, J. N., Harris, T. J. R., and Brown, F., 1978, The structure and replication of calicivirus RNA. *Nature* 274:614–615.

Blackhall, J., Fuentes, A., and Magnusson, G., 1996, Genetic stability of a porcine rotavirus RNA segment during repeated plaque isolation. *Virology* 225:181–190.

Blacklow, N. R., Dolin, R., Fedson, D. S., DuPont, H. L., Northrup, R. S., Hornick, R. B., and Chanock, R. M., 1972, Acute infectious nonbaterial gastroenteritis: etiology and pathogenesis. *Ann. Intern. Med.* 76:993–1008.

Blutt, S. E., Kirkwood, C. D., Parreño, V., Warfield, K. L., Ciarlet, M., Estes, M. K., Bok, K., Bishop, R. F., and Conner, M. E., 2003, Rotavirus antigenaemia and viraemia: a common event? *Lancet* 362:1445–1449.

Blutt, S. E., Crawfird, S. E., Warfield, K. L., Lewis, D. E., Estes, M. K., and Conner, M. E., 2004, The VP7 outer capsid protein of rotavirus induces polyclonal B-cell activation. *J. Virol.* 78:6974–6981.

Bon, F., Fascia, P., Dauvergne, M., Tenenbaum, D., Planson, H., Petion, A. M., Pothier, P., and Kohli, E., 1999, Prevalence of group A rotavirus, human calicivirus, astrovirus, and adenovirus type 40 and 41 infections among children with acute gastroenteritis in Dijon, France. *J. Clin. Microbiol.* 37:3055–3058.

Borman, A. M., Michel, Y. M., and Kean, K. M., 2001, Detailed analysis of the requirements of hepatitis A virus internal ribosome entry segment for the eukaryotic initiation factor complex eIF4F. *J. Virol.* 75:7864–7871.

Borovec, S. V., and Andersen, D. A., 1993, Synthesis and assembly of hepatitis A virus-specific proteins in BS-C-1 cells. *J. Virol.* 67:3095–3102.

Brown, E. A., Zajac, A. J., and Lemon, S. M., 1994, In vitro characterization of an internal ribosomal entry site (IRES) present within the 5′ nontranslated region of hepatitis A virus RNA—comparison with the IRES of encephalomyocarditis virus. *J. Virol.* 68:1066–1074.

Browne, E. P., Bellamy, R., and Taylor, J. A., 2000, Membrane-destabilizing activity of rotavirus NSP4 is mediated by a membrane-proximal amphipatic domain. *J. Gen. Virol.* 81:1955–1959.

Brunet, J.-P., Cotte-Laffitte, J., Linxe, C., Quero, A.-M., Géniteau-Legendre, M., and Servin, A., 2000, Rotavirus infection induces an increase in intracellular calcium concentration in human intestinal epithelial cells: role in microvillar actin alteration. *J. Virol.* 74:2323–2332.

Bryan, J., Tsarev, S., Iqbal, M., Ticehurst, J., Emerson, S., Ahmed, A., Duncan, J., Rafiqui, A. R., Malik, I. A., and Purcell, R. H., 1994, Epidemic hepatitis E in Pakistan: patterns of serologic response and evidence that antibody to hepatitis E virus protects against disease. *J. Infect. Dis.* 170:517–521.

Buesa, J., de Souza, C. O., Asensi, M., Martínez, C., Prat, J., Gil, M. T., 2000, VP7 and VP4 genotypes among rotavirus strains recovered from children with gastroenteritis over a 3-year period in Valencia, Spain. *Eur. J. Epidemiol.* 16:501–506.

Buesa, J., Collado, B., López-Andújar, P., Abu-Mallouh, R., Rodríguez Díaz, J., García Díaz, A., Prat, J., Guix, S., Llovet, T., Prats, G., and Bosch, A., 2002, Molecular epidemiology of caliciviruses causing outbreaks and sporadic cases of acute gastroenteritis in Spain. *J. Clin. Microbiol.* 40:2854–2859.

Burke, B., and Desselberger, U., 1996, Rotavirus pathogenicity. *Virology* 218:299–305.

Caballero, S., Guix, S., El-Senousy, W. M., Calicó, I., Pintó, R. M., and Bosch, A., 2003, Persistent gastroenteritis in children infected with astrovirus: association with serotype-3 strains. *J. Med. Virol.* 71:245–250.

Carter, M. J., 1990, Transcription of feline calicivirus RNA. *Arch. Virol.* 114:143–152.

Cascio, A., Bosco, M., Vizzi, E., Giammanco, A., Ferraro, D., and Arista, S., 1996, Identification of picobirnavirus from faeces of Italian children suffering from acute diarrhea. *Eur. J. Epidemiol.* 12:545–547.

Castello, A. A., Argüelles, M. H., Villegas, G. A., López, N., Ghiringhelli, D. P., Semorile, L., and Glikmann, G., 2000, Characterization of human group C rotavirus in Argentina. *J. Med. Virol.* 62:199–207.

Caul, E. O., and Appleton, H., 1982, The electron microscopical and physical characteristics of small round human fecal viruses: an interim scheme for classification. *J. Med. Virol.* 9:257–265.
Caul, E. O., and Clarke, S. K. R., 1975, Coronavirus propagated from patient with nonbacterial gastroenteritis. *Lancet* 2:953–954.
Caul, E. O., and Egglestone, S. I., 1977, Further studies on human enteric coronaviruses. *Arch. Virol.* 54:107–117.
Chandra, R., 1997, Picobirnavirus, a novel group of undescribed viruses of mammals and birds: a minireview. *Acta Virol.* 41:59–62.
Charpilienne, A., Lepault, J., Rey, F., and Cohen, J., 2002, Identification of rotavirus VP6 residues located at the interface with VP2 that are essential for capsid assembly and transcriptase activity. *J. Virol.* 76:7822–7831.
Chen, D., and Patton, J. T., 1998, Rotavirus RNA replication requires a single-stranded 3′ end for efficient minus-strand synthesis. *J. Virol.* 72:7387–7396.
Chen, D., Estes, M. K., and Ramig, R. F., 1992, Specific interactions between rotavirus outer capsid proteins VP4 and VP7 determine expression of a cross-reactive, neutralizing VP4-specific epitope. *J. Virol.* 66:432–439.
Chen, D. Y., Zeng, C. Q.-Y., Wentz, M. J., Gorziglia, M., Estes, M. K., and Ramig, R. F., 1994, Template-dependent, *in vitro* replication of rotavirus RNA, *J. Virol.* 68:7030–7039.
Chen, D., Barros, M., Spence, E., and Patton, J. T., 2001, Features of the 3′-consensus sequence of rotavirus mRNAs critical to minus strand synthesis. *Virology* 281:221–229.
Chen, R., Neill, J. D., Noel, J. S., Hutson, A. M., Glass, R. I., Estes, M. K., and Prasad, B. V., 2004, Inter- and intragenus structural variations in caliciviruses and their functional implications. *J. Virol.* 78:6469–6479.
Chiba, S., Sakuma, Y., Kogasaka, R., Akihara, M., Horino, K., Nakao, T., and Fukui, S., 1979, An outbreak of gastroenteritis associated with calicivirus in an infant home. *J. Med. Virol.* 4:249–254.
Ciarlet, M., and Estes, M. K., 1999, Human and most animal rotavirus strains do not require the presence of sialic acid on the cell surface for efficient infectivity. *J. Gen. Virol.* 80:943–948.
Ciarlet, M., and Estes, M., 2001, Interactions between rotavirus and gastrointestinal cells. *Curr. Opin. Microbiol.* 4:435–441.
Ciarlet, M., Crawford, S. E., and Estes, M., 2001, Differential infection of polarized epithelial cell lines by sialic acid-dependent and sialic acid-independent rotavirus strains. *J. Virol.* 75:11834–11850.
Ciarlet, M., Crawford, S. E., Cheng, E., Blutt, S. E., Rice, D. A., Bergelson, J. M., and Estes, M. K., 2002a, VLA-2 (α2β1) integrin promotes rotavirus entry into cells but is not necessary for rotavirus attachment. *J. Virol.* 76:1109–1123.
Ciarlet, M., Ludert, J. E., Iturriza-Gómara, M., Liprandi, F., Gray, J. J., Desselberger, U., and Estes, M. K., 2002b, Initial interaction of rotavirus strains with N-acetylneuraminic (sialic) acid residues on the cell surface correlates with VP4 genotype, not species of origin. *J. Virol.* 76:4087–4095.
Clarke, I. N., and Lambden, P. R., 1997, Viral zoonoses and food of animal origin: caliciviruses and human disease. *Arch. Virol.* Suppl 13:141–152.
Clarke, I. N., and Lambden, P. R., 2000, Organization and expression of calicivirus genes. *J. Infect. Dis.* 181(Suppl 2):S309–316.
Clarke, I. N., and Lambden, P. R., 2001, The molecular biology of human caliciviruses. *Novartis Found. Symp.* 238:180–191; discussion 191–196.

Cohen, J. I., Ticehurst, J. R., Feinstone, S. M., Rosenblum, B., and Purcell, R. H., 1987a, Hepatitis A virus cDNA and its RNA transcripts are infectious in cell culture. *J. Virol.* 61:3035–3039.

Cohen, J. I., Ticehurst, J. R., Purcell, R. H., Buckler-White, A., and Baroudy, B. M., 1987b, Complete nucleotide sequence of wild-type hepatitis A virus: comparison with different strains of hepatitis A virus and other picornaviruses. *J. Virol.* 61:50–59.

Coppo, P., Scieux, C., Ferchal, F., Clauvel, J., and Lassoued, K., 2000, Astrovirus enteritis in a chronic lymphocytic leukemia patient treated with fludarabine monophosphate. *Ann. Hematol.* 79:43–45.

Costa-Mattioli, M., Ferre, V., Monpoeho, S., Garcia, L., Colina, R., Billaudel, S., Vega, I., Perez-Bercoff, R., and Cristina, J., 2001, Genetic variability of hepatitis A virus in South America reveals heterogeneity and co-circulation during epidemic outbreaks. *J. Gen. Virol.* 82:2647–2652.

Coulepis, A. G., Locarnini, S. A., Westaway, E. G., Tannock, G. A., and Gust, I. D., 1982, Biophysical and biochemical characterization of hepatitis A virus. *Intervirology* 18:107–127.

Coulson, B. S., Londrigan, S. L., and Lee, D. J., 1997, Rotavirus contains integrin ligand sequences and a disintegrin-like domain that are implicated in virus entry into cells. *Proc. Natl. Acad. Sci. U.S.A.* 94:5389–5394.

Crawford, S. E., Mukherjee, S. K., Estes, M. K., Lawton, J. A., Shaw, A. L., Ramig, R. F., and Prasad, V. V., 2001, Trypsin cleavage stabilizes the rotavirus VP4 spike. *J. Virol.* 75:6052–6061.

Cubitt, W. D., Mitchell, D. K., Carter, M. J., Willcocks, M. M., and Holzel, H., 1999, Application of electronmicroscopy, enzyme immunoassay, and RT-PCR to monitor an outbreak of astrovirus type 1 in a paediatric bone marrow transplant unit. *J. Med. Virol.* 57:313–321.

Cunliffe, N. A., Woods, P. A., Leite, J. P. G., Das, B. K., Ramachandran, M., Bhan, M. K., Hart, C. A., Glass, R. I., and Gentsch, J. R., 1997, Sequence analysis of NSP4 gene of human rotavirus allows classification into two main genetic groups. *J. Med. Virol.* 53:41–50.

Cunliffe, N. A., Gondwe, J. S., Broadhead, R. L., Molyneaux, M. E., Woods, P. A., Bresee, J. S., Glass, R. I., Gentsch, J. R., and Hart, C. A., 1999, Rotavirus G and P types in children with acute diarrhea in Blantyre, Malawi, from 1997 to 1998: predominance of novel P[6]G8 strains. *J. Med. Virol.* 57:308–312.

Dalton, R. M., Pastrana, E. P., and Sanchez-Fauquier, A., 2003, Vaccinia virus recombinant expressing an 87-kilodalton polyprotein that is sufficient to form astrovirus-like particles. *J. Virol.* 77:9094–9098.

Dastjerdi, A. M., Green, J., Gallimore, C. I., Brown, D. W., and Bridger, J. C., 1999, The bovine Newbury agent-2 is genetically more closely related to human SRSVs than to animal caliciviruses. *Virology* 254:1–5.

Daughenbaugh, K. F., Fraser, C. S., Hershey, J. W., and Hardy, M. E., 2003, The genome-linked protein VPg of the Norwalk virus binds eIF3, suggesting its role in translation initiation complex recruitment. *EMBO J.* 22:2852–2859.

Davidson, A. J., Telford, E. A. R., Watson, M. S., McBride, K., and Mautner, V., 1993, The DNA sequence of adenovirus type 40. *J. Mol. Biol.* 234:1308–1316.

de Chastonay, J., and Siegl, G., 1987, Replicative events in hepatitis A virus-infected MRC-5 cells. *Virology* 157:268–275.

de Jong, J. C., Wigand, R., Kidd, A. H., Wadell, G., Kapsenberg, J. G., Muzerie, C. J., Wermenbol, A. G., and Firtzlaff, R. G., 1983, Candidate adenoviruses 40 and 41: fastidious adenoviruses from human infant stool. *J. Med. Virol.* 11:215–231.

De Serres, G., Cromeans, T. L., Levesque, B., Brassard, N., Barthe, C., Dionne, M., Prud'homme, H., Paradis, D., Shapiro, C. N., Nainan, O. V., and Margolis, H. S., 1999, Molecular confirmation of hepatitis A virus from well water: epidemiology and public health implications. *J. Infect. Dis.* 179:37–43.

Denisova, E., Dowling, W., LaMonica, R., Shaw, R., Scarlata, S., Ruggeri, F., and Mackow, E. R., 1999, Rotavirus capsid protein VP5* permeabilizes membranes. *J. Virol.* 73:3147–3153.

Desselberger, U., 1996, Genome rearrangements of rotaviruses. *Adv. Virus Res.* 46:69–95.

Dolin, R., Levy, A. G., Wyatt, T. S., Thornhill, T. S., and Gardner, J. D., 1975, Viral gastroenteritis induced by the Hawaii agent. Jejunal histopathology and serologic response. *Am. J. Med.* 59:761–768.

Dormitzer, P. R., Greenberg, H. B., and Harrison, S. C., 2001, Proteolysis of monomeric recombinant rotavirus VP4 yields an oligomeric VP5* core. *J. Virol.* 75:7339–7350.

Dowdle, W. R., Cochi, S. L., Oberste, S., and Sutter, R., 2001, Preventing polio from becoming a reemerging disease. *Emerg. Infect. Dis.* 7:549–550.

Duckmanton, L., Luan, B., Devenish, J., Tellier, R., and Petric, M., 1997, Characterization of torovirus from human fecal specimens. *Virology* 239:158–168.

Duizer, E., Schwab, K. J., Neill, F. H., Atmar, R. L., Koopmans, M. P., and Estes, M. K., 2004, Laboratory efforts to cultivate noroviruses. *J. Gen. Virol.* 85:79–87.

Dunham, D. M., Jiang, X., Berke, T., Smith, A. W., and Matson, D. O., 1998, Genomic mapping of a calicivirus VPg. *Arch. Virol.* 143:2421–2430.

Enjuanes, L., Brian, D., Cavanagh, D., Holmes, K., Lai, M. M. C., Laude, H., Masters, P., Rottier, P. J. M., Siddell, S. G., Spaan, W. J. M., Taguchi, F., and Talbot, P., 2000, *Coronaviridae*, in: *Virus Taxonomy: The Classification and Nomenclature of Viruses. The Seventh Report of the International Committee on Taxonomy of Viruses*, Academic Press, San Diego, pp. 835–849.

Emerson, S. U., and Purcell, R. H., 2003, Hepatitis E virus. *Rev. Med. Virol.* 13:145–154.

Emerson, S. U., Zhang, M., Meng, X. J., Nguyen, H., St Claire, M., Govindarajan, S., Huang, Y. K., and Purcell, R. H., 2001, Recombinant hepatitis E virus genomes infectious for primates: importance of capping and discovery of a cis-reactive element. *Proc. Natl. Acad. Sci. U.S.A.* 98:15270–15275.

Estes, M. K., 2001, Rotaviruses and their replication, in: *Fields Virology* (D. M. Knipe, P. M. Howley et al., eds.), Lippincott Williams & Wilkins, Philadelphia, pp. 1747–1785.

Estes, M., and Cohen, J., 1989, Rotavirus gene structure and function. *Microbiol. Rev.* 53:410–449.

Ettayebi, K., and Hardy, M. E., 2003, Norwalk virus nonstructural protein p48 forms a complex with the SNARE regulator VAP-A and prevents cell surface expression of vesicular stomatitis virus G protein. *J. Virol.* 77:11790–11797.

Fankhauser, R. L., Monroe, S. S., Noel, J. S., Humphrey, C. D., Bresee, J. S., Parashar, U. D., Ando, T., and Glass, R. I., 2002, Epidemiologic and molecular trends of "Norwalk-like viruses" associated with outbreaks of gastroenteritis in the United States. *J. Infect. Dis.* 186:1–7.

Farkas, T., Zhong, W. M., Jing, Y., Huang, P. W., Espinosa, S. M., Martinez, N., Morrow, A. L., Ruiz-Palacios, G. M., Pickering, L. K., and Jiang, X., 2004, Genetic diversity among sapoviruses. *Arch. Virol.* 149:1309–1323.

Fiore, A. E., 2004, Hepatitis A transmitted by food. *Clin. Infect. Dis.* 38:705–715.

Fiore, L., Greenberg, H. B., and Mackow, E. R., 1991, The VP8 fragment of VP4 is the rhesus rotavirus hemagglutinin. *Virology* 181:553–563.

Flewett, T. H., Bryden, A. S., Davies, H. A., and Morris, C. A., 1973, Epidemic viral enteritis in a long-stay childrens ward. *Lancet* 1:4–5.

Flynn, W. T., and Saif, L. J., 1988, Serial propagation of porcine enteric calicivirus-like virus in porcine kindney cells. *J. Clin. Microbiol.* 26:203–212.

Fujiwara, K., Yokosuka, O., Imazeki, F., Saisho, H., Saotome, N., Suzuki, K., Okita, K., Tanaka, E., and Omata, M., 2003, Analysis of the genotype-determining region of hepatitis A viral RNA in relation to disease severities. *Hepatol. Res.* 25:124–134.

Gaggero, A., O'Ryan, M., Noel, J. S., Glass, R. I., Monroe, S. S., Mamani, N., Prado, V., and Avendano, L. F., 1998, Prevalence of astrovirus infection among Chilean children with acute gastroenteritis. *J. Clin. Microbiol.* 36:3691–3693.

Gallimore, C. I., Appleton, H., Lewis, D., Green, J., and Brown, D. W., 1995, Detection and characterisation of bisegmented double-stranded RNA viruses (picobirnaviruses) in human faecal specimens. *J. Med. Virol.* 45:135–140.

Garkavenko, O., Obriadina, A., Meng, J., Anderson, D. A., Benard, H. J., Schroeder, B. A., Khudyakov, Y. E., Fields, H. A., and Croxson, M. C., 2001, Detection and characterisation of swine hepatitis E virus in New Zealand. *J. Med. Virol.* 65: 525–529.

Geigenmuller, U., Ginzton, N. H., and Matsui, S. M., 1997, Construction of a genome-length cDNA clone for human astrovirus serotype 1 and synthesis of infectious RNA transcripts. *J. Virol.* 71:1713–1717.

Geigenmuller, U., Chew, T., Ginzton, N., and Matsui, S. M., 2002a, Processing of non-structural protein 1a of human astrovirus. *J. Virol.* 76:2003–2008.

Geigenmuller, U., Ginzton, N. H., and Matsui, S. M., 2002b, Studies on intracellular processing of the capsid protein of human astrovirus serotype 1 in infected cells. *J. Gen. Virol.* 83:1691–1695.

Gilbert, J. M., and Greenberg, H. B., 1998, Cleavage of rhesus rotavirus VP4 after arginine 247 is essential for rotavirus like particle-induced fusion from without. *J. Virol.* 72:5323–5327.

Glass, M. J., Jia, X. Y., and Summers, D. F., 1993, Identification of the hepatitis A virus internal ribosome entry site: in vivo and in vitro analysis of bicistronic RNAs containing the HAV 5′ noncoding region. *Virology* 193:842–852.

Glass, P. J., White, L. J., Ball, J. M., Leparc-Goffart, I., Hardy, M. E., and Estes, M. K., 2000a, Norwalk virus open reading frame 3 encodes a minor structural protein. *J. Virol.* 74:6581–6591.

Glass, R. I., Noel, J., Ando, T., Fankhauser, R., Belliot, G., Mounts, A., Parashar, U. D., Bresee, J. S., and Monroe, S. S., 2000b, The epidemiology of enteric caliciviruses from humans: a reassessment using new diagnostics. *J. Infect. Dis.* 181(Suppl 2): S254–S261.

Gonzalez, G. G., Pujol, F. H., Liprandi, F., Deibis, L., and Ludert, J. E., 1998, Prevalence of enteric viruses in human immunodeficiency virus seropositive patients in Venezuela. *J. Med. Virol.* 55:288–292.

Gosert, R., Cassinotti, P., Siegl, G., and Weitz, M., 1996, Identification of hepatitis A virus non-structural protein 2B and its release by the major virus protease 3C. *J. Gen. Virol.* 77:247–255.

Grabow, W. O. K., Puttergill, D. L., and Bosch, A., 1992, Propagation of adenoviruses types 40 and 41 in the PLC/PRF/5 primary liver carcinoma cell line. *J. Virol. Methods* 37:201–208.

Graham, F. L., Smiley, J., Russell, W. C., and Nairn, R., 1977, Characteristics of a human cell line transformed by DNA from human adenovirus type 5. *J. Gen. Virol.* 36: 59–72.

Graham, K. L., Halasz, P., Tan, Y., Hewish, M. J., Takada, Y., Mackow, E., Robinson, M. K., and Coulson, B. S., 2003, Integrin-using rotaviruses bind α2β1 integrin α2 I domain via VP4 DGE sequence and recognize αXβ2 and αVβ3 by using VP7 during cell entry. *J. Virol.* 77:9969–9978.

Green, K. Y., Kapikian, A. Z., Valdesuso, J., Sosnovtsev, S., Treanor, J. J., and Lew, J. F., 1997, Expression and self-assembly of recombinant capsid protein from the antigenically distinct Hawaii human calicivirus. *J. Clin. Microbiol.* 35:1909–1914.

Green, J., Gallimore, C. I., Clewley, J. P., and Brown, D. W., 1999, Genomic characterisation of the large segment of a rabbit picobirnavirus and comparison with the atypical picobirnavirus of Cryptosporidium parvum. *Arch. Virol.* 144:2457–2465.

Green, J., Vinje, J., Gallimore, C. I., Koopmans, M., Hale, A., Brown, D. W., Clegg, J. C., and Chamberlain, J., 2000a, Capsid protein diversity among Norwalk-like viruses. *Virus Genes* 20:227–236.

Green, K. Y., Ando, T., Balayan, M. S., Clarke, I. N., Estes, M. K., Matson, D. O., Nakata, S., Neill, J. D., Studdert, M. J., and Thiel, H. J., 2000b, Taxonomy of the caliciviruses. *J. Infect. Dis.* 181:S322–S330.

Green, K. Y., Chanock, R. M., and Kapikian, A. Z., 2001, Human caliciviruses, in: *Fields Virology*, 4th ed. (D. M. Knipe, P. M. Howley et al., eds.), Lippincott Williams & Wilkins, Philadelphia, pp. 841–874.

Green, K. Y., Mory, A., Fogg, M. H., Weisberg, A., Belliot, G., Wagner, M., Mitra, T., Ehrenfeld, E., Cameron, C. E., and Sosnovtsev, S. V., 2002, Isolation of enzymatically active replication complexes from feline calicivirus-infected cells. *J. Virol.* 76:8582–8595.

Griffin, D. D., Fletcher, M., Levy, M. E., Ching-Lee, M., Nogami, R., Edwards, L., Peters, H., Montague, L., Gentsch, J. R., and Glass, R. I., 2002, Outbreaks of adult gastroenteritis traced to a single genotype of rotavirus. *J. Infect. Dis.* 185: 1502–1505.

Grohmann, G. S., Glass, R. I., Pereira, H. G., Monroe, S. S., Hightower, A. W., Weber, R., and Bryan, R. T., 1993, Enteric viruses and diarrhea in HIV-infected patients. Enteric Opportunistic Infections Working Group. *N. Engl. J. Med.* 329:14–20.

Guerrero, C. A., Zárate, S., Corkidi, G., López, S., and Arias, C., 2000, Biochemical characterization of rotavirus receptors in MA104 cells. *J. Virol.* 74:9362–9371.

Guerrero, C. A., Bouyssounade, D., Zárate, S., Isa, P., López, T., Espinosa, R., Romero, P., Méndez, E., López, S., and Arias, C. F., 2002, Heat shock cognate protein 70 is involved in rotavirus cell entry. *J. Virol.* 76:4096–4102.

Guix, S., Caballero, S., Villena, C., Bartolomé, R., Latorre, C., Rabella, N., Simó, M., Bosch, A., and Pintó, R. M., 2002, Molecular epidemiology of astrovirus infection in Barcelona, Spain. *J. Clin. Microbiol.* 40:133–139.

Guo, M., and Saif, L. J., 2003, Pathogenesis of enteric calicivirus infections, in: *Viral Gastroenteritis* (U. Desselberger and J. Gray, eds.), Elsevier, Amsterdam, pp. 489–503.

Guo, C., Nakagomi, M., Mochizuki, M., Ishida, H., Kiso, M., Ohta, Y., Suzuki, T., Miyamoto, D., Hidari, K., and Suzuki, Y., 1999a, Ganglioside GM(1a) on the cell surface is involved in the infection by human rotavirus KUN and MO strains. *J. Biochem. (Tokyo)* 126:683–688.

Guo, M., Chang, K. O., Hardy, M. E., Zhang, Q., Parwani, A. V., and Saif, L. J., 1999b, Molecular characterization of a porcine enteric calicivirus genetically related to Sapporo-like human caliciviruses. *J. Virol.* 73:9625–9631.

Guo, M., Hayes, J., Cho, K. O., Parwani, A. V., Lucas, L. M., and Saif, L. J., 2001, Comparative pathogenesis of tissue culture adapted and wild type Cowden porcine enteric calicivirus (PEC) in gnotobiotic pigs and induction of diarrhea by intravenous inoculation of wild type PEC. *J. Virol.* 75:9239–9251.

Halaihel, N., Liévin, V., Ball, J., Estes, M. K., Alvarado, F., and Vasseur, M., 2000, Direct inhibitory effect of rotavirus NSP4(114-135) peptide on the Na+-D-glucose symporter of rabbit intestinal brush border membrane. *J. Virol.* 74:9464–9470.

Hale, A. D., Crawford, S. E., Ciarlet, M., Green, J., Gallimore, C., Brown, D. W., Jiang, X., and Estes, M. K., 1999, Expression and self-assembly of Grimsby virus: antigenic distinction from Norwalk and Mexico viruses. *Clin. Diagn. Lab. Immunol.* 6:142–145.

Hardy, M. E., Tanaka, T. N., Kitamoto, N., White, L. J., Ball, J. M., Jiang, X., and Estes, M. K., 1996, Antigenic mapping of the recombinant Norwalk virus capsid protein using monoclonal antibodies. *Virology* 217:252–261.

Hardy, M. E., Kramer, S. F., Treanor, J. J., and Estes, M. K., 1997, Human calicivirus genogroup II capsid sequence diversity revealed by analyses of the prototype Snow Mountain agent. *Arch. Virol.* 142:1469–1479.

Herbert, T. P., Brierley, I., and Brown, T. D., 1996, Detection of the ORF3 polypeptide of feline calicivirus in infected cells and evidence for its expression from a single, functionally bicistronic, subgenomic mRNA. *J. Gen. Virol.* 77:123–127.

Herbert, T. P., Brierley, I., and Brown, T. D., 1997, Identification of a protein linked to the genomic and subgenomic mRNAs of feline calicivirus and its role in translation. *J. Gen. Virol.* 78:1033–1040.

Hewish, M. J., Takada, Y., and Coulson, B. S., 2000, Integrins alfa2beta1 and alfa4beta1 can mediate SA11 rotavirus attachment and entry into cells. *J. Virol.* 74:228–236.

Hollinger, F. B., and Emerson, S. U., 2001, Hepatitis A virus, in: *Fields Virology* (D. M. Knipe, P. M. Howley et al., eds.), Lippincott Williams & Wilkins, Philadelphia, pp. 779–840.

Horie, Y., Nakagomi, O., Koshimura, Y., Nakagomi, Y., Suzuki, Y., Oka, T., Sasaki, S., Matsuda, Y., and Watanabe, S., 1999, Diarrhea induction by rotavirus NSP4 in the homologous mouse model system. *Virology* 398–407.

Horwitz, M. S., 2001, Adenoviruses, in: *Fields Virology*, 4th ed. (D. M. Knipe, P. M. Howley et al., eds.), Lippincott Williams & Wilkins, Philadelphia, pp. 2301–2326.

Hoshino, Y., Nishikawa, K., Benfield, D. A., and Gorziglia, M., 1994, Mapping of antigenic sites involved in serotype-cross-reactive neutralization on group A rotavirus outercapsid glycoprotein. *Virology* 199:233–237.

Hsieh, S. Y., Meng, X. J., Wu, Y. H., Liu, S. T., Tam, A. W., Lin, D. Y., and Liaw, Y. F., 1999, Identity of a novel swine hepatitis E virus in Taiwan forming a monophyletic group with Taiwan isolates of human hepatitis E virus. *J. Clin. Microbiol.* 37:3828–3834.

Hughes, A. L., 2004, Phylogeny of the *Picornaviridae* and differential evolutionary divergence of picornavirus proteins. *Infect. Genet. Evol.* 4:143–152.

Hughes, P. J., and Stanway, G., 2000, The 2A proteins of three diverse picornaviruses are related to each other and to the H-rev107 family of proteins involved in the control of cell proliferation. *J. Gen. Virol.* 81:201–207.

Hutin, Y. J., Pool, V., Cramer, E. H., Nainan, O. V., Weth, J., Williams, I. T., Goldstein, S. T., Gensheimer, K. F., Bell, B. P., Shapiro, C. N., Alter, M. J., and Margolis, H. S., 1999, A multistate, foodborne oubreak of hepatitis A. National Hepatitis A Investigation Team. *N. Engl. J. Med.* 340:595–602.

Hutson, A. M., Atmar, R. L., Graham, D. Y., and Estes, M. K., 2002, Norwalk virus infection and disease is associated with ABO histo-blood group type. *J. Infect. Dis.* 185:1335–1337.

Hutson, A. M., Atmar, R. L., and Estes, M. K., 2004, Norovirus disease: changing epidemiology and host susceptibility factors. *Trends Microbiol.* 12:279–287.

Isa, P., Realpe, M., Romero, P., S., L., and Arias, C. F., 2004, Rotavirus RRV associates with lipid membrane microdomains during cell entry. *Virology* 322:370–381.

Ishino, M., Ohashi, Y., Emoto, T., Sawada, Y., and Fujinaga, K., 1988, Characterisation of adenovirus type 40 E1 region. *Virology* 165:95–102.

Iturriza-Gomara, M., Green, J., Brown, D. W. G., Ramsay, M., Desselberger, U., and Gray, J., 2000, Molecular epidemiology of human group A rotavirus infections in the United Kingdom between 1995 and 1998. *J. Clin. Microbiol.* 38:4394–4401.

Iturriza-Gómara, M., Wong, C., Blome, S., Desselberger, U., and Gray, J., 2002, Molecular characterization of VP6 genes of human rotavirus isolates: correlation of genogroups with subgroups and evidence of independent segregation. *J. Virol.* 76:6596–6601.

Jayaram, H., Estes, M. K., and Prasad, B. V. V., 2004, Emerging themes in rotavirus cell entry, genome organization, transcription and replication. *Virus. Res.* 101: 67–81.

Jiang, X., Graham, D. Y., Wang, K., and Estes, M. K., 1990, Norwalk virus genome cloning and characterization. *Science* 250:1580–1583.

Jiang, X., Wang, M., Graham, D. Y., and Estes, M. K., 1992, Expression, self-assembly, and antigenicity of the Norwalk virus capsid protein. *J. Virol.* 66:6527–6532.

Jiang, B., Monroe, S. S., Koonin, E. V., Stine, S. E., and Glass, R. I., 1993a, RNA sequence of astrovirus: distinctive genomic organization and a putative retrovirus-like ribosomal frameshifting signal that directs the viral replicase synthesis. *Proc. Natl. Acad. Sci. U. S. A.* 90:10539–10543.

Jiang, X., Wang, M., Wang, K., and Estes, M. K., 1993b, Sequence and genomic organization of Norwalk virus. *Virology* 195:51–61.

Jiang, X., Turf, E., Hu, J., Barrett, E., Dai, X. M., Monroe, S., Humphrey, C., Pickering, L. K., and Matson, D. O., 1996, Outbreaks of gastroenteritis in elderly nursing homes and retirement facilities associated with human caliciviruses. *J. Med. Virol.* 50:335–341.

Jiang, X., Espul, C., Zhong, W. M., Cuello, H., and Matson, D. O., 1999a, Characterization of a novel human calicivirus that may be a naturally occurring recombinant. *Arch. Virol.* 144:2377–2387.

Jiang, X., Huang, P. W., Zhong, W. M., Farkas, T., Cubitt, D. W., and Matson, D. O., 1999b, Design and evaluation of a primer pair that detects both Norwalk- and Sapporo-like caliciviruses by RT-PCR. *J. Virol. Methods* 83:145–154.

Johansen, K., Hinkula, J., Espinoza, F., Levi, M., Zeng, C., Rudén, U., Vesikari, T., Estes, M., and Svensson, L., 1999, Humoral and cell-mediated immune responses in humans to the NSP4 enterotoxin of rotavirus. *J. Med. Virol.* 59:369–377.

Jourdan, N., Brunet, J. P., Sapin, C., Blais, A., Cotte-Laffitte, J., Forestier, F., Quero, A.-M., Trugnan, G., and Servin, A. L., 1998, Rotavirus infection reduces sucrase-isomaltase expression in human intestinal epithelial cells by perturbing protein targeting and organization of microvillar cytoskeleton. *J. Virol.* 72: 7228–7236.

Kabrane-Lazizi, Y., Fine, J. B., Elm, J., Glass, G. E., Higa, H., Diwan, A., Gibbs, C. J., Jr., Meng, X. J., Emerson, S. U., and Purcell, R. H., 1999a, Evidence for widespread infection of wild rats with hepatitis E virus in the United States. *Am. J. Trop. Med. Hyg.* 61:331–335.

Kabrane-Lazizi, Y., Meng, X. J., Purcell, R. H., and Emerson, S. U., 1999b, Evidence that the genomic RNA of hepatitis E virus is capped. *J. Virol.* 73:8848–8850.

Kamer, G., and Argos, P., 1984, Primary structural comparison of RNA-dependent polymerases from plant, animal and bacterial viruses. *Nucleic. Acids. Res.* 12:7269–7282.

Kang, J. A., and Funkhouser, A. W., 2002, A proposed vestigial translation initiation motif in VP1 of hepatitis A virus. *Virus. Res.* 87:11–19.

Kapikian, A. Z., Wyatt, R. G., Dolin, R., Thornhill, T. S., Kalica, A. R., and Chanock, R. M., 1972, Visualization by immune electron microscopy of a 27 nm particle associated with acute infectious nonbacterial gastroenteritis. *J. Virol.* 10:1075–1081.

Kapikian, A. Z., Hoshino, Y., and Chanock, R. M., 2001, Rotaviruses, in: *Fields. Virology* (D. M. Knipe, P. M. Howley et al., eds.), Lippincott Williams & Wilkins, Philadelphia, pp. 1787–1833.

Karst, S. M., Wobus, C. E., Lay, M., Davidson, J., and Virgin, H. W., 2003, STAT1-dependent innate immunity to a Norwalk-like virus. *Science* 299:1575–1578.

Khudyakov, Y. E., Lopareva, E. N., Jue, D. L., Crews, T. K., Thyagarajan, S. P., and Fields, H. A., 1999, Antigenic domains of the open reading frame 2-encoded protein of hepatitis E virus. *J. Clin. Microbiol.* 37:2863–2871.

Kiang, D., and Matsui, S. M., 2002, Proteolytic processing of a human astrovirus nonstructural protein. *J. Gen. Virol.* 83:25–34.

King, A. M. Q., Brown, F., Christian, P., Hovi, T., Hyypia, T., Knowles, N. J., Lemon, S. M., Minor, P. D., Palmenberg, A. C., Skern, T., and Stanway, G., 1999, Picornavirus taxonomy: a modified species definition and proposal for three new genera. XIth International Congress of Virology, Sydney, Australia.

King, A. M. Q., Brown, F., Christian, P., Hovi, T., Hyypiä, T., Knowles, N. J., Lemon, S. M., Minor, P. D., Palmenburg, A. C., Skern, T., and Stanway, G., 2000, Family *Picornaviridae*, in: *Virus Taxonomy. The Classification and Nomenclature of Viruses. The Seventh Report of the International Committee on Taxonomy of Viruses*. Academic Press, San Diego, pp. 657–678.

Kobayashi, S., Sakae, K., Suzuki, Y., Ishiko, H., Kamata, K., Suzuki, K., Natori, K., Miyamura, T., and Takeda, N., 2000, Expression of recombinant capsid proteins of chitta virus, a genogroup II Norwalk virus, and development of an ELISA to detect the viral antigen. *Microbiol. Immunol.* 44:687–693.

Konno, T., Suzuki, H., Ishida, N., Chiba, R., Mochizuki, K., and Tsunoda, A., 1982, Astrovirus-associated epidemic gastroenteritis in Japan. *J. Med. Virol.* 9:11–17.

Koonin, E. V., Gorbalenya, A. E., Purdy, M. A., Rozanov, M. N., Reyes, G. R., and Bradley, D. W., 1992, Computer-assisted assignment of functional domains in the nonstructural polyprotein of hepatitis E virus: delineation of an additional group of positive-strand RNA plant and animal viruses. *Proc. Natl. Acad. Sci. U. S. A.* 89:8259–8263.

Koopmans, M., and Horzinek, M. C., 1994, Toroviruses of animals and humans: a review. *Adv. Virus Res.* 43:233–273.

Koopmans, M. P., Goosen, E. S., Lima, A. A., McAuliffe, I. T., Nataro, J. P., Barrett, L. J., Glass, R. I., and Guerrant, R. L., 1997, Association of torovirus with acute and persistent diarrhea in children. *Pediatr. Infect. Dis. J.* 16:504–507.

Koopmans, M., Vinjé, J., de Wit, M., Leenen, I., van der Poel, W., and van Duynhoven, Y., 2000, Molecular epidemiology of human enteric caliciviruses in The Netherlands. *J. Infect. Dis.* 181(Suppl 2):S262–S269.

Koopmans, M., Vinje, J., Duizer, E., de Wit, M., and van Duijnhoven, Y., 2001, Molecular epidemiology of human enteric caliciviruses in The Netherlands. *Novartis Found. Symp.* 238:197–214; discussion 214–218.

Koopmans, M., von Bonsdorff, C.-H., Vinjé, J., de Medici, D., and Monroe, S., 2002, Foodborne viruses. *FEMS Microbiol. Rev.* 26:187–205.

Koopmans, M., van Strien, E., and Vennema, H., 2003, Molecular epidemiology of human caliciviruses, in: *Viral Gastroenteritis* (U. Desselberger and J. Gray, eds.), Elsevier, Amsterdam, pp. 523–554.

Korkaya, H., Jameel, S., Gupta, D., Tyagi, S., Kumar, R., Zafrullah, M., Mazumdar, M., Lal, S. K., Xiaofang, L., Sehgal, D., Das, S. R., and Sahal, D., 2001, The ORF3 protein of hepatitis E virus binds to Src homology 3 domains and activates MAPK. *J. Biol. Chem.* 276:42389–42400.

Kotloff, K. L., Losonsky, G. A., Morris, J. G., Jr., Wasserman, S. S., Singh-Naz, N., and Levine, M. M., 1989, Enteric adenovirus infection and childhood diarrhea: an epidemiologic study in three clinical settings. *Pediatrics* 84:219–225.

Kusov, Y. Y., Weitz, M., Dollenmaier, G., et al., 1996, RNA-protein interactions at the 3′-end of the hepatitis A virus RNA. *J. Virol.* 70:1890–1897.

Kuyumcu-Martinez, M., Belliot, G., Sosnovtsev, S. V., Chang, K. O., Green, K. Y., and Lloyd, R. E., 2004, Calicivirus 3C-Like proteinase inhibits cellular translation by cleavage of poly(A)-binding protein. *J. Virol.* 78:8172–8182.

Labbé, M., Charpilienne, A., Crawford, S. E., Estes, M. K., and Cohen, J., 1991, Expression of rotavirus VP2 produces empty corelike particles. *J. Virol.* 65:2946–2952.

Lai, M. M. C., and Holmes, K. V., 2001, *Coronaviridae*: The viruses and their replication, in: *Fields Virology*, 4th ed. (D. M. Knipe, P. M. Howley et al., eds.), Lippincott Williams & Wilkins, Philadelphia, pp. 1163–1185.

Lambden, P. R., Caul, E. O., Ashley, C. R., and Clarke, I. N., 1993, Sequence and genome organization of a human small round-structured (Norwalk-like) virus. *Science* 259:516–519.

Lambden, P. R., Caul, E. O., Ashley, C. R., and Clarke, I. N., 1994, Human enteric caliciviruses are genetically distinct from small round structured viruses. *Lancet* 343:666–667.

Lawton, J. A., Estes, M. K., and Prasad, B. V., 1997, Three-dimensional visualization of mRNA release from actively transcribing rotavirus particles. *Nat. Struct. Biol.* 4:118–121.

Lawton, J. A., Estes, M. K., and Prasad, B. V., 2000, Mechanisms of genome transcription in segmented dsRNA viruses. *Adv. Virus Res.* 55:185–229.

Lazdins, I., Coulson, B., Kirkwood, C., Dyall-Smith, M., Masendycz, P. J., Sonza, S., and Holmes, I. H., 1995, Rotavirus antigenicity is affected by the genetic context and glycosylation of VP7. *Virology* 209:80–89.

Lee, T. W., and Kurtz, J. B., 1982, Human astrovirus serotypes. *J. Hyg. (Lond.)* 89:539–540.

Lee, T. W., and Kurtz, J. B., 1994, Prevalence of human astrovirus serotypes in the Oxford region 1976–92, with evidence for two new serotypes. *Epidemiol. Infect.* 112:187–193.

Lemon, S. M., and Binn, L. N., 1983, Antigenic relatedness of two strains of hepatitis A virus determinant by cross-neutralization. *Infect. Immunol.* 42:418–420.

Lemon, S. M., and Robertson, B. H., 1993, Current perspectives in the virology and molecular biology of hepatitis A virus. *Semin. Virol.* 4:285–295.

Lemon, S. M., Jansen, R. W., and Newbold, J. E., 1985, Infectious hepatitis A virus particles produced in cell culture consist of three distinct types with different buoyant densities in CsCl. *J. Virol.* 54:78–85.

Leong, J. C., Brown, D., Dobos, P., Kibenge, F. S. B., Lufert, J. E., Muller, H., Mundt, E., and Nicholson, B., 2000, Family *Birnaviridae*, in: *Virus Taxonomy. Classifica-*

tion and Nomenclature of Viruses. Seventh Report of the International Committee on Taxonomy of Viruses. Academic Press, San Diego, pp. 481–490.

Lewis, D. C., Lightfoot, N. F., Cubitt, W. D., and Wilson, S. A., 1989, Outbreaks of astrovirus type 1 and rotavirus gastroenteritis in a geriatric in-patient population. *J. Hosp. Infect.* 14:9–14.

Lewis, T. L., Greenberg, H. B., Herrmann, J. E., Smith, L. S., and Matsui, S. M., 1994, Analysis of astrovirus serotype 1 RNA, identification of the viral RNA-dependent RNA polymerase motif, and expression of a viral structural protein. *J. Virol.* 68:77–83.

Li, F., Riddell, M. A., Seow, H. F., Takeda, N., Miyamura, T., and Anderson, D. A., 2000, Recombinant subunit ORF2.1 antigen and induction of antibody against immunodominant epitopes in the hepatitis E virus capsid protein. *J. Med. Virol.* 60: 379–386.

Lindesmith, L., Moe, C., Marionneau, S., Ruvoen, N., Jiang, X., Lindblad, L., Stewart, P., LePendu, J., and Baric, R., 2003, Human susceptibility and resistance to Norwalk virus infection. *Nat. Med.* 9:548–553.

Liste, M. B., Natera, I., Suarez, J. A., Pujol, F. H., Liprandi, F., and Ludert, J. E., 2000, Enteric virus infections and diarrhea in healthy and human immunodeficiency virus-infected children. *J. Clin. Microbiol.* 38:2873–2877.

Liu, M., Mattion, N. M., and Estes, M. K., 1992, Rotavirus VP3 expressed in insect cells possesses guanylyltransferase activity. *Virology* 188:77–84.

Liu, B. L., Clarke, I. N., Caul, E. O., and Lambden, P. R., 1995, Human enteric caliciviruses have a unique genome structure and are distinct from the Norwalk-like viruses. *Arch. Virol.* 140:1345–1356.

Liu, B., Clarke, I. N., and Lambden, P. R., 1996, Polyprotein processing in Southampton virus: identification of 3C-like protease cleavage sites by in vitro mutagenesis. *J. Virol.* 70:2605–2610.

Liu, B. L., Lambden, P. R., Gunther, H., Otto, P., Elschner, M., and Clarke, I. N., 1999a, Molecular characterization of a bovine enteric calicivirus: relationship to the Norwalk-like viruses. *J. Virol.* 73:819–825.

Liu, B. L., Viljoen, G. J., Clarke, I. N., and Lambden, P. R., 1999b, Identification of further proteolytic cleavage sites in the Southampton calicivirus polyprotein by expression of the viral protease in *E. coli*. *J. Gen. Virol.* 80:291–296.

López, S., and Arias, C. F., 2004, Multistep entry of rotavirus into cells: a Versaillesque dance. *Trends Microbiol.* 12:271–278.

López-Vazquez, A., Martin Alonso, J. M., Casais, R., Boga, J. A., and Parra, F., 1998, Expression of enzymatically active rabbit hemorrhagic disease virus RNA-dependent RNA polymerase in *Escherichia coli*. *J. Virol.* 72:2999–3004.

Lopman, B. A., Brown, D. W., and Koopmans, M., 2002, Human caliciviruses in Europe. *J. Clin. Virol.* 24:137–160.

Lopman, B. A., Reacher, M. H., Van Duijnhoven, Y., Hanon, F. X., Brown, D., and Koopmans, M., 2003, Viral gastroenteritis outbreaks in Europe, 1995–2000. *Emerg. Infect. Dis.* 9:90–96.

Lopman, B., Vennema, H., Kohli, E., Pothier, P., Sanchez, A., Negredo, A., Buesa, J., Schreier, E., Reacher, M., Brown, D., Gray, J., Iturriza, M., Gallimore, C., Bottiger, B., Hedlund, K. O., Torven, M., von Bonsdorff, C. H., Maunula, L., Poljsak-Prijatelj, M., Zimsek, J., Reuter, G., Szucs, G., Melegh, B., Svennson, L., van Duijnhoven, Y., and Koopmans, M., 2004, Increase in viral gastroenteritis outbreaks in Europe and epidemic spread of new norovirus variant. *Lancet* 363:682–688.

Love, D. N., and Sabine, M., 1975, Electron microscopic observation of feline kidney cells infected with a feline calicivirus. *Arch. Virol.* 48:213–228.

Luby, J. P., Clinton, R., and Kurtz, S., 1999, Adaptation of human enteric coronavirus to growth in cell lines. *J. Clin. Virol.* 12:43–51.

Ludert, J. E., and Liprandi, F., 1993, Identification of viruses with bi- and trisegmented double-stranded RNA genome in faeces of children with gastroenteritis. *Res. Virol.* 144:219–224.

Ludert, J. E., Hidalgo, M., Gil, F., and Liprandi, F., 1991, Identification in porcine faeces of a novel virus with a bisegmented double stranded RNA genome. *Arch. Virol.* 117:97–107.

Ludert, J. E., Abdul-Latiff, L., Liprandi, A., and Liprandi, F., 1995, Identification of picobirnavirus, viruses with bisegmented double stranded RNA, in rabbit faeces. *Res. Vet. Sci.* 59:222–225.

Macnaughton, M. R., and Davies, H. A., 1981, Human enteric coronaviruses. Brief review. *Arch. Virol.* 70:301–313.

Madeley, C. R., 1979, Comparison of the features of astroviruses and caliciviruses seen in samples of feces by electron microscopy. *J. Infect. Dis.* 139:519–523.

Madore, H. P., Treanor, J. J., and Dolin, R., 1986, Characterization of the Snow Mountain agent of viral gastroenteritis. *J. Virol.* 58:487–492.

Maguire, A. J., Green, J., Brown, D. W., Desselberger, U., and Gray, J. J., 1999, Molecular epidemiology of outbreaks of gastroenteritis associated with small round-structured viruses in East Anglia, United Kingdom, during the 1996–1997 season. *J. Clin. Microbiol.* 37:81–89.

Malcolm, B. A., Chin, S. M., Jewell, D. A., Stratton-Thomas, J. R., Thudium, K. B., Ralston, R., and Rosenberg, S., 1992, Expression and characterization of recombinant hepatitis A virus 3C proteinase. *Biochemistry* 31:3358–3363.

Manes, S., del Real, G., and Martinez, A. C., 2003, Pathogens: raft hijackers. *Nat. Rev. Immunol.* 3:557–568.

Marin, M. S., Casais, R., Alonso, J. M., and Parra, F., 2000, ATP binding and ATPase activities associated with recombinant rabbit hemorrhagic disease virus 2C-like polypeptide. *J. Virol.* 74:10846–10851.

Marionneau, S., Cailleau-Thomas, A., Rocher, J., Le Moullac-Vaidye, B., Ruvoen, N., Clément, M., and Le Pendu, J., 2001, ABH and Lewis histo-blood group antigens, a model for the meaning of oligosaccharide diversity in the face of a changing world. *Biochimie* 83:565–573.

Marionneau, S., Ruvoen, N., Le Moullac-Vaidye, B., Clement, M., Cailleau-Thomas, A., Ruiz-Palacios, G., Huang, P., Jiang, X., and Le Pendu, J., 2002, Norwalk virus binds to histo-blood group antigens present on gastroduodenal epithelial cells of secretor individuals. *Gastroenterology* 122:1967–1977.

Martin, A., Escriou, N., Chao, S. F., Girard, M., Lemon, S. M., and Wychowski, C., 1995, Identification and site-directed mutagenesis of the primary (2A/2B) cleavage site of the hepatitis A virus polyprotein: functional impact on the infectivity of HAV RNA transcripts. *Virology* 213:213–222.

Martin, A., Bénichou, D., Chao, S.-F., Cohen, L. M., and Lemon, S. M., 1999, Maturation of the hepatitis A virus capsid protein VP1 is not dependent on processing by the $3C^{pro}$ proteinase. *J. Virol.* 73:6220–6227.

Mast, E. E., Alter, M. J., Holland, P. V., and Purcell, R. H., 1998, Evaluation of assays for antibody to hepatitis E virus by a serum panel. Hepatitis E Virus Antibody Serum Panel Evaluation Group. *Hepatology* 27:857–861.

Mathan, M., Mathan, V. I., Swaminathan, S. P., Yesudoss, S., and Baker, S. J., 1975, Pleomorphic virus-like particles in human faeces. *Lancet* 1:1068–1069.

Mathieu, M., Petitpas, I., Navaza, J., Lepault, J., Kohli, E., Pothier, P., Prasad, B. V., Cohen, J., and Rey, F. A., 2001, Atomic structure of the major capsid protein of rotavirus: implications for the architecture of the virion. *EMBO J.* 20:1485–1497.

Matsui, M., 1997, Astrovirus, in: *Clinical Virology* (D. D. Richman, R. J. Whitley, and F. G. Hayden, eds.), Churchill Livingstone, New York, pp. 1111–1121.

Matsui, S. M., and Greenberg, H. B., 2000, Immunity to calicivirus infection. *J. Infect. Dis.* 181(Suppl 2):S331–S335.

Matsui, M., and Greenberg, H. B., 2001, Astroviruses, in: *Fields Virology* (D. M. Knipe, P. M. Howley et al., eds.), Lippincott Williams & Wilkins, Philadelphia, pp. 875–893.

Maunula, L., and von Bonsdorf, C.-H., 1998, Short sequences define genetic lineages: phylogenetic analysis of group A rotaviruses based on partial sequences of genome segments 4 and 9. *J. Gen. Virol.* 79:321–332.

Mautner, V., MacKay, N., and Steinthorsdottir, V., 1989, Complementation of enteric adenovirus type 40 for lytic growth in tissue culture by E1B 55K function of adenovirus types 5 and 12. *Virology* 171:619–622.

Mautner, V., Steinthorsdottir, V., and Bailey, A., 1995, Enteric adenoviruses. *Curr. Top. Microbiol. Immunol.* 179:229–282.

Mayo, M. A., 2002, Virus taxonomy—Houston 2002. *Arch. Virol.* 147:1071–1076.

Melnick, J. L., 1996, Enteroviruses: polioviruses, coxsackieviruses, echoviruses, and newer enteroviruses, in: *Fields Virology* (B. N. Fields, D. M. Knipe, P. M. Howley et al., eds.), Lippincott-Raven, Philadelphia, pp. 655–712.

Méndez, E., Fernández-Luna, T., López, S., Mendez-Toss, M., and Arias, C. F., 2002, Proteolytic processing of a serotype 8 human astrovirus ORF2 polyprotein. *J. Virol.* 76:7996–8002.

Meng, X. J., 2000, Novel strains of hepatitis E virus identified from humans and other animal species: is hepatitis E a zoonosis? *J. Hepatol.* 33:842–845.

Meng, X. J., Purcell, R. H., Halbur, P. G., Lehman, J. R., Webb, D. M., Tsareva, T. S., Haynes, J. S., Thacker, B. J., and Emerson, S. U., 1997, A novel virus in swine is closely related to the human hepatitis E virus. *Proc. Natl. Acad. Sci. U. S. A.* 94:9860–9865.

Meng, X. J., Halbur, P. G., Haynes, J. S., Tsareva, T. S., Bruna, J. D., Royer, R. L., Purcell, R. H., and Emerson, S. U., 1998a, Experimental infection of pigs with the newly identified swine hepatitis E virus (swine HEV), but not with human strains of HEV. *Arch. Virol.* 143:1405–1415.

Meng, X. J., Halbur, P. G., Shapiro, M. S., Govindarajan, S., Bruna, J. D., Mushahwar, I. K., Purcell, R. H., and Emerson, S. U., 1998b, Genetic and experimental evidence for cross-species infection by swine hepatitis E virus. *J. Virol.* 72:9714–9721.

Meng, J., Dai, X., Chang, J. C., Lopareva, E., Pillot, J., Fields, H. A., and Khudyakov, Y. E., 2001, Identification and characterization of the neutralization epitope(s) of the hepatitis E virus. *Virology* 288:203–211.

Midthun, K., Greenberg, H. B., Kurtz, J. B., Gary, G. W., Lin, F. Y., and Kapikian, A. Z., 1993, Characterization and seroepidemiology of a type 5 astrovirus associated with an outbreak of gastroenteritis in Marin County, California. *J. Clin. Microbiol.* 31:955–962.

Mikami, T., Nakagomi, T., Tsutsui, R., Ishikawa, K., Onodera, Y., Arisawa, K., and Nakagomi, O., 2004, An outbreak of gastroenteritis during school trip caused by serotype G2 group A rotavirus. *J. Med. Virol.* 73:460–464.

Minor, P., 1991, *Picornaviridae*, in: *Classification and nomenclature of viruses* (R. I. B. Francki, C. M. Fauquet, and D. L. Knudson, eds.), Springer-Verlag, Wien, pp. 320–326.

Mirazimi, A., Magnusson, K.-E., and Svensson, L., 2003, A citoplasmic region of the NSP4 enterotoxin of rotavirus is involved in retention in the endoplasmic reticulum. *J. Gen. Virol.* 84:875–883.

Monroe, S. S., Stine, S. E., Gorelkin, L., Herrmann, J. E., Blacklow, N. R., and Glass, R. I., 1991, Temporal synthesis of proteins and RNAs during human astrovirus infection of cultured cells. *J. Virol.* 65:641–648.

Monroe, S. S., Holmes, J. L., and Belliot, G. M., 2001, Molecular epidemiology of human astroviruses. *Novartis Found. Symp.* 238:237–245; discussion 245–249.

Morales, M., Barcena, J., Ramirez, M. A., Boga, J. A., Parra, F., and Torres, J. M., 2004, Synthesis in vitro of rabbit hemorrhagic disease virus subgenomic RNA by internal initiation on (-)sense genomic RNA: mapping of a subgenomic promoter. *J. Biol. Chem.* 279:17013–17018.

Morris, A. P., and Estes, M. K., 2001, Microbes and microbial toxins: Paradigms for microbial-mucosal interactions. VIII. Pathological consequences of rotavirus infection and its enterotoxin. *Am. J. Physiol. Gastrointest. Liver Physiol.* 281: G303–G310.

Morris, A. P., Scott, J. K., Ball, J. M., Zeng, C. Q.-Y., O'Neal, W. K., and Estes, M. K., 1999, NSP4 elicits age-dependent diarrhea and Ca^{2+}-mediated I^{-}Influx into intestinal crypts of CF mice. *Am. J. Physiol. Gastrointest. Liver Physiol.* 277: G431–G444.

Mossel, E. C., and Ramig, R. F., 2003, A lymphatic mechanism of rotavirus extraintestinal spread in the neonatal mouse. *J. Virol.* 77:12352–12356.

Mustafa, H., Palombo, E. A., and Bishop, R. F., 2000, Epidemiology of astrovirus infection in young children hospitalized with acute gastroenteritis in Melbourne, Australia, over a period of four consecutive years, 1995 to 1998. *J. Clin. Microbiol.* 38:1058–1062.

Nakata, S., Kogawa, K., Numata, K., Ukae, S., Adachi, N., Matson, D. O., Estes, M. K., and Chiba, S., 1996, The epidemiology of human calicivirus/Sapporo/82/Japan. *Arch. Virol.* Suppl 12:263–270.

Neill, J. D., 1990, Nucleotide sequence of a region of the feline calicivirus genome which encodes picornavirus-like RNA-dependent RNA polymerase, cysteine protease and 2C polypeptides. *Virus Res.* 17:145–160.

Nilsson, M., Hedlund, K. O., Thorhagen, M., Larson, G., Johansen, K., Ekspong, A., and Svensson, L., 2003, Evolution of human calicivirus RNA in vivo: accumulation of mutations in the protruding P2 domain of the capsid leads to structural changes and possibly a new phenotype. *J. Virol.* 77:13117–13124.

Nishikawa, K., Hoshino, Y., Taniguchi, K., Green, J., Greenberg, H. B., Kapikian, A. Z., Chanock, R. M., and Gorziglia, M., 1989, Rotavirus VP7 neutralization epitopes of serotype 3 strains. *Virology* 171:503–515.

Noel, J., and Cubitt, D., 1994, Identification of astrovirus serotypes from children treated at the Hospitals for Sick Children, London 1981–93. *Epidemiol. Infect.* 113:153–159.

Noel, J. S., Lee, T. W., Kurtz, J. B., Glass, R. I., and Monroe, S. S., 1995, Typing of human astroviruses from clinical isolates by enzyme immunoassay and nucleotide sequencing. *J. Clin. Microbiol.* 33:797–801.

Noel, J. S., Liu, B. L., Humphrey, C. D., Rodriguez, E. M., Lambden, P. R., Clarke, I. N., Dwyer, D. M., Ando, T., Glass, R. I., and Monroe, S. S., 1997, Parkville virus: a novel genetic variant of human calicivirus in the Sapporo virus clade, associated with an outbreak of gastroenteritis in adults. *J. Med. Virol.* 52:173–178.

Numata, K., Hardy, M. E., Nakata, S., Chiba, S., and Estes, M. K., 1997, Molecular characterization of morphologically typical human calicivirus Sapporo. *Arch. Virol.* 142:1537–1552.

Oberste, M. S., Maher, K., Kilpatrick, D. R., and Pallansch, M. A., 1999, Molecular evolution of the human enteroviruses: correlation of serotype with VP1 sequence and application to picornavirus classification. *J. Virol.* 73:1941–1948.

Obert, G., Peiffer, I., and Servin, A., 2000, Rotavirus-induced structural and functional alterations in tight junctions of polarized intestinal Caco-2 cell monolayers. *J. Virol.* 74:4645–4651.

O'Brien, J. A., Taylor, J. A., and Bellamy, A. R., 2000, Probing the structure of rotavirus NSP4: a short sequence at the extreme C terminus mediates binding to the inner capsid particle. *J. Virol.* 74:5388–5394.

Oliver, S. L., Dastjerdi, A. M., Wong, S., El-Attar, L., Gallimore, C., Brown, D. W., Green, J., and Bridger, J. C., 2003, Molecular characterization of bovine enteric caliciviruses: a distinct third genogroup of noroviruses (Norwalk-like viruses) unlikely to be of risk to humans. *J. Virol.* 77:2789–2798.

Palombo, E. A., Bishop, R. F., and Cotton, R. G. H., 1993, Sequence conservation within neutralization epitope regions of VP7 and VP4 proteins of human serotype G4 rotavirus isolates. *Arch. Virol.* 133:323–334.

Panda, S. K., Ansari, I. H., Durgapal, H., Agrawal, S., and Jameel, S., 2000, The in vitro-synthesized RNA from a cDNA clone of hepatitis E virus is infectious. *J. Virol.* 74:2430–2437.

Parashar, U. D., Hummelman, E. G., Bresee, J. S., Miller, M. A., and Glass, R. I., 2003, Global illness and deaths caused by rotavirus disease in children. *Emerg. Infect. Dis.* 9:565–572.

Parwani, A. V., Flynn, W. T., Gadfield, K. L., and Saif, L. J., 1991, Serial propagation of porcine enteric calicivirus: effects of medium supplementation with intestinal contents or enzymes. *Arch. Virol.* 120:115–122.

Patton, J. T., and Gallegos, C. O., 1990, Rotavirus RNA replication: single-strand RNA extends from the replicase particle. *J. Gen. Virol.* 71:1087–1094.

Patton, J. T., Kearney, K., and Taraporewala, Z., 2003, Rotavirus genome replication: role of the RNA-binding proteins, in: *Viral Gastroenteritis* (U. Desselberger and J. Gray, eds.), Elsevier, Amsterdam, pp. 165–183.

Pereira, H. G., Fialho, A. M., Flewett, T. H., Candeias, J. A. N., and Andrade, Z. P., 1988a, Novel viruses in human faeces. *Lancet* 2:103–104.

Pereira, H. G., Flewett, T. H., Candeias, J. A., and Barth, O. M., 1988b, A virus with bisegmented double-stranded RNA genome in rat (*Oryzomys nigripes*). *J. Gen. Virol.* 69:2749–2754.

Perron-Henry, D. M., Herrmann, J. E., and Blacklow, N. R., 1988, Isolation and propagation of enteric adenoviruses in HEp-2 cells. *J. Clin. Microbiol.* 26:1445–1447.

Pesavento, J. B., Lawton, J. A., Estes, M. K., and Prasad, B. V. V., 2001, The reversible condensation and expansion of the rotavirus genome. *Proc. Natl. Acad. Sci. U. S. A.* 98:1381–1386.

Pesavento, J. B., Estes, M. K., and Prasad, B. V. V., 2003, Structural organization of the genome in rotavirus, in: *Viral Gastroenteritis* (U. Desselberger and J. Gray, eds.), Elsevier, Amsterdam.

Pfister, T., and Wimmer, E., 2001, Polypeptide p41 of a Norwalk-like virus is a nucleic acid-independent nucleoside triphosphatase. *J. Virol.* 75:1611–1619.

Pintó, R. M., Díez, J. M., and Bosch, A., 1994, Use of the colonic carcinoma cell line CaCo-2 for in vivo amplification and detection of enteric viruses. *J. Med. Virol.* 44:310–315.

Pintó, R. M., Abad, F. X., Gajardo, R., and Bosch, A., 1996, Detection of infectious astroviruses in water. *Appl. Environ. Microbiol.* 62:1811–1813.

Pintó, R. M., Villena, C., Le Guyader, F., Guix, S., Caballero, S., Pommepuy, M., and Bosch, A., 2001, Astrovirus detection in wastewater samples. *Water Sci. Technol.* 43:73–76.

Piron, M., Vende, P., Cohen, J., and Poncet, D., 1998, Rotavirus RNA-binding protein NSP3 interacts with eIF4GI and evicts the poly(A) binding protein from eIF4F. *EMBO J.* 17:5811–5821.

Poncet, D., Laurent, S., and Cohen, J., 1994, Four nucleotides are the minimal requirement for RNA recognition by rotavirus non-structural protein NSP3. *EMBO J.* 13:4165–4173.

Prasad, B. V., Wang, G. J., Clerx, J. P. M., and Chiu, W., 1988, Three-dimensional structure of rotavirus. *J. Mol. Biol.* 199:269–275.

Prasad, B. V., Rothnagel, R., Jiang, X., and Estes, M. K., 1994, Three-dimensional structure of baculovirus-expressed Norwalk virus capsids. *J. Virol.* 68:5117–5125.

Prasad, B. V., Hardy, M. E., Jiang, X., and Estes, M. K., 1996a, Structure of Norwalk virus. *Arch. Virol.* Suppl 12:237–242.

Prasad, B. V., Rothnagel, R., Zeng, C. Q., Jakana, J., Lawton, J. A., Chiu, W., and Estes, M. K., 1996b, Visualization of ordered genomic RNA and localization of transcriptional complexes in rotavirus. *Nature* 382:471–473.

Prasad, B. V., Hardy, M. E., Dokland, T., Bella, J., Rossmann, M. G., and Estes, M. K., 1999, X-ray crystallographic structure of the Norwalk virus capsid. *Science* 286:287–290.

Probst, C., Jecht, M., and Gauss-Mueller, V., 1998, Processing of proteinase precursors and their effect on hepatitis A virus particle formation. *Virology* 72:8013–8020.

Purcell, R. H., and Emerson, S. U., 2001, Hepatitis E virus, in: *Fields Virology* (D. M. Knipe, P. M. Howley et al., eds.), Lippincott Williams & Wilkins, Philadelphia, pp. 3051–3061.

Racaniello, V. R., 2001, *Picornaviridae*: the viruses and their replication, in: *Fields Virology*, 4th ed. (D. M. Knipe, P. M. Howley et al., eds.), Lippincott Williams & Wilkins, Philadelphia, pp. 685–722.

Ramig, R., 2000, Mixed infections with rotaviruses: protocols for reassortment, complementation and other assays, in: *Rotaviruses: Methods and Protocols* (J. Gray and U. Desselberger, eds.), Humana Press, Totowa, NJ, pp. 79–100.

Ramig, R. F., and Estes, M. K., 2003, The dsRNA segments and proteins of simian rotavirus A / SA11 (genus *Rotavirus*: family *Reoviridae*), in: *The RNAs and Proteins of dsRNA Viruses* (P. C. Mertens and D. H. Bamford, eds.). Available at http://www.iah.bbsrc.ac.uk/dsRNA_virus_proteins/Rotavirus.htm.

Ray, P., Malik, J., Singh, R. K., Bhatnagar, S., Kumar, R., and Bhan, M. K., 2003, Rotavirus nonstructural protein NSP4 induces heterotypic antibody responses during natural infection in children. *J. Infect. Dis.* 187:1786–1793.

Resta, S., Luby, J. P., Rosenfeld, C. R., and Siegel, J. D., 1985, Isolation and propagation of a human enteric coronavirus. *Science* 229:978–981.

Riddell, M. A., Li, F., and Anderson, D. A., 2000, Identification of immunodominant and conformational epitopes in the capsid protein of hepatitis E virus by using monoclonal antibodies. *J. Virol.* 74:8011–8017.

Risco, C., Carrascosa, J. L., Pedregosa, A. M., Humphrey, C. D., and Sanchez-Fauquier, A., 1995, Ultrastructure of human astrovirus serotype 2. *J. Gen. Virol.* 76:2075–2080.

Robertson, B. H., Jansen, R. W., Khanna, B., Totsuka, A., Nainan, O. V., Siegl, G., Widell, A., Margolis, H. S., Isomura, S., Ito, K., Ishizu, T., Moritsugu, Y., and Lemon, S. M., 1992, Genetic relatedness of hepatitis A virus strains recovered from different geographical regions. *J. Gen. Virol.* 73:1365–1377.

Robinson, R. A., Burgess, W. H., Emerson, S. U., Leibowitz, R. S., Sosnovtseva, S. A., Tsarev, S., and Purcell, R. H., 1998, Structural characterization of recombinant hepatitis E virus ORF2 proteins in baculovirus-infected insect cells. *Protein Expr. Purif.* 12:75–84.

Rolsma, M. D., Kuhlenschmidt, T. B., Gelberg, H. B., and Kuhlenschmidt, M. S., 1998, Structure and function of a ganglioside receptor for porcine rotavirus. *J. Virol.* 72:9079–9091.

Rosen, B., 2003, Molecular characterization and epidemiology of picobirnaviruses, in: *Viral Gastroenteritis* (U. Desselberger and J. Gray, eds.), Elsevier, Amsterdam, pp. 633–644.

Rosen, B. I., Fang, Z. Y., Glass, R. I., and Monroe, S. S., 2000, Cloning of human picobirnavirus genomic segments and development of an RT-PCR detection assay. *Virology* 277:316–329.

Rossmann, M. G., and Johnson, J. E., 1989, Icosahedral RNA virus structure. *Annu. Rev. Biochem.* 58:533–573.

Ruvöen-Clouet, N., Ganière, J. P., André-Fontain, G., Blanchard, D., and Le Pendu, J., 2000, Binding of rabbit haemorrhagic disease virus to antigens of the ABH blood group family. *J. Virol.* 74:11950–11954.

Sánchez, G., Bosch, A., and Pintó, R. M., 2003, Genome variability and capsid structural constraints of hepatitis A virus. *J. Virol.* 77:452–459.

Sanekata, T., Ahmed, M. U., Kader, A., Taniguchi, K., and Kobayashi, N., 2003, Human group B rotavirus infections cause severe diarrhea in children and adults in Bangladesh. *J. Clin. Microbiol.* 41:2187–2190.

Sasaki, J., Kusuhara, Y., Maeno, Y., Kobayashi, N., Yamashita, T., Sakae, K., Takeda, N., and Taniguchi, K., 2001, Construction of an infectious cDNA clone of Aichi virus (a new member of the family *Picornaviridae*) and mutational analysis of a stem-loop structure at the 5′ end of the genome. *J. Virol.* 75:8021–8030.

Schlauder, G. G., Dawson, G. J., Erker, J. C., Kwo, P. Y., Knigge, M. F., Smalley, D. L., Rosenblatt, J. E., Desai, S. M., and Mushahwar, I. K., 1998, The sequence and phylogenetic analysis of a novel hepatitis E virus isolated from a patient with acute hepatitis reported in the United States. *J. Gen. Virol.* 79:447–456.

Schlauder, G. G., Desai, S. M., Zanetti, A. R., Tassopoulos, N. C., and Mushahwar, I. K., 1999, Novel hepatitis E virus (HEV) isolates from Europe: evidence for additional genotypes of HEV. *J. Med. Virol.* 57:243–251.

Schuffenecker, I., Ando, T., Thouvenot, D., Lina, B., and Aymard, M., 2001, Genetic classification of "Sapporo-like viruses." *Arch. Virol.* 146:2115–2132.

Servin, A. L., 2003, Effects of rotavirus infection on the structure and functions of intestinal cells, in: *Viral Gastroenteritis* (U. Desselberger and J. Gray, eds.), Elsevier, Amsterdam, pp. 237–254.

Shenk, E. S., 2001, Adenoviridae: the viruses and their replication, in: *Fields Virology*, 4th ed. (D. M. Knipe and P. M. Howley, eds.), Lippincott Williams & Wilkins, Philadelphia, pp. 2265–2300.

Siegl, G., de Chastonay, J., and Kronauer, G., 1984, Propagation and assay of hepatitis A virus in vitro. *J. Virol. Methods* 9:53–67.

Smiley, J. R., Hoet, A. E., Traven, M., Tsunemitsu, H., and Saif, L. J., 2003, Reverse transcription-PCR assays for detection of bovine enteric caliciviruses (BEC) and analysis of the genetic relationships among BEC and human caliciviruses. *J. Clin. Microbiol.* 41:3089–3099.

Smith, A. W., Akers, T. G., Madin, S. H., and Vedros, N. A., 1973, San Miguel sea lion virus isolation, preliminary characterization and relationship to vesicular exanthema of swine virus. *Nature* 244:108–110.

Smith, A. W., Skilling, D. E., Ensley, P. K., Benirschke, K., and Lester, T. L., 1983, Calicivirus isolation and persistence in a pygmy chimpanzee (*Pan paniscus*). *Science* 221:79–81.

Smits, S. L., Lavazza, A., Matiz, K., Horzinek, M. C., Koopmans, M. P., and de Groot, R. J., 2003, Phylogenetic and evolutionary relationships among torovirus field variants: evidence for multiple intertypic recombination events. *J. Virol.* 77: 9567–9577.

Smyth, M., Pettitt, T., Symonds, A., and Martin, J., 2003, Identification of the pocket factors in a picornavirus. *Arch. Virol.* 148:1225–1233.

Snijder, E. J., and Horzinek, M. C., 1993, Toroviruses: replication, evolution and comparison with other members of the coronavirus-like superfamily. *J. Gen. Virol.* 74:2305–2316.

Snijder, E. J., den Boon, J. A., Bredenbeek, P. J., Horzinek, M. C., Rijnbrand, R., and Spaan, W. J., 1990a, The carboxyl-terminal part of the putative Berne virus polymerase is expressed by ribosomal frameshifting and contains sequence motifs which indicate that toro- and coronaviruses are evolutionarily related. *Nucleic Acids Res.* 18:4535–4542.

Snijder, E. J., Horzinek, M. C., and Spaan, W. J., 1990b, A 3′-coterminal nested set of independently transcribed mRNAs is generated during Berne virus replication. *J. Virol.* 64:331–338.

Sosnovtsev, S., and Green, K. Y., 2003, Feline calicivirus as a model for the study of calicivirus replication, in: *Viral Gastroenteritis* (U. Desselberger and J. Gray, eds.), Elsevier, Amsterdam, pp. 467–503.

Stevenson, F., and Mautner, V., 2003, Aspects of the molecular biology of enteric adenoviruses, in: *Viral Gastroenteritis* (U. Desselberger and J. Gray, eds.), Elsevier, Amsterdam, pp. 389–406.

Sugieda, M., and Nakajima, S., 2002, Viruses detected in the caecum contents of healthy pigs representing a new genetic cluster in genogroup II of the genus "Norwalk-like viruses." *Virus Res.* 87:165–172.

Tafazoli, F., Zeng, C. Q., Estes, M., Magnusson, K.-E., and Svensson, L., 2001, NSP4 enterotoxin of rotavirus induces paracellular leakage in polarized epithelial cells. *J. Virol.* 75:1540–1546.

Takahashi, K., Iwata, K., Watanabe, N., Hatahara, T., Ohta, Y., Baba, K., and Mishiro, S., 2001, Full-genome nucleotide sequence of a hepatitis E virus strain that may be indigenous to Japan. *Virology* 287:9–12.

Takiff, H. E., Strauss, S. E., and Garon, C. F., 1981, Propagation and in vitro studies of previously non-cultivable enteral adenovirus in 293 cells. *Lancet* 2:832–834.

Takiff, H. E., Reinhold, W., Garon, C. F., and Straus, S. E., 1984, Cloning and physical mapping of enteric adenoviruses (candidate types 40 and 41). *J. Virol.* 51:131–136.

Tam, A. W., Smith, M. M., Guerra, M. E., Huang, C. C., Bradley, D. W., Fry, K. E., and Reyes, G. R., 1991, Hepatitis E virus (HEV): molecular cloning and sequencing of the full-length viral genome. *Virology* 185:120–131.

Tan, M., Huang, P., Meller, J., Zhong, W., Farkas, T., and Jiang, X., 2003, Mutations within the P2 domain of norovirus capsid affect binding to human histo-blood group antigens: evidence for a binding pocket. *J. Virol.* 77:12562–12571.

Tan, M., Hegde, R. S., and Jiang, X., 2004, The P domain of norovirus capsid protein forms dimer and binds to histo-blood group antigen receptors. *J. Virol.* 78:6233–6242.

Taylor, J. A., O'Brien, J. A., Lord, V. J., Meyer, J. C., and Bellamy, A. R., 1993, The RER-localized rotavirus intracellular receptor: a truncated purified soluble form is multivalent and bind virus particles. *Virology* 194:807–814.

Taylor, M. B., Walter, J., Berke, T., Cubitt, W. D., Mitchell, D. K., and Matson, D. O., 2001, Characterisation of a South African human astrovirus as type 8 by antigenic and genetic analyses. *J. Med. Virol.* 64:256–261.

Terashima, H., Chiba, S., Sakuma, Y., Kogasaka, R., Nakata, S., Minami, R., Horino, K., and Nakao, T., 1983, The polypeptide of a human calicivirus. *Arch. Virol.* 78:1–7.

Tian, P., Hu, Y., Schilling, W. P., Lindsay, D. A., Eiden, J., and Estes, M. K., 1994, The nonstructural glycoprotein of rotavirus affects intracellular calcium levels. *J. Virol.* 68:251–257.

Tian, P., Ball, J. M., Zeng, C. Q. Y., and Estes, M. K., 1996, The rotavirus nonstructural glycoprotein NSP4 possesses membrane destabilization activity. *J. Virol.* 70:6973–6981.

Tiemessen, C. T., Nel, M. J., and Kidd, A. H., 1996, Adenovirus 41 replication: cell-related differences in viral gene transcription. *Mol. Cell Probes* 10:279–287.

Tsarev, S., A., Tsareva, T. S., Emerson, S. U., Govindarajan, S., Shapiro, M., Gerin, J. L., and Purcell, R. H., 1994, Successful passive and active immunization of cynomolgus monkeys against hepatitis E. *Proc. Natl. Acad. Sci. U. S. A.* 91:10198–10202.

Tyagi, S., Korkaya, H., Zafrullah, M., Jameel, S., and Lal, S. K., 2002, The phosphorylated form of the ORF3 protein of hepatitis E virus interacts with its non-glycosylated form of the major capsid protein, ORF2. *J. Biol. Chem.* 277: 22759–22767.

Uhnoo, I., Wadell, G., Svensson, L., and Johansson, M., 1983, Two new serotypes of enteric adenovirus causing infantile diarrhoea. *Dev. Biol. Stand.* 53:311–318.

Uhnoo, I., Wadell, G., Svensson, L., and Johansson, M. E., 1984, Importance of enteric adenoviruses 40 and 41 in acute gastroenteritis in infants and young children. *J. Clin. Microbiol.* 20:365–372.

Uhnoo, I., Svensson, L., and Wadell, G., 1990, Enteric adenoviruses. *Baillieres Clin. Gastroenterol.* 4:627–642.

Unicomb, L. E., Banu, N. N., Azim, T., Islam, A., Bardhan, P. K., Faruque, A. S., Hall, A., Moe, C. L., Noel, J. S., Monroe, S. S., Albert, M. J., and Glass, R. I., 1998, Astrovirus infection in association with acute, persistent and nosocomial diarrhea in Bangladesh. *Pediatr. Infect. Dis. J.* 17:611–614.

van der Poel, W. H., Vinje, J., van Der Heide, R., Herrera, M. I., Vivo, A., and Koopmans, M. P., 2000, Norwalk-like calicivirus genes in farm animals. *Emerg. Infect. Dis.* 6:36–41.

van Loon, A. E., Maas, R., Vaessen, R. T. M., Reemst, A. M. C. B., Sussenbach, J. M., and Rozijn, T. H., 1985a, Cell transformation by the left terminal regions of the adenovirus 40 and 41 genomes. *Virology* 147:227–230.

van Loon, A. E., Rozijn, T. H., de Jong, J. C., and Sussenbach, J. S., 1985b, Physicochemical properties of the DNAs of the fastidious adenovirus species 40 and 41. *Virology* 140:197–200.

van Loon, A. E., Gilardi, P., Perricaudet, M., Rozijn, T. H., and Sussenbach, J. S., 1987, Transcriptional activation by the E1A regions of adenovirus types 40 and 41. *Virology* 160:305–307.

van Vliet, A. L., Smits, S. L., Rottier, P. J., and de Groot, R. J., 2002, Discontinuous and non-discontinuous subgenomic RNA transcription in a nidovirus. *EMBO J.* 21:6571–6580.

Vinje, J., and Koopmans, M. P., 1996, Molecular detection and epidemiology of small round-structured viruses in outbreaks of gastroenteritis in the Netherlands. *J. Infect. Dis.* 174:610–615.

Vinje, J., Green, J., Lewis, D. C., Gallimore, C. I., Brown, D. W., and Koopmans, M. P., 2000, Genetic polymorphism across regions of the three open reading frames of "Norwalk-like viruses." *Arch. Virol.* 145:223–241.

Vinje, J., Hamidjaja, R. A., and Sobsey, M. D., 2004, Development and application of a capsid VP1 (region D) based reverse transcription PCR assay for genotyping of genogroup I and II noroviruses. *J. Virol. Methods* 116:109–117.

Wadell, G., 1984, Molecular epidemiology of human adenoviruses. *Curr. Top. Microbiol. Immunol.* 110:191–220.

Weitz, M., Baroudy, B. M., Maloy, W. L., Ticehurst, J. R., and Purcell, R. H., 1986, Detection of a genome-linked protein (VPg) of hepatitis A virus and its comparison with other picornaviral VPgs. *J. Virol.* 60:124–130.

Weitz, M., and Siegl, G., 1998, Hepatitis A virus: structure and molecular virology, in: *Viral Hepatitis* (A. J. Zuckerman and H. C. Thomas, eds.), Churchill Livingstone, London, pp. 15–27.

White, L. J., Ball, J. M., Hardy, M. E., Tanaka, T. N., Kitamoto, N., and Estes, M. K., 1996, Attachment and entry of recombinant Norwalk virus capsids to cultured human and animal cell lines. *J. Virol.* 70:6589–6597.

White, L. J., Hardy, M. E., and Estes, M. K., 1997, Biochemical characterization of a smaller form of recombinant Norwalk virus capsids assembled in insect cells. *J. Virol.* 71:8066–8072.

Willcocks, M. M., Carter, M. J., Laidler, F. R., and Madeley, C. R., 1990, Growth and characterisation of human faecal astrovirus in a continuous cell line. *Arch. Virol.* 113:73–81.

Willcocks, M. M., Brown, T. D., Madeley, C. R., and Carter, M. J., 1994, The complete sequence of a human astrovirus. *J. Gen. Virol.* 75:1785–1788.

Willcocks, M. M., Boxall, A. S., and Carter, M. J., 1999, Processing and intracellular location of human astrovirus non-structural proteins. *J. Gen. Virol.* 80:2607–2611.

Wirblich, C., Thiel, H. J., and Meyers, G., 1996, Genetic map of the calicivirus rabbit hemorrhagic disease virus as deduced from the *in vitro* translation studies. *J. Virol.* 70:7974–7983.

Woode, G. N., Reed, D. E., Runnels, P. L., Herrig, M. A., and Hill, H. T., 1982, Studies with an unclassified virus isolated from diarrheic calves. *Vet. Microbiol.* 7:221–240.

Woode, G. N., Pohlenz, J. F., Gourley, N. E., and Fagerland, J. A., 1984, Astrovirus and Breda virus infections of dome cell epithelium of bovine ileum. *J. Clin. Microbiol.* 19:623–630.

Woode, G. N., Saif, L. J., Quesada, M., Winand, N. J., Pohlenz, J. F., and Gourley, N. K., 1985, Comparative studies on three isolates of Breda virus of calves. *Am. J. Vet. Res.* 46:1003–1010.

Yamashita, T., and Sakae, K., 2003, Molecular biology and epidemiology of Aichi virusand other diarrhoeogenic enteroviruses, in: *Viral Gastroenteritis* (U. Desselberger and J. Gray, eds.), Elsevier, Amsterdam, pp. 645–657.

Yamashita, T., Kobayashi, S., Sakae, K., Nakata, S., Chiba, S., Ishihara, Y., and Isomura, S., 1991, Isolation of cytopathic small round viruses with BS-C-1 cells from patients with gastroenteritis. *J. Infect. Dis.* 164:954–957.

Yamashita, T., Sakae, K., Kobayashi, S., Ishihara, Y., Miyake, T., Mubina, A., and Isomura, S., 1995, Isolation of cytopathic small round virus (Aichi virus) from Pakistani children and Japanese travelers from Southeast Asia. *Microbiol. Immunol.* 39:433–435.

Yamashita, T., Sakae, K., Tsuzuki, H., Suzuki, Y., Ishikawa, N., Takeda, N., Miyamura, T., and Yamazaki, S., 1998, Complete nucleotide sequence and genetic organization of Aichi virus, a distinct member of the Picornaviridae associated with acute gastroenteritis in humans. *J. Virol.* 72:8408–8412.

Yamashita, T., Ito, M., Kabashima, Y., Tsuzuki, H., Fujiura, A., and Sakae, K., 2003, Isolation and characterization of a new species of kobuvirus associated with cattle. *J. Gen. Virol.* 84:3069–3077.

Yeager, M., Dryden, K. A., Olson, N. H., Greenberg, H. B., and Baker, T. S., 1990, Three-dimensional structure of rhesus rotavirus by cryoelectron microscopy and image reconstruction. *J. Cell Biol.* 110:2133–2144.

Yokosuka, O., 2000, Molecular biology of hepatitis A virus: significance of various substitutions in the hepatitis A virus genome. *J. Gastroenterol. Hepatol.* 15(Suppl.):D91–D97.

Zafrullah, M., Ozdener, M. H., Panda, S. K., and Jameel, S., 1997, The ORF3 protein of hepatitis E virus is a phosphoprotein that associates with the cytoskeleton. *J. Virol.* 71:9045–9053.

Zárate, S., Espinosa, R., Romero, P., Guerrero, C. A., Arias, C. F., and López, S., 2000a, Integrin $\alpha 2\beta 1$ mediates the cell atachment of the rotavirus neuraminidase-resistant variant nar3. *Virology* 278:50–54.

Zárate, S., Espinosa, R., Romero, P., Méndez, E., Arias, C., and López, S., 2000b, The VP5 domain of VP4 can mediate attachment of rotaviruses to cells. *J. Virol.* 74:593–599.

Zhang, Y., McAtee, P., Yarbough, P. O., Tam, A. W., and Fuerst, T., 1997, Expression, characterization, and immunoreactivities of a soluble hepatitis E virus putative capsid protein species expressed in insect cells. *Clin. Diagn. Lab. Immunol.* 4:423–428.

Zhang, M., Zeng, C. Q.-Y., Morris, M. P., and Estes, M. K., 2000, A functional NSP4 enterotoxin peptide secreted from rotavirus-infected cells. *J. Virol.* 74:11663–11670.

Zhang, M., Purcell, R. H., and Emerson, S. U., 2001, Identification of the 5′ terminal sequence of the SAR-55 and MEX-14 strains of hepatitis E virus and confirmation that the genome is capped. *J. Med. Virol.* 65:293–295.

Zhou, Y., Li, L., Okitsu, S., Maneekarn, N., and Ushijima, H., 2003, Distribution of human rotaviruses, especially G9 strains, in Japan from 1996 to 2000. *Microbiol. Immunol.* 47:591–599.

CHAPTER 4

Methods of Virus Detection in Foods

Sagar M. Goyal

1.0. INTRODUCTION

Viral contamination of food and water represents a significant threat to human health. Many different types of foods are implicated in food-borne outbreaks but shellfish (oysters, clams, mussels), cold foods, and fresh produce (fruits and vegetables) are considered to be the most important vehicles. In recent years, viral food-borne outbreaks have been traced to raspberries (Ponka et al., 1999), strawberries (Gaulin et al., 1999), well water (Beller et al., 1997), sandwiches (Daniels et al., 2000), and oysters (Kohn et al., 1995). The source of viral contamination of shellfish is fecal contamination of water in which they reside whereas produce may be contaminated through the use of contaminated irrigation or wash water, infected food handlers involved in the preparation and processing of food, and contact of produce with contaminated surfaces.

The cases of produce-associated outbreaks are on the rise because the consumption of such foods has increased due to health reasons and because produce may often be imported from areas lacking in strict hygienic measures. In addition, produce is usually eaten uncooked thereby eliminating the added safety factor provided by cooking. Produce-associated outbreaks attributed to a single food source have occurred in several countries simultaneously (Koopmans et al., 2003). In addition, food is also subject to intentional contamination with highly infectious pathogens including category A and B pathogens such as *Bacillus anthracis*, *Yersinia pestis*, *Francisella tularensis*, Brucella spp., smallpox virus, filoviruses, arenaviruses, and alphaviruses.

Food-borne outbreaks are believed to cause an estimated 76 million illnesses, 5,000 deaths, and 325,000 hospitalizations annually in the United States (Mead et al., 1999). In many outbreaks the causative agent cannot be confirmed but they are suspected to be caused by viruses. It is generally believed that the number of viral food-borne outbreaks far exceeds the number currently being reported. One reason for the failure to confirm a viral etiology is the lack of sensitive and reliable methods for the detection of viruses in the implicated food. Svensson (2000) estimated that at least half of the viral food-borne disease outbreaks are not recognized because of inadequate sampling and detection methods.

Many viruses are associated with food-borne illnesses including norovirus (NV), hepatitis A virus (HAV), hepatitis E virus (HEV), and rotavirus (RV). Most of these viruses originate from the human gastrointestinal tract. Of these, caliciviruses (NVs) are the most important nonbacterial cause of food and waterborne disease outbreaks causing diarrhea, nausea and vomiting. Persons of all ages are affected and reinfections can occur because prior infection results only in a short term immune response. The problem of viral food-borne outbreaks has been neglected until recently probably because the diseases caused by such viruses are less severe and seldom fatal except in very young and elderly, pregnant women, and immunocompromised hosts (Gerba et al., 1996). However, viral food-borne outbreaks do cause high morbidity and suffering and hence should be investigated thoroughly.

Microbial monitoring is a useful tool in risk assessment of various food products. Simple, rapid, and sensitive methods for the detection of viruses in food and water can be used to help establish the cause and source of outbreaks (Jaykus, 1997) and to understand the epidemiologic features of the outbreak (Bresee et al., 2002). Existing methods to detect viruses in patients and in victims of food-borne outbreaks are relatively robust because infected individuals shed large number of viruses which can be easily detected by the current methods such as enzyme immunoassay (Fleissner et al., 1989), RT-PCR (reverse transcription-polymerase chain reaction; Anderson et al., 2001), and solid phase immune electron microscopy (SPIEM) using convalescent serum (Cunney et al., 2000; Girish et al., 2002). In addition, seroconversion (greater than fourfold antibody rise from acute to convalescent phase serum), as measured by enzyme immunoassay, can also be used for indirect evidence of viral infection (Gordon et al., 1990). Unfortunately, foods are rarely tested for viral contamination because simple and rapid methods for the detection of viruses in foods (except for shellfish) are not available (Leggitt and Jaykus, 2000; Koopmans et al., 2002).

The lack of sensitive surveillance systems to detect food contamination and the lack of available laboratory methods to concentrate and detect viruses in food has limited the ability of public health officials to identify or investigate outbreaks associated with widely distributed commodities or food products. One reason for the lack of these methods is that the number of viruses present in food is too small to be detected by methods used in clinical virology, although this low level viral contamination can still cause infection in a susceptible host. In addition, the direct detection and identification of viruses in food is difficult because of a large variety and complexity of foods, heterogeneous distribution of contaminating viruses in the food milieu, and the presence of substances in food that may inhibit or interfere with virus detection methods.

Another problem is that two of the most important food-borne viruses either do not grow in cell cultures (norovirus) or their primary isolation in cell cultures is very inefficient (hepatitis A virus). It is difficult to develop methods using these viruses thus necessitating the use of surrogates of these viruses for experimental studies. Although no validated model is available for these viruses, many investigators have utilized feline calicivirus (FCV) as a surrogate of NV (Slomka and Appleton, 1998; Gulati et al., 2001; Taku et al., 2002; Duizer et al., 2004a, 2004b) because FCV is easily propagated and titrated in CRFK (Crandell-Reese feline kidney) cells.

The need to develop simple, rapid, and accurate methods for the concentration of viruses from large amounts of food cannot be overemphasized. An ideal method would produce a final sample that does not interfere with conventional or molecular virology techniques used for virus detection. In addition, the method should be able to concentrate viruses from many different types of foods so that it can be used in situations where the integrity and safety of food is in question and to help develop laboratory-based surveillance for the early and rapid detection of large, common-source outbreaks. These methods will also be helpful in the event of an actual or suspected act of agro-terrorism thereby maintaining the confidence of the public in public health authorities.

Considerable progress has been made in the development of sensitive methods for the detection of viruses in shellfish. The method consists basically of two steps. The first step is 'sample processing' in which viruses (or their nucleic acids) are removed and/or concentrated from shellfish tissues. The second 'detection' step uses either conventional virus isolation in cell culture or molecular techniques for the detection of viruses or their nucleic acids, respectively. Broadly speaking, there are two types of 'sample processing' methods that have been used for virus detection in shellfish; the whole virus concentration-detection method and nucleic acid extraction-detection method.

The nucleic acid extraction-detection method is relatively new. In this procedure total RNA (not whole virus) is extracted from oyster meat followed by RT-PCR (Coelho et al., 2003a). The direct RNA isolation protocol is simple because no fastidious concentration steps are involved as in the whole virus concentration-detection method. The disadvantage of this procedure is that no opportunity exists for the detection of infectious/viable virus particles. Legeay et al. (2000) described a simple procedure in which viral RNA was isolated directly from the shellfish extract by a guanidium thiocyanate-silica extraction method. Viral RNA was detected by RT-PCR followed by confirmation of the amplicons by hybridization with DIG-labeled specific probes. Using this procedure, as little as 20 PFU (plaque forming units) of HAV per g of shellfish tissues could be detected.

Goswami et al. (2002) described a method by which HAV RNA was detected in spiked samples of shellfish and cilantro. Total RNA was first isolated followed by isolation of poly(A)-containing RNA because HAV genomic RNA contains a poly(A) tail. The isolated RNA was amplified by RT-PCR and then re-amplified with internal primers to improve the quality and the quantity of amplified DNA. With this procedure, 0.15 $TCID_{50}$ of HAV could be detected in 0.62 g of tissue. In addition, this procedure was used to successfully detect naturally occurring HAV in clams involved in an outbreak of gastroenteritis.

In the whole virus procedure, on the other hand, the virus is extracted (eluted) from shellfish tissues in an alkaline buffer solution (Sobsey et al.,

1975; Sobsey et al., 1978; Seidel et al., 1983; Bouchriti and Goyal, 1992, 1993). This step is based on the fact that viruses can be adsorbed to, or eluted from, tissues and other solids by regulating the pH and ionic conditions of the suspending medium thus effectively separating viruses from solids. The viruses can then be concentrated from these extracts (eluates) by a concentration step involving acid precipitation, polyethylene glycol precipitation, or organic flocculation (Katzenelson et al., 1976; Bouchriti and Goyal, 1992; Atmar et al., 1995). During concentration, the volume of the extract (eluate) is reduced resulting in a small final sample that can be conveniently assayed for viruses by conventional or molecular methods. The most commonly used molecular diagnostic method is RT-PCR since a large majority of viruses in food happen to be RNA viruses. Ideally the final sample should not contain cytotoxic agents or PCR inhibitors when tested by cell culture inoculation or RT-PCR, respectively. A few methods that have been used for the detection of viruses in non-shellfish foods are a modification of methods used for shellfish. It is important, therefore, to review methods that have been developed for virus detection in shellfish.

2.0. METHODS FOR THE DETECTION OF VIRUSES IN SHELLFISH

Viruses are usually found at low levels in shellfish and cannot be detected by direct examination of shellfish extracts for viruses. Several methods have been developed over the last 30 years for the concentration of small amounts of viruses from large amounts of shellfish tissues. Some of these methods have been further modified to increase the efficiency of virus recovery and to reduce cytotoxic agents and PCR inhibitors in shellfish extracts. The first step involved in these procedures is extraction of virions by homogenization of shellfish tissues in a buffer solution followed by low speed centrifugation to remove solids. The eluted viruses can then be concentrated from eluates by e.g., acid precipitation, polyelectrolyte flocculation, adsorption-elution-ultrafiltration, elution-adsorption-precipitation, or elution-precipitation methods (Katzenelson et al., 1976; Bouchriti and Goyal, 1992; Atmar et al., 1995).

In a classical study, Sobsey et al. (1975) manipulated the pH and ionic conditions of shellfish homogenate to adsorb viruses to, or elute from, shellfish meat (adsorption-elution-precipitation method). The shellfish tissues were homogenized in 7 volumes of distilled water followed by the adjustment of pH and salinity to 5.0 and ≤1,500 mg NaCl/L, respectively. Under the conditions of acidic pH and low conductivity, the viruses adsorbed to oyster solids, which were collected by centrifugation. The viruses were then eluted from these solids by resuspending them in 0.05 M glycine buffer (pH 7.5, conductivity of ≥8,000 ppm NaCl). After centrifugation, the virus-containing supernatant was adjusted to pH 4.5, the precipitate collected by centrifugation, and then resuspended in a small volume of buffer. Using this method, poliovirus from 100g of oyster tissue was concentrated in a final volume of 15ml with a recovery efficiency of 48%.

Over the years, many different modifications of the Sobsey method have been reported (Gerba and Goyal, 1978; Bouchriti and Goyal, 1993; Muniain-Mujika et al., 2000, 2003). In the elution-precipitation method, the initial adsorption step is eliminated (Richards et al., 1982). In some cases, an organic compound (such as beef extract powder) is incorporated during acid precipitation to provide a matrix to which viruses can be adsorbed (Vaughn et al., 1979). In yet another modification (elution-adsorption-elution), Goyal et al. (1979) eluted viruses from oyster tissues by homogenizing them in 0.05M glycine buffer (pH 9.0). The virus containing supernatant was adjusted to pH 5.5 and conductivity of ≤1,500ppm NaCl. After centrifugation, the supernatant was discarded and viruses were re-eluted from oyster solids by re-suspending them in glycine saline (pH 11.5). The overall virus recovery averaged 60%.

The choice of extraction and concentration methods depends on efficiency of virus recovery, ease and rapidity of the method, small final volume, and absence of interfering substances in the final sample. In their quest to increase virus recovery and eliminate PCR inhibitors, Dix and Jaykus (1998) used a protein-precipitating agent Pro-Cipitate for the concentration of NV from hard-shelled clams (*Mercenaria mercenaria*) in an adsorption-elution-precipitation scheme. Using this procedure along with RT-PCR and oligoprobe hybridization, they were able to detect as low as 450 RT-PCR amplifiable units of NV.

Traore et al. (1998) compared four methods of extraction and three methods of concentration. Mussel tissues in 60 gram amounts were spiked with astrovirus, HAV, or poliovirus and then extracted with borate buffer, glycine solution, saline beef, or saline beef-Freon. The viruses were then concentrated by precipitation with polyethylene glycol 6000 (PEG 6000) or PEG 8000 or by organic flocculation. Extraction with glycine solution and borate buffer resulted in significantly more RT-PCR-positive samples than the saline beef extraction method. Of the 20 different combinations of extraction and concentration methods tested, the borate buffer-organic flocculation, borate buffer-PEG 6000, and glycine solution-PEG 6000 were found to be the most efficient.

A modified procedure described by Mullendore et al. (2001) consisted of acid adsorption at pH 4.8, first elution with 0.05M glycine, second elution with 0.5M threonine, PEG-precipitation twice, chloroform-extraction twice, RNA-extraction, and a single round of RT-PCR. Using this procedure, HAV was detected at a seeding density of ≥1 plaque forming unit (PFU)/g of oyster. Kingsley and Richards (2001) developed a rapid extraction method for the detection of HAV and NV from shellfish using a pH 9.5 glycine buffer, PEG precipitation, Tri-reagent treatment, and purification of viral poly(A) RNA by using magnetic poly(dT) beads. When coupled with RT-PCR-based detection, this method could detect as low as 0.015 PFU of HAV and 22.4 RT-PCR units of NV in hard-shelled clams and oysters. Homogenization in glycine/NaCl buffer (pH 9.5) followed by PEG 8000 precipitation and DNA extraction by proteinase K and phenol/chloroform treatment was used successfully by Karamoko et al. (2005) to detect adenoviruses in mussels (Mytilus sp.) harvested from Moroccan waters.

As is clear from the above discussion, methods for the detection of viruses in shellfish have been in use for a long time. Several studies have compared different methods and have documented the advantages and disadvantages of each. For example, Sunen et al. (2004) compared two processing procedures for the detection of HAV in clams. The first method involved acid adsorption, elution, PEG precipitation, chloroform extraction, and PEG precipitation while the second method was based on virus elution with glycine buffer (pH 10), chloroform extraction, and concentration by ultracentrifugation. Final clam concentrates were processed by RNA extraction or immunomagnetic capture of viruses (IMC) followed by RT-nested PCR. Although both methods of sample processing were effective in detecting HAV, the first method was more effective in removal of PCR inhibitors whereas the second method was simpler and faster.

3.0. DETECTION OF VIRUSES BY CONVENTIONAL VIRUS ISOLATION

Classical methods for detecting viral contamination of foods by inoculation of cell cultures are costly and time consuming. In addition, food extracts may be cytotoxic to the indicator cells and viruses commonly found in food either do not grow in cell cultures or grow very poorly. For example, Duizer et al. (2004b) made concerted efforts to grow NV in 27 different cell culture systems. Insulin, DMSO, and butyric acid were used as cell culture supplements to induce differentiation. In some cases, the cells and the NV-containing stool samples were treated with bioactive digestive additives. Even after five blind passages, no reproducible viral growth was observed. Similarly, Malik et al. (2005) evaluated 19 different cell types from 11 different animal species for the propagation of NV but were unsuccessful in propagating the virus.

Because of the above difficulties, molecular methods such as PCR and RT-PCR are commonly used for the detection of such viruses (Greiser-Wilke and Fries, 1994). However, these methods detect both infectious and non-infectious viruses and hence a sample positive by these techniques may or may not contain live infectious virus (Olsvik et al., 1994; Richards, 1999). Other limitations of these techniques include lack of sensitivity and specificity, high assay costs, and a level of technical expertise not available in most food-testing laboratories (Richards, 1999). To overcome this problem, integrated cell culture/strand-specific RT-PCR and integrated cell culture (ICC)/RT-PCR assays are available, that detect negative-strand RNA replicative intermediate, thus distinguishing infectious from non-infectious virus (Jiang et al., 2004). In these procedures, limited virus propagation occurs in cell cultures, which increases the amount of target material and hence the sensitivity of the immunological or molecular method (Chironna et al., 2002; Bosch et al., 2004).

4.0. DETECTION OF VIRUSES BY MOLECULAR DIAGNOSTIC TECHNIQUES

The application of molecular techniques to diagnose and investigate disease outbreaks during recent years has led to a growing appreciation of the importance of these techniques (Koopmans et al., 2004). Such methods can also be used for molecular tracing of virus strains (Koopmans et al., 2002). However, the advantage of the conventional virus isolation procedure is that a live, infectious virus is detected whereas molecular procedures detect nucleic acid from both infectious and non-infectious virus particles. To determine food safety, it is important to know if the virus is capable of causing infection or not. In addition, direct isolation and purification of intact (whole) virions from foods prior to the application of nucleic acid amplification methods may remove PCR inhibitors. Many modifications of molecular procedures have been described as summarized below. More detailed description is provided in chapter 5.

4.1. PCR

Schwab et al. (2001) developed an RT-PCR-oligoprobe amplification and detection method using rTth polymerase, a heat-stable enzyme that functions as both a reverse transcriptase and DNA polymerase, in a single-tube, single-buffer, elevated temperature reaction. An internal standard NV RNA control was added to each RT-PCR to identify sample inhibition, and thermolabile uracil N-glycosylase was incorporated into the reaction to prevent PCR product carryover contamination. The amplicons were detected by ELISA using virus-specific biotinylated oligoprobes. Low levels of NV were detected in stools and bivalve mollusks following bioaccumulation. In addition, this method successfully detected NV in oysters implicated in an outbreak of NV gastroenteritis.

Many different variations of RT-PCR reaction have been described. For example, semi-nested or nested PCR has been used to increase the sensitivity of virus detection (Abad et al., 1997). Di Pinto et al. (2003) described an RT-PCR for the detection of HAV in shellfish (*Mytilus galloprovincialis*). The virus particles were first concentrated by polyethylene glycol followed by RT-PCR detection in a single step using primers specific for the VP3-VP1 region of the genome. The specificity of the PCR products was determined by heminested PCR. Using this procedure, 0.6 PFU/25 g of shellfish homogenate could be detected.

Rigotto et al. (2005) compared conventional-PCR, nested-PCR (nPCR), and integrated cell culture PCR (ICC/PCR) to detect adenovirus in oysters seeded with known amounts of adenovirus serotype 5 (Ad5) and found that the nPCR was more sensitive (limit of detection: 1.2 PFU/g of tissue) than conventional-PCR and ICC-PCR. Jothikumar et al. (2005) developed two broadly reactive one-step TaqMan RT-PCR assays for the detection of genogroup I (GI) and II (GII) of NV in fecal and shellfish samples. The sensitivity of these assays was found to be similar to that of an nPCR.

Loisy et al. (2005) developed a real-time RT-PCR based on one-step detection using single primer sets and probes for NV genogroups I and II. Using this method, they were able to detect 70 and 7 RT-PCR units of genogroup I and II norovirus strains, respectively, in artificially contaminated oysters. Burkhardt et al. (2002) compared a single, compartmentalized tube-within-a-tube (TWT) device for nPCR with conventional protocol of nPCR. The TWT device decreased the calicivirus assay detection limit 10-fold over that of conventional nPCR.

4.2. mPCR

Multiplex RT-PCR (mRT-PCR) can be used for the detection of several viruses in a single reaction tube (Rosenfield and Jaykus, 1999; Coelho et al., 2003b; Beuret, 2004). Rosenfield and Jaykus (1999) described an mRT-PCR for the simultaneous detection of human poliovirus type 1 (PV1), HAV and NV using three different sets of primers to produce three size-specific amplicons. Detection limits of ≤ 1 infectious unit (PV1 and HAV) or RT-PCR-amplifiable unit (NV) were achieved. Formiga-Cruz et al. (2005) developed a nested mRT-PCR for the detection of adenovirus, enterovirus and HAV in urban sewage and shellfish, which was able to detect all three viruses simultaneously when the concentration of each virus was equal to or lower than 1,000 copies per PCR reaction.

4.3. NASBA

Nucleic acid sequence-based amplification (NASBA) uses three enzymes (reverse transcriptase, RNaseH, and RNA polymerase) and is designed to detect single stranded RNA. The product of NASBA is also ssRNA which can be detected by gel electrophoresis followed by ethidium bromide staining. Jean et al. (2002) developed NASBA for HAV targeting the capsid protein gene VP2. The assay was able to detect 10^6 PFU of HAV artificially inoculated onto the surfaces of lettuce and blueberries. However, the method suffers from the low amount of food that can be processed and tested. In a later study, Jean et al. (2004) developed a multiplex format of NASBA method to simultaneously detect HAV and NV (genogroups I and II). The amplicons were detected and confirmed by agarose gel electrophoresis, electrochemiluminescence, and Northern hybridization. Using this method, they were able to detect all three viruses inoculated into two ready-to-eat foods (deli sliced turkey and lettuce) at 10^0 to 10^2 RT-PCR-detectable units in both food commodities.

4.4. PCR Inhibitors and Their Removal

Although PCR and RT-PCR assays provide rapid virus detection, their use in food samples may be hampered by the presence of PCR inhibitors. These inhibitors are either present in the sample or are introduced during the concentration procedure (Atmar 1993; Abbaszadegan, 1993; Le Guyader et al., 1994). Naturally occurring substances such as clay, humic acid, and mussel tissues can act as PCR inhibitors (Lewis et al., 2000). In shellfish extracts, glycogen and acidic polysaccharides have been found to inhibit PCR (Schwab et al., 1998). The presence of endogenous inhibitors in sample concentrates can be detected by spiking a control reaction with a known amplifiable target and its respective primers.

Some of these problems can be resolved by using a processing method that effectively concentrates low number of viruses from large amounts of sample and in doing so gets rid of PCR inhibitors also. Often, re-extraction of the nucleic acid or ethanol precipitation and/or centrifugal ultrafiltration is sufficient to remove PCR inhibitors. Other methods that have been tried include PEG precipitation (Lewis and Metcalf, 1988; Shieh et al., 1999, 2000), freon extraction, ultrafiltration, silica gel adsorption-elution (Shieh et al., 2000), aluminum hydroxide precipitation (Farrah et al., 1978), hydroextraction (Farrah et al., 1977), membrane adsorption-elution (Goyal and Gerba, 1983), Sephadex columns, protein-precipitating agent (Jaykus et al., 1996), and beef extract flocculation (Gerba and Goyal, 1982; Traore et al., 1998). Kingsley et al. (2002) used digestive tissues of clams instead of the whole clam. In this method, they extracted virus by glycine extraction, PEG treatment, Tri-reagent treatment, and purification of poly(A) RNA with magnetic beads coupled to poly(dT) oligonucleotides. Le Guyader et al. (1994) suspended the final pellet of the concentrated sample in distilled water instead of phosphate buffered saline, thus effectively eliminating PCR inhibitors.

5.0. METHODS FOR VIRUS DETECTION ON ENVIRONMENTAL SURFACES

Food contact surfaces on which raw foods are processed often become contaminated with pathogens, which can subsequently be transferred to other foods prepared on those surfaces. Outbreaks of NV have often originated in food service establishments. Methods that can detect viral contamination on food contact surfaces should be helpful in efforts to control food-borne disease outbreaks. Taku et al. (2002) described a simple method for elution and detection of NV from stainless steel surfaces using feline calicivirus (FCV) as a model. Stainless steel surfaces were artificially contaminated with known amounts of FCV, followed by its elution in a buffer solution. Three methods of virus elution were compared. In the first method, moistened cotton swabs or pieces of positively charged filter (1MDS) were used to elute the contaminating virus. The second method consisted of flooding the contaminated surface with eluting buffer, allowing it to stay in contact for 15 min, followed by aspiration of the buffer (aspiration method). The third method, the scraping-aspiration method, was similar to the aspiration method, except that the surfaces were scraped with a cell scraper before buffer aspiration. Maximum virus recovery (32% to 71%) was obtained with the scraping-aspiration method using 0.05 M glycine buffer at pH 6.5. Two methods (organic flocculation and filter adsorption elution) were compared to reduce the volume of the eluate recovered from larger surfaces. The organic flocculation method gave an average overall recovery of 55% compared to the filter-adsorption-elution method, which yielded an average recovery of only 8%. The newly developed method was validated for the detection of NV by artificial contamination of 929-cm^2 stainless steel sheets with NV-positive stool samples followed by RT-PCR for the detection of the recovered virus.

6.0. METHODS FOR VIRUS DETECTION IN NON-SHELLFISH FOODS

As stated earlier, one problem with food contamination is that viruses would be present in food in very small amounts even in the event of a deliberate contamination because of the large quantities of food involved. Although minimal contamination of food items may go undetected by direct detection methods, they remain hazardous to human health because of the low infectious dose of viruses. It is important, therefore, that any proposed method for virus detection in food should be capable of detecting small numbers of pathogens in large amounts/volumes of food. To do so, it is necessary to separate and concentrate viruses from food matrices followed by their detection by conventional and/or molecular methods (Sair et al., 2002). It is also important that the final concentrate should not be cytotoxic to cell cultures used in infectivity assays and be free of PCR inhibitors that may be co-extracted or co-concentrated from food. Alternately, the final sample can be treated in some manner to remove PCR inhibitors or the nucleic acid extraction method can be modified specifically to remove these inhibitors.

The development of robust, simple and sensitive methods to recover pathogens from produce (and other foods) will facilitate prevalence studies that are useful in risk assessment and for developing food safety guidelines. They can also be used to detect pathogens in "suspect" foods, will permit the identification of contaminated food, and improve our understanding of the modes of food contamination and pathogen transmission thereby assisting state and federal agricultural and health agencies to design methods for the prevention and control of pathogen contamination of foods.

Methods for the detection of viruses in non-shellfish foods are in their infancy. Daniels et al. (2000) used RT-PCR and sequence analysis for the first time to confirm the presence of viral nucleic acid in deli ham. The sequence of RT-PCR product was similar to that found in a stool specimen from an infant whose mother had prepared implicated sandwiches. Leggitt and Jaykus (2000) developed a method to extract and detect PV, HAV, and NV from lettuce and hamburger using an elution-concentration approach followed by detection with RT-PCR. Samples of lettuce or hamburger were artificially inoculated with one of the three viruses and then processed by the sequential steps of homogenization, filtration, Freon extraction (hamburger), and PEG precipitation. To further reduce sample volume and to remove PCR inhibitors, a secondary PEG precipitation was added. Using this method, 50 g samples were reduced to a final volume of 3 to 5 ml with recovery efficiency of 10% to 70% for PV and 2% to 4% for HAV. Total RNA from the final sample was extracted in a small volume (30 to 40 microl) and subjected to RT-PCR amplification of viral RNA sequences. Viral RNA was consistently detected by RT-PCR at initial inoculum of $\geq 10^2$ PFU/50 g of food for PV and $\geq 10^3$ PFU/50 g for HAV.

Bidawid et al. (2000) used immunomagnetic beads-PCR (IM-PCR), positively-charged virosorb filters (F), or a combination of both methods (F-IM-PCR) to capture, concentrate and rapidly detect HAV in experimentally contaminated samples of lettuce and strawberries. Direct RT-PCR of the collected HAV-bead complex showed a detection limit of 0.5 PFU of the virus in 1-ml of wash solution from the produce. In the F-IM-PCR method, virus-containing washes from produce were passed through positively-charged virosorb filters and the captured virus was eluted with 10 ml volumes of 1% beef extract. Of the 62% filter-captured HAV, an average of 35% was eluted by the 1% beef extract but PCR amplification of 2 µl from this eluate failed to produce a clear positive signal. However, considering the large volumes used in F-IM-PCR, the sensitivity of detection could be much greater than that of the IM-PCR.

Schwab et al. (2000) developed a method to recover NV and HAV from food samples. The method involves washing of food samples with a guanidinium-phenol-based reagent, extraction with chloroform, and precipitation in isopropanol. Recovered viral RNA is amplified with HAV- or NV-specific primers in RT-PCR, using a viral RNA internal standard control to identify potential sample inhibition. By this method, 10 to 100 PCR units of HAV and NV seeded onto ham, turkey, and roast beef were detected. The method was applied to food samples implicated in an NV-associated outbreak at a university cafeteria. Sliced deli ham was positive for a genogroup II NV. Sequence analysis of the PCR-amplified capsid region of the genome indicated that the sequence was identical to that from virus detected in the stools of ill students. D'Souza and Jaykus (2002) used zirconium hydroxide to concentrate PV, HAV, and NV from food. Recovery of PV1 ranged from 16% to 59% with minimal loss to the supernatant. For both HAV and NV, RT-PCR amplicons of appropriate sizes were detected and confirmed in the pellet fraction with no visible amplicons from the supernatant.

Dubois et al. (2002) modified an elution-concentration method based on PEG precipitation to detect PV, HAV, and NV in fresh and frozen berries and fresh vegetables. The surface of produce was washed with a buffer containing 100 mM Tris-HCl, 50 mM glycine, 50 mM $MgCl_2$, and 3% beef extract (pH 9.5). PCR inhibitors and cytotoxic compounds were removed from viral concentrates by chloroform-butanol extraction. Viruses from 100 g of vegetal products could be recovered in volumes of 3 to 5 ml. The presence of virus was detected by RT-PCR and cell culture inoculation. Using the latter method, 15% to 20% of PV and HAV were recovered from frozen raspberry surfaces. By RT-PCR, the recovery was estimated to be 13% for NV, 17% for HAV, and 45% to 100% for PV.

Sair et al. (2002) compared four methods of RNA extraction for optimizing the detection of viruses in food. Hamburger and lettuce samples, processed for virus concentration using a previously reported filtration-extraction-precipitation procedure, were inoculated with HAV or NV. Several RNA extraction methods (guanidinium isothiocyanate, microspin column, QIAshredder Homogenizer, and TRIzol) and primer pairs were compared for overall RNA yield (μg/ml), purity (A(260)/A(280)), and RT-PCR limits of detection. The use of TRIzol with the QIAshredder homogenizer (TRIzol/Shred) yielded the best RT-PCR detection, and the NVp110/NVp36 primer set was the most efficient for detecting NV from seeded food samples. A one-step RT-PCR protocol using the TRIzol/Shred extraction method and the NVp110/NVp36 or HAV3/HAV5 primer sets demonstrated improved sensitivity over the routinely used two-step method. Residual RT-PCR inhibitors were effectively removed as evidenced by the ability to detect viral RNA in food concentrates without prior dilution.

Recently, Kobayashi et al. (2004) used magnetic beads coated with an antibody to the baculovirus-expressed recombinant capsid proteins of the Chiba virus (rCV) to facilitate the capture of NV from food items implicated in an outbreak. Following immunomagnetic capture, NV bound to the beads was detected by RT-PCR. Two of the nine food items were positive for genogroup I NV, the nucleotide sequence of which was almost identical to those of NV strains detected in stool samples of ill patients. The immunocapture RT-PCR method seems simple and easy to perform and may be helpful in the detection of NV from outbreak-implicated foods.

Le Guyader et al. (2004) conducted a round-robin study to compare five different methods for the detection of three different viruses (PV, NV, and canine calicivirus as a surrogate of NV) in artificially contaminated lettuce. All five methods consisted of virus elution followed by concentration, and RNA extraction. The methods were compared for efficiency of virus recovery and removal of PCR inhibitors from the final samples. The first method (method A) consisted of virus elution in phosphate buffered saline (PBS) and Vertrel (1,1,1,2,3,4,4,5,5,5-decafluoropentane) and virus concentration by PEG precipitation. In the second method (method B), the viruses were eluted with 3% beef extract solution (pH 9.5) followed by ultracentrifugation to concentrate viruses. The eluent in the third method (method C) was 0.05 M glycine-NaCl buffer (pH 9.5) followed by chloroform-butanol (1:1, vol/vol) extraction and PEG precipitation. The fourth method (method D) consisted of PBS-Vertrel elution and ultracentrifugation. In the fifth method (method E), viruses were eluted with glycine buffer (pH 8.5) and concentrated by ultrafiltration. Methods C and E were found to result in a concentrate that was free of PCR inhibitors and yielded good virus recoveries (approximately 10 RT-PCR units of viruses per gram of lettuce).

We have recently developed a unified method for the concentration of feline calicivirus (FCV; a surrogate for human NV) from strawberries, lettuce, green onions, and cabbage (Goyal et al., unpublished data). In this study, produce was experimentally contaminated with known amounts of FCV followed by its elution in 7 volumes of beef extract-0.05 M glycine buffer (pH 8.5) by shaking the produce and eluent for 30 min. The volume of the eluate was reduced to 3–5 ml by precipitation with polyethylene glycol. Average virus recovery using this procedure was 70% and the final sample was not inhibitory to PCR. We believe that this method can be modified to test larger quantities of produce, other types of food, and other types of pathogens including bacteria.

7.0. COMPARISON OF METHODS

The choice of extraction/processing method will depend on per cent virus recovery and absence of cytotoxic agents and PCR inhibitors in the final concentrate (Arnal et al., 1999). It is important that a newly developed method be subjected to inter- and intralaboratory standardization and validation before recommending it for routine use (Romalde et al., 2002, 2004). In a multicenter, collaborative study to evaluate a method for the detection of NV in shellfish tissues, replicate samples of stomach and hepatopancreas of oysters or hard-shell clams were seeded with NV and then shipped to several laboratories, where viral nucleic acids were extracted followed by their detection by RT-PCR (Atmar et al., 1996). The sensitivity and specificity of the assay were 87% and 100%, respectively, when results were determined by ethidium bromide-staining of agarose gels followed by confirmation with hybridization with a digoxigenin-labeled, virus-specific probe.

Arnal et al. (1999) compared seven methods for detecting HAV in stool and shellfish samples. The protocols tested were either techniques for the recovery and purification of total RNA (RNAzol, PEG-CETAB, GTC-silica and Chelex) or techniques for isolating specifically HAV using a nucleotide probe or a monoclonal antibody. For stool samples, RNAzol, PEG-CETAB, and magnetic beads with antibody allowed efficient virus detection. For shellfish samples, three protocols (RNAzol, PEG-CETAB, and GTC-silica) allowed RNA to be extracted in 90% of cases. The authors suggested that the rapidity and low cost of RNAzol and GTC-silica made them the most suitable methods for routine diagnostic testing.

Ribao et al. (2004) compared several nucleic acid extraction and RT-PCR commercial kits for the detection of HAV from seeded mussel tissues and found that Total Quick RNA Cells & Tissues version mini (Talent) for RNA extraction and Superscript One-Step RT-PCR System (Life Technologies) for the RT-PCR reaction were the best. Di Pinto et al. (2004) compared two RT-PCR based techniques for the detection of HAV in shellfish. Both techniques involved virus extraction in glycine buffer followed by concentration of eluted virus by one or two PEG precipitation steps. RNA extraction was done by the use of oligo (dT) cellulose to select poly (A) RNA or by another method in which total RNA is bound on silica membrane. The first approach was found to be less time-consuming and less technically demanding than the second method.

8.0. CONCLUSIONS

Methods for the detection of small amounts of viruses in large amounts of shellfish meat are available and have been used for surveillance and epidemiological studies. Such methods for the detection of viruses in non-shellfish food are not available because of the low number of viruses present in large amounts of food and because of complex and varied food matrices. To be successful, a virus detection method for foods needs to be simple, sensitive, and robust; the final sample should not contain cytotoxic agents and/or PCR inhibitors; and the method should be applicable to a large variety of food items. Because of the complexity of food matrices, it may often be necessary to use two different methods to maximize the validity of diagnosis (Rabenau et al., 2003).

9.0. REFERENCES

Abad, F. X., Pintó, R. M., Villena, C., Gajardo, R., and Bosch, A., 1997, Astrovirus survival in drinking water, *Appl. Environ. Microbiol.* 63:3119–3122.

Abbaszadegan, M., Huber, M. S., Gerba, C. P., and Pepper, I. L., 1993, Detection of enteroviruses in groundwater with the polymerase chain reaction, *Appl. Environ. Microbiol.* 59:1318–1324.

Anderson, A. D., Garrett. V. D., and Sobel, J., 2001, Multistate outbreak of Norwalk-like virus gastroenteritis associated with a common caterer, *Am. J. Epidemiol.* 154:1013–1019.

Arnal, C., Ferre-Aubineau, V., Besse, B., Mignotte, B., Schwartzbrod, L., and Billaudel, S., 1999, Comparison of seven RNA extraction methods on stool and shellfish samples prior to hepatitis A virus amplification, *J. Virol. Methods* 77:17–26.

Atmar, R. L., Metcalf, T. G., Neill, F. H., and Estes, M. K., 1993, Detection of enteric viruses in oysters by using the polymerase chain reaction, *Appl. Environ. Microbiol.* 59:631–635.

Atmar, R. L., Neill, F. H., Romalde, J. L., Le Guyader, F., Woodley, C. M., Metcalf, T. G., and Estes, M. K., 1995, Detection of Norwalk virus and hepatitis A virus in shellfish tissues with the PCR, *Appl. Environ. Microbiol.* 61:3014–3018.

Atmar, R. L., Neill, F. H., Woodley, C. M., Manger, R., Fout, G. S., Burkhardt, W., Leja, L., McGovern, E. R., Le Guyader, F., Metcalf, T. G., and Estes, M. K., 1996, Collaborative evaluation of a method for the detection of Norwalk virus in shellfish tissues by PCR, *Appl. Environ. Microbiol.* 62:254–258.

Beller, M., Ellis, A., Lee, S. H., Drebot, M. A., Jenkerson, S. A., Funk, E., Sobsey, M. D., Simmons III, O. D., Monroe, S. S., Ando, T., Noel, J., Petric, M., Hockin, J., Middaugh, J. P., and Spika, J. S., 1997, Outbreak of viral gastroenteritis due to a contaminated well—International consequences, *J. Am. Med. Assn.* 278:563–568.

Beuret, C., 2004, Simultaneous detection of enteric viruses by multiplex real-time RT-PCR, *J. Virol. Methods* 115:1–8.

Bidawid, S., Farber, J. M., and Sattar, S. A., 2000, Rapid concentration and detection of hepatitis A virus from lettuce and strawberries, *J. Virol. Methods* 88:175–185.

Bosch, A., Pinto, R. M., Comas, J., and Abad, F. X., 2004, Detection of infectious rotaviruses by flow cytometry, *Methods Mol. Biol.* 268:61–68.

Bouchriti, N., and Goyal, S. M., 1992, Evaluation of three methods for the concentration of poliovirus from oysters, *Microbiologica.* 15:403–408.

Bouchriti, N., and Goyal, S. M., 1993, Methods for the concentration and detection of human enteric viruses in shellfish: a review, *New Microbiol.* 16:105–103.

Bresee, J. S., Widdowson, M. A., Monroe, S. S., and Glass, R. I., 2002, Foodborne viral gastroenteritis: challenges and opportunities, *Clin. Infect. Dis.* 35:748–753.

Burkhardt, W., Blackstone, G. M., Skilling, D., and Smith, A. W., 2002, Applied technique for increasing calicivirus detection in shellfish extracts, *J. Appl. Microbiol.* 93:235–240.

Chironna, M., Germinario, C., De Medici, D., Fiore, A., Di Pasquale, S., Quarto, M., and Barbuti, S., 2002, Detection of hepatitis A virus in mussels from different sources marketed in Puglia region (South Italy), *Int. J. Food Microbiol.* 75:11–18.

Coelho, C., Heinert, A. P., Simoes, C. M., and Barardi, C. R., 2003a, Hepatitis A virus detection in oysters (Crassostrea gigas) in Santa Catarina State, Brazil, by reverse transcription-polymerase chain reaction, *J. Food Prot.* 66:507–511.

Coelho, C., Vinatea, C. E., Heinert, A. P., Simoes, C. M., and Barardi, C. R., 2003b, Comparison between specific and multiplex reverse transcription-polymerase chain reaction for detection of hepatitis A virus, poliovirus and rotavirus in experimentally seeded oysters, *Memorias do Instituto Oswaldo Cruz.* 98:465–468.

Cunney, R. J., Costigan, P., and McNamara, E. B., 2000, Investigation of an outbreak of gastroenteritis caused by Norwalk-like virus, using solid phase immune electron microscopy, *J. Hosp. Infect.* 44:113–118.

Daniels, N. A., Bergmire-Sweat, D. A., and Schwab, K. J., 2000, A foodborne outbreak of gastroenteritis associated with Norwalk-like viruses: first molecular traceback to deli sandwiches contaminated during preparation, *J. Infect. Dis.* 181:1467–1470.

Di Pinto, A., Forte, V. T., and Tantillo, G. M., 2003, Detection of hepatitis A virus in shellfish (*Mytilus galloprovincialis*) with RT-PCR, *J. Food Prot.* 66:1681–1685.

Di Pinto, A., Conversano, M. C., Forte, V. T., La Balandra, G., Montervino, C., and Tantillo G. M., 2004, A comparison of RT-PCR-based assays for the detection of HAV from shellfish, *New Microbiologica* 27:119–124.

Dix, A. B., and Jaykus, L. A., 1998, Virion concentration method for the detection of human enteric viruses in extracts of hard-shelled clams, *J. Food Prot.* 61:458–465.

D'Souza, D. H., and Jaykus, L. A., 2002, Zirconium hydroxide effectively immobilizes and concentrates human enteric viruses, *Letters Appl. Microbiol.* 35:414–418.

Dubois, E., Agier, C., and Traore, O., 2002, Modified concentration method for the detection of enteric viruses on fruits and vegetables by reverse transcriptase-polymerase chain reaction or cell culture, *J. Food Prot.* 65:1962–1969.

Duizer, E., Bijkerk, P., Rockx, B., De Groot, A., Twisk, F., and Koopmans, M., 2004a, Inactivation of caliciviruses, *Appl. Environ. Microbiol.* 70:4538–4543.

Duizer, E., Schwab, K. J., and Neill, F. H., 2004b, Laboratory efforts to cultivate noroviruses, *J. Gen. Virol.* 85:79–87.

Farrah, S. R., Goyal, S. M., Gerba, C. P., Wallis, C., and Melnick, J. L., 1977, Concentration of enteroviruses form estuarine water, *Appl. Environ. Microbiol.* 33: 1192–1196.

Farrah, S. R., Goyal, S. M., Gerba, C. P., Conklin, R. H., and Smith, E. M., 1978, Comparison between adsorption of poliovirus and rotavirus by aluminum hydroxide and activated sludge flocs, *Appl. Environ. Microbiol.* 25:360–363.

Fleissner, M. L., Herrmann, J. E., and Booth, J. W., 1989, Role of Norwalk virus in two foodborne outbreaks of gastroenteritis: definitive virus association, *Am. J. Epidemiol.* 129:165–172.

Formiga-Cruz, M., Hundesa, A., Clemente-Casares, P., Albinana-Gimenez, N., Allard, A., and Girones, R., 2005, Nested multiplex PCR assay for detection of human enteric viruses in shellfish and sewage, *J. Virol. Methods* 125:111–118.

Gaulin, C., Frigon, M., Poirier, D., and Fournier, C., 1999, Transmission of calicivirus by a foodhandler in the pre-symptomatic phase of illness, *Epidemiol. Infect.* 123:475–478.

Gerba, C. P., and Goyal, S. M., 1978, Detection and occurrence of enteric viruses in shellfish: a review, *J. Food Protect.* 41:743–754.

Gerba, C. P., and Goyal, S. M. (Eds.), 1982, Methods in Environmental Virology, Marcel Dekker, New York, NY, 378 pp.

Gerba, C. P., Rose, J. B., and Haas, C. N. 1996, Sensitive populations-who is at the greatest risk? *Int. J. Food Microbiol.* 30:113–123.

Goswami, B. B., Kulka, M., Ngo, D., Istafanos, P., and Cebula, T. A., 2002, A polymerase chain reaction-based method for the detection of hepatitis A virus in produce and shellfish, *J. Food Prot.* 65:393–402.

Girish, R., Broor, S., Dar, L., and Ghosh, D., 2002, Foodborne outbreak caused by a Norwalk-like virus in India, *J. Med. Virol.* 67:603–607.

Gordon, S. M., Oshiro, L. S., and Jarvis, W. R., 1990, Foodborne Snow Mountain agent gastroenteritis with secondary person-to-person spread in a retirement community, *Am. J. Epidemiol.* 131:702–710.

Goyal, S. M., Gerba, C. P., and Melnick, J. L., 1979, Human enteroviruses in oysters and their overlying waters, *Appl. Environ. Microbiol.* 37:572–581.

Goyal, S. M., and Gerba, C. P., 1983, Viradel method for detection of rotavirus from seawater, *J. Virol. Methods* 7:279–285.

Greiser-Wilke, I., and Fries, R., 1994, Methods for the detection of viral contamination in food of animal origin, *Dtsch. Tierarztl. Wochensch.* 101:284–290.

Gulati, B. R., Allwood, P. B., Hedberg, C. W., and Goyal, S. M., 2001, Efficacy of commonly used disinfectants for the inactivation of calicivirus on strawberry, lettuce, and a food-contact surface, *J. Food Prot.* 64:1430–1434.

Jaykus, L. A., 1997, Epidemiology and detection as options for control of viral and parasitic foodborne disease, *Emerg. Infect. Dis.* 3:529–539.

Jaykus, L. A., De Leon, R., and Sobsey, M. D., 1996, A virion concentration method for detection of human enteric viruses in oysters by PCR and oligoprobe hybridization, *Appl. Environ. Microbiol.* 62:2074–2080.

Jean, J., Blais, B., Darveau, A., and Fliss, I., 2002, Simultaneous detection and identification of hepatitis A virus and rotavirus by multiplex nucleic acid sequence-based amplification (NASBA) and microtiter plate hybridization system, *J. Virol. Methods* 105:123–132.

Jean, J., D'Souza, D. H., and Jaykus, L. A., 2004, Multiplex nucleic acid sequence-based amplification for simultaneous detection of several enteric viruses in model ready-to-eat foods, *Appl. Environ. Microbiol.* 70:6603–6610.

Jiang, Y. J., Liao, G. Y., Zhao, W., Sun, M. B., Qian, Y., Bian, C. X., and Jiang, S. D., 2004, Detection of infectious hepatitis A virus by integrated cell culture/

strand-specific reverse transcriptase-polymerase chain reaction, *J. Appl. Microbiol.* 97:1105–1112.

Jothikumar, N., Lowther, J. A., Henshilwood, K., Lees, D. N., Hill, V. R., and Vinje, J., 2005, Rapid and sensitive detection of noroviruses by using TaqMan-based one-step reverse transcription-PCR assays and application to naturally contaminated shellfish samples, *Appl. Environ. Microbiol.* 71:1870–1875.

Karamoko, Y., Ibenyassine, K., Aitmhand, R., Idaomar, M., and Ennaji, M. M., 2005, Adenovirus detection in shellfish and urban sewage in Morocco (Casablanca region) by the polymerase chain reaction, *J. Virol. Methods* 126:135–137.

Katzenelson, E., Fattal, B., and Hostovesky, T., 1976, Organic flocculation: an efficient second-step concentration method for the detection of viruses in tap water, *Appl. Environ. Microbiol.* 32:638–639.

Kingsley, D. H., and Richards, G. P., 2001, Rapid and efficient extraction method for reverse transcription-PCR detection of hepatitis A and Norwalk-like viruses in shellfish, *Appl. Environ. Microbiol.* 67:4152–4157.

Kingsley, D. H., Meade, G. K., and Richards, G. P., 2002, Detection of both hepatitis A virus and Norwalk-like virus in imported clams associated with food-borne illness, *Appl. Environ. Microbiol.* 68:3914–3918.

Kobayashi, S., Natori, K., Takeda, N., and Sakae, K., 2004, Immunomagnetic capture RT-PCR for detection of norovirus from foods implicated in a foodborne outbreak, *Microbiol. Immunol.* 48:201–204.

Kohn, M. A., Farley, T. A., Ando, T., Curtis, M., Wilson, S. A., Jin, Q., Monroe, S. S., Baron, R. C., McFarland, L. M., and Glass, R. I., 1995, An outbreak of Norwalk virus gastroenteritis associated with eating raw oysters. Implications for maintaining safe oyster beds, *J. Am. Med. Assn.* 273:1492.

Koopmans, M., von Bonsdorff, C. H., Vinje, J., de Medici, D., and Monroe, S., 2002, Foodborne viruses, *FEMS Microbiol. Rev.* 26:187–205.

Koopmans, M., Vennema, H., and Heersma, H., 2003, Early identification of common-source foodborne virus outbreaks in Europe, *Emerg. Infect. Dis.* 9:1136–1142.

Koopmans, M., and Duizer, E., 2004, Foodborne viruses: an emerging problem. *Inter. J. Food Microbiol.* 90:23–41.

Legeay, O., Caudrelier, Y., Cordevant, C., Rigottier-Gois, L., and Lange, M., 2000, Simplified procedure for detection of enteric pathogenic viruses in shellfish by RT-PCR, *J. Virol. Methods* 90:1–14.

Le Guyader, F., Dubois, E., Menard, D., and Pommepuy, M., 1994, Detection of hepatitis A virus, rotavirus, and enterovirus in naturally contaminated shellfish and sediment by reverse transcription-seminested PCR, *Appl. Environ. Microbiol.* 60:3665–3671.

Le Guyader, F. S., Schultz, A. C., Haugarreau, L., Croci, L., Maunula, L., Duizer, E., Lodder-Verschoor, F., von Bonsdorff, C. H., Suffredini, E., van der Poel, W. M., Reymundo, R., and Koopmans, M., 2004, Round-robin comparison of methods for the detection of human enteric viruses in lettuce, *J. Food Prot.* 67:2315–2319.

Leggitt, P. R., and Jaykus, L. A., 2000, Detection methods for human enteric viruses in representative foods, *J. Food Prot.* 63:1738–1744.

Lewis, G. D., and Metcalf, T. G., 1988, Polyethylene glycol precipitation for recovery of pathogenic viruses, including hepatitis A virus and human rotavirus, from oyster, water, and sediment samples, *Appl. Environ. Microbiol.* 54:1983–1988.

Lewis, G. D., Molloy, S. L., Greening, G. E., and Dawson, J., 2000. Influence of environmental factors on virus detection by RT-PCR and cell culture, *J. Appl. Microbiol.* 88:633–640.

Loisy, F., Atmar, R. L., Guillén, P., Le Cann, P., Pommepuy, M., and Le Guyader, F. S., 2005, Real-time RT-PCR for norovirus screening in shellfish, *J. Virol. Methods* 123:1–7.

Malik, Y. S., Maherchandani, S., Allwood, P. B., and Goyal, S. M., 2005, Evaluation of animal origin cell cultures for in vitro cultivation of Noroviruses. *J. Appl. Res. Clin. Expt. Therapeutics*. 5:312–317.

Mead, P. S., Slutsker, L., Dietz, V., McCaig, L. F., Bresee, J. S., Shapiro, C., Griffin, P. M., and Tauxe, R. V., 1999, Food-related illness and death in the United States, *Emerg. Infect. Dis*. 5:607–625.

Mullendore, J. L., Sobsey, M. D., and Shieh, Y. C., 2001. Improved method for the recovery of hepatitis A virus from oysters, *J. Virol. Methods* 94:25–35.

Muniain-Mujika, I., Girones, R., and Lucena, F., 2000, Viral contamination of shellfish: evaluation of methods and analysis of bacteriophages and human viruses, *J. Virol. Methods* 89:109–118.

Muniain-Mujika, I., Calvo, M., Lucena, F., and Girones, R., 2003, Comparative analysis of viral pathogens and potential indicators in shellfish, *Inter. J. Food Microbiol.* 83:75–85.

Olsvik, O., Popovic, T., and Skjerve, E., 1994, Magnetic separation techniques in diagnostic microbiology, *Clin. Microbiol. Rev*. 7:43–54.

Ponka, A., Maunula, L., von Bonsdorff, C. H., and Lyytikainen, O., 1999, An outbreak of calicivirus associated with consumption of frozen raspberries, *Epidemiol Infect.* 123:469–474.

Rabenau, H. F., Sturmer, M., Buxbaum, S., Walczok A., Preiser, W., and Doerr, H. W., 2003, Laboratory diagnosis of norovirus: which method is the best? *Intervirology* 46:232–238.

Ribao, C., Torrado, I., Vilarino, M. L., and Romalde, J. L., 2004, Assessment of different commercial RNA-extraction and RT-PCR kits for detection of hepatitis A virus in mussel tissues, *J. Virol. Methods* 115:177–182.

Richards, G. P., 1999, Limitations of molecular biological techniques for assessing the virological safety of foods, *J. Food Prot.* 62:691–697.

Richards, G. P., Goldmintz, D., and Babinchak, L., 1982, Rapid method for extraction and concentration of poliovirus from oyster tissues, *J. Virol. Methods* 5:285–291.

Rigotto, C., Sincero, T. C., Simoes, C. M., and Barardi, C. R., 2005, Detection of adenoviruses in shellfish by means of conventional-PCR, nested-PCR, and integrated cell culture PCR (ICC/PCR), *Water Res.* 39:297–304.

Romalde, J. L., Area, E., Sanchez, G., Ribao, C., Torrado, I., Abad, X., Pinto, R. M., Barja, J. L., and Bosch, A., 2002. Prevalence of enterovirus and hepatitis A virus in bivalve molluscs from Galicia (NW Spain): inadequacy of the EU standards of microbiological quality, *Int. J. Food Microbiol.* 74:119–130.

Romalde, J. L., Ribao, C., Luz Vilarino, M., and Barja, J. L., 2004, Comparison of different primer sets for the RT-PCR detection of hepatitis A virus and astrovirus in mussel tissues, *Water Sci. Technol.* 50:131–136.

Rosenfield, S. I., and Jaykus, L. A., 1999, A multiplex reverse transcription polymerase chain reaction method for the detection of foodborne viruses, *J. Food Prot.* 62:1210–1214.

Sair, A. I., D'Souza, D. H., Moe, C. L., and Jaykus, L. A., 2002, Improved detection of human enteric viruses in foods by RT-PCR, *J. Virol Methods* 100:57–69.

Schwab, K. J., Neill, F. H., Estes, K. K., Metcalf, T. G., and Atmar, R. L., 1998, Distribution of Norwalk virus within shellfish following bioaccumulation and subsequent depuration by detection using RT-PCR, *J. Food Prot.* 61:1674–1680.

Schwab, K. J., Neill, F. H., Fankhauser, R. L., Daniels, N. A., Monroe, S. S., Bergmire-Sweat, D. A., Estes, M. K., and Atmar, R. L., 2000, Development of methods to detect "Norwalk-like viruses" (NLVs) and hepatitis A virus in delicatessen foods: application to a food-borne NLV outbreak, *Appl. Environ. Microbiol* 66:213–218.

Schwab, K. J., Neill, F. H., Le Guyader, F., Estes, M. K., and Atmar, R. L., 2001, Development of a reverse transcription-PCR-DNA enzyme immunoassay for detection of "Norwalk-like" viruses and hepatitis A virus in stool and shellfish, *Appl. Environ. Microbiol.* 67:742–749.

Seidel, K. M., Goyal, S. M., Rao, V. C., and Melnick, J. L., 1983, Concentration of rotaviruses and enteroviruses from blue crabs (Callinectes sapidus), *Appl. Environ. Microbiol.* 46:1293–1296.

Shieh, Y. C., Calci, K. R., and Baric, R. S., 1999, A method to detect low levels of enteric viruses in contaminated oysters. *Appl. Environ. Microbiol.* 65:4709–4714.

Shieh, Y. C., Monroe, S. S., Fankhauser, R. L., Langlois, G. W., and Burkhardt, W., Baric, R. S., 2000, Detection of norwalk-like virus in shellfish implicated in illness, *J. Infect Dis.* 181 Suppl 2:S360–S366.

Slomka, M. J., and Appleton, H., 1998, Feline calicivirus as a model system for heat inactivation studies of small round structured viruses in shellfish, *Epidemiol. Infect.* 121:401–407.

Sobsey, M. D., Wallis, C., and Melnick, J. L., 1975, Development of a simple method for concentrating enteroviruses from oysters, *Appl. Microbiol.* 29:21–26.

Sobsey, M. D., Carrick, R. J., and Jensen, H. R., 1978, Improved methods for detecting enteric viruses in oysters, *Appl. Environ. Microbiol.* 36:121–128.

Sunen, E., Casas, N., Moreno, B., and Zigorraga, C., 2004, Comparison of two methods for the detection of hepatitis A virus in clam samples (Tapes spp.) by reverse transcription-nested PCR, *Int. J. Food Microbiol.* 91:147–154.

Svensson, L., 2000, Diagnosis of foodborne viral infections in patients, *Int. J. Food Microbiol.* 59:117–126.

Taku, A., Gulati, B. R., Allwood, P. B., Palazzi, K., Hedberg, C. W., and Goyal, S. M., 2002, Concentration and detection of caliciviruses from food-contact surfaces, *J. Food Prot.* 65:999–1004.

Traore, O., Arnal, C., Mignotte, B., Maul, A., Laveran, H., Billaudel, S., and Schwartzbrod, L. 1998, Reverse transcriptase PCR detection of astrovirus, hepatitis A virus, and poliovirus in experimentally contaminated mussels: comparison of several extraction and concentration methods, *Appl. Environ. Microbiol.* 64:3118–3122.

Vaughn, J. M., Landry, E. F., Vicale, T. J., and Dahl, M. C., 1979, Modified procedure for the recovery of naturally accumulated poliovirus from oysters, *Appl. Environ. Microbiol.* 38:594–598.

CHAPTER 5

Molecular Methods of Virus Detection in Foods

Robert L. Atmar

1.0. INTRODUCTION

Food-borne transmission of virus infections has been recognized for more than 5 decades (Svensson, 2000). The principal clinical syndromes associated with food-borne viruses are hepatitis and gastroenteritis, but not all enteric viruses (Table 5.1) have been linked to food-borne illness either epidemiologically or by direct pathogen detection. The primary means of identifying viruses as causes of food-borne outbreaks has been through the recognition of a common viral pathogen in consumers and the use of epidemiologic methods to identify a particular food as the vector. More direct methods, such as the detection of viruses in food, have largely been unsuccessful because only small quantities of viruses are generally present in food, and these viruses are either difficult to grow in cell cultures or are noncultivable.

In the past two decades, molecular assays have been developed for the detection of a number of pathogens, including food-borne viruses. In the past decade, the sensitivity of these assays has improved to the point that their application to virus detection in foods can be considered. This chapter will provide an overview of molecular assays that are available for pathogen detection, examples of application of these methods for the detection of viruses associated with food-borne illness, and the limitations of these assays.

2.0. NONAMPLIFICATION METHODS (PROBE HYBRIDIZATION)

Probe hybridization assays were the first molecular assays applied to the detection of enteric viruses. In these assays, single-stranded RNA or DNA probes that are complementary to a viral genomic sequence are linked to a reporter (radioisotope, enzyme, chemiluminescent agent) and hybridized with the target. The probes can range in size from 15 to 20 to several hundred nucleotides. Detection of signal from the reporter after the hybridization reaction indicates the presence of the target nucleic acid.

Several different hybridization formats can be used, including solid-phase hybridization, liquid hybridization, and *in situ* hybridization. In solid-phase hybridization, the target nucleic acid is fixed to a nylon or nitrocellulose membrane and a solution containing the labeled probe is applied. After hybridization, the unbound probe is washed away and the bound probe is

Table 5.1 Enteric Viruses and Their Association with Food-borne Illness

Virus	*Family*	*Disease*	*Food-borne Transmission Demonstrated*
Enteric adenovirus	Adenoviridae	Gastroenteritis	No
Astrovirus	Astroviridae	Gastroenteritis	Yes
Human caliciviruses	Caliciviridae		
Norovirus		Gastroenteritis	Yes
Sapovirus		Gastroenteritis	Yes
Hepatitis A virus	Picornaviridae	Hepatitis	Yes
Hepatitis E virus	Unclassified	Hepatitis	Yes
Rotavirus	Reoviridae	Gastroenteritis	Yes

detected by fluorescence, radioactivity, or color development. In liquid-phase hybridization, both the target and probe are in solution at the time of hybridization. Probe signal can then be detected by fluorescence or color change. *In situ* hybridization is used to detect target nucleic acids within an infected cell. The principal application of this method for the detection of viruses in foods would be to couple the hybridization assay to a cell culture system (Jiang et al., 1989). Because the method is more cumbersome than antigen detection methods, it is rarely used for this purpose.

Probe hybridization assays have been described for the detection of a number of enteric viral pathogens (Dimitrov et al., 1985; Jansen et al., 1985; Takiff et al., 1985; Willcocks et al., 1991). However, their sensitivity is no better than that of antigen detection methods (approximately 10,000 genomic copies) (Jiang et al., 1992). Thus, this is not a practical method for the detection of enteric viruses in foods but can be incorporated into some of the other molecular methods described below to confirm their specificity.

3.0. AMPLIFICATION METHODS

The era of molecular diagnostics began with the development of methods to detect low numbers of pathogens in clinical and environmental samples. Since the initial description of the polymerase chain reaction (PCR) assay (Saiki et al., 1985), a number of different strategies have been developed for use in molecular assays (Table 5.2). These methods are based on amplification of the target nucleic acids, amplification of the signal generated after probe hybridization, and amplification of a probe sequence (Nolte and Caliendo, 2003). One or more examples of each are described in the following sections.

3.1. Target Amplification

Several target amplification systems have been described, with PCR-based assays being the best known and most commonly used. In each of these systems, enzymatic reactions are used to amplify a portion of the target nucleic acid 1 million–fold or more to the point that the amplified products can be easily detected and analyzed. One of the major pitfalls of these strategies is that the products generated can serve as a contaminating template for subsequent assays and lead to false-positive results. Strategies to prevent carryover contamination are addressed later in this chapter.

3.1.1. PCR

In its simplest form, the PCR uses a DNA polymerase to amplify a DNA template. The core components of this chemical reaction are the DNA polymerase, equimolar concentrations of deoxyribonucleotide triphosphates (dATP, dCTP, dGTP, dTTP), molar excess of two oligonucleotide primers, and an appropriate buffer. The oligonucleotide primers are complementary to sequences on opposite strands of the target template, and these primers flank the region to be amplified. There are three basic steps that are repeated through a variable number of cycles: (1) heat denaturation, (2) primer annealing, and (3) primer extension. The initial step denatures double-stranded DNA into single-stranded DNA using heat (92°C and 95°C). The reaction mix is then rapidly cooled to 40–60°C when oligonucleotide primers preferentially bind to the single-stranded DNA template because they are present in a much higher concentration than the template. In the third step, the DNA polymerase adds nucleotides to the 3′ end of the primer that are complementary to the sequence of the template (Fig. 5.1). This step occurs

Table 5.2 Common Nucleic Acid Amplification Methods Used for Pathogen Detection and Application to Enteric Viruses

Method	*Amplification Strategy*	*Potential for Carryover Contamination*	*Assay Described for Enteric Viruses*	*Assay Applied to Detect Enteric Viruses in Food*
Polymerase chain reaction (PCR)	Target amplification	Yes	Yes	Yes
Nucleic acid sequence-based amplification (NASBA)	Target amplification	Yes	Yes	Yes
Strand displacement amplification (SDA)	Target amplification	Yes	No	No
Branched DNA (bDNA) amplifcation	Signal amplification	No	No	No
Ligase chain reaction (LCR)	Probe amplification	Yes	No	No

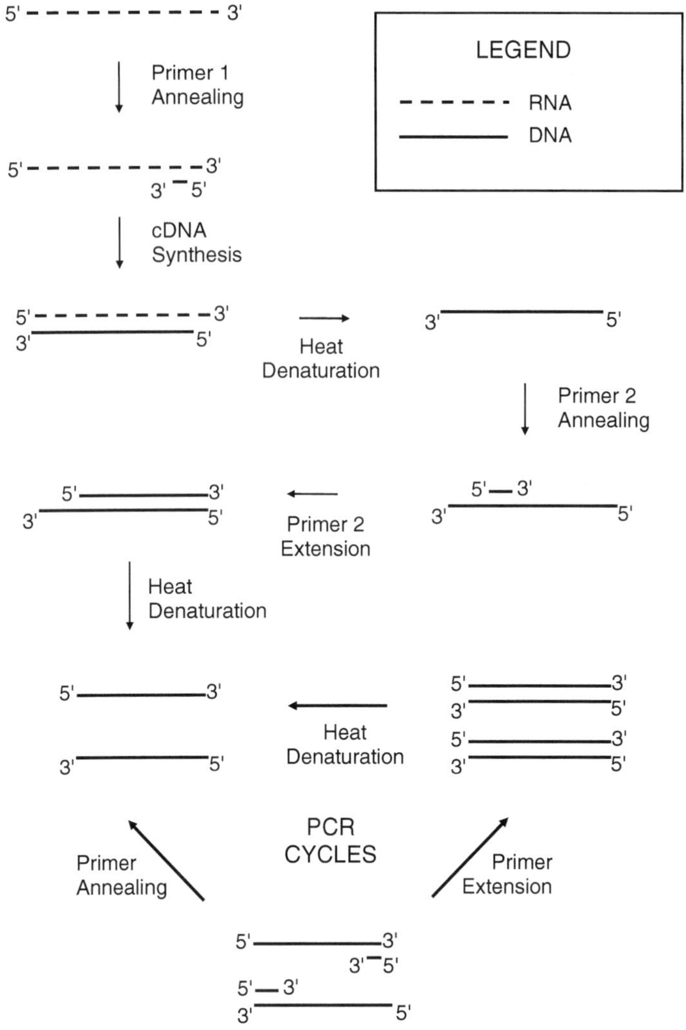

Figure 5.1 Schematic of RT-PCR reaction. Primer 1 anneals to the RNA, and cDNA is synthesized by reverse transcriptase. After denaturation of the RNA:cDNA hybrid, primer 2 anneals to the cDNA and is extended to form a second strand. This double-stranded DNA then enters cycles of heat denaturation, primer annealing, and primer extension. The number of copies of amplicons can double with each cycle if the reaction occurs at maximal efficiency.

at a temperature (70–75°C) that is optimal for the enzymatic activity of the DNA polymerase.

The three steps of heat denaturation, primer annealing, and primer extension are repeated in each cycle. If the reaction proceeds with absolute efficiency, the amount of amplified DNA doubles after each cycle. Thus, the target DNA sequence can be amplified approximately 1 million (2^{20})-fold after 20 cycles. In practice, the efficiency of PCR amplification is less than ideal especially with increases in the number of cycles.

The initial description of the PCR assay used a thermolabile DNA polymerase (Klenow fragment of *Escherichia coli* DNA polymerase), and hence the enzyme had to be replaced after each heat denaturation step (Saiki et al., 1985). The identification of a thermostable DNA polymerase prevented heat inactivation of the enzyme after each cycle and allowed the automation of the amplification reaction using programmable thermal cyclers (Saiki et al., 1988). Since the initial description of Taq polymerase derived from *Thermus aquaticus*, additional thermostable DNA polymerases have been described. Some of the newer thermostable DNA polymerases have higher fidelity (lower error rates) than the Taq polymerase (Cline et al., 1996).

3.1.2. RT-PCR

Reverse transcription–PCR (RT-PCR), also called RNA-PCR, is a modification of the PCR reaction that allows amplification of an RNA template. In the initial step, a complementary DNA (cDNA) is synthesized, which is then amplified by PCR using the same steps as described above for a DNA template. The cDNA synthesis step also requires deoxyribonucleotide triphosphates, an oligonucleotide primer, an appropriate buffer, and a DNA polymerase with reverse transcriptase activity. The oligonucleotide primer can be template-specific (as used in the PCR reaction), or it can be random hexamers or oligo-dT (if the genomic region to be amplified is near a polyadenylated site). A two-step RT-PCR assay is the one in which the cDNA is synthesized in a separate reaction and all or a part of the reaction mix is subsequently added to the PCR reaction mix. In a one-step RT-PCR assay, all reagents necessary for both cDNA synthesis and PCR amplification are added at the same time.

Most RT-PCR assays use a heat-labile reverse transcriptase, such as avian myelobastosis virus (AMV) reverse transcriptase or Moloney murine leukemia virus (MMLV) reverse transcriptase, and the cDNA synthesis step is carried out at a lower temperature (<50°C). The development of a thermostable DNA polymerase from *Thermus thermophilus* that has both reverse transcriptase and DNA polymerase activities has allowed the use of a single enzyme for RT-PCR assay (Myers and Gelfand, 1991). This enzyme has been successfully incorporated into assays used to detect noroviruses in shellfish, and the limits of virus detection are similar to assays using the Taq polymerase (Schwab et al., 2001). However, subsequent studies have demonstrated that this enzyme fails to adequately amplify viral genome with certain primer pairs that can be used successfully with Taq polymerase.

3.1.3. Nested PCR

Nested PCR is the serial amplification of a target sequence using two different primer pairs (Haqqi et al., 1988). The initial amplification is performed using an outer primer pair and 15–30 cycles of amplification. A second round of amplification is then performed using primers that anneal to a region located (or nested) between the initial two primers. Hemi-nested PCR is a variant of nested PCR in which one of the primers used in the second round of amplification is the same as that used in the first round of amplification and the second primer anneals to a region on the opposite strand that is nested between the initial two primers. Nested and hemi-nested PCR have been used to detect sequences that cannot be detected after a single round of PCR, thus increasing the sensitivity of the assay.

Nested PCR also has the potential to increase the specificity of the assay because both primer pairs must amplify the target sequence. However, the major problem with this approach is the potential for carryover contamination. PCR products from the first round of amplification may contaminate the laboratory or cross-contaminate other tubes during their transfer to a second tube for the second round of amplification. The cross-contamination causes false-positive results and thus can decrease the specificity of the assay. Although carryover contamination is a potential problem for any PCR assay, the methodologic controls (see Section 6.0) used to prevent this problem are more likely to fail for nested assays. Nevertheless, strategies that do not require reopening the tube after the first round of amplification have been developed (Ratcliff et al., 2002). In the "hanging drop" method, reagents (inner primer pair, additional DNA polymerase) for the nested PCR step are placed in the cap of the tube and are added to the overall reaction mix by centrifugation after the first round of PCR is complete.

3.1.4. Multiplex PCR

Multiplex PCR assays use two or more primer pairs to amplify different target sequences in a single tube (Chamberlain et al., 1988). This strategy allows the evaluation of a sample for more than one virus at a time and can also identify the presence of multiple viruses in a single reaction. However, the primers for different targets should have similar annealing temperatures and lack complementarity so that each target can be efficiently amplified. Even when such steps are taken, the multiplex PCR assays usually have decreased sensitivity as compared with the standard PCR assays due to competition for reagents. A high concentration of one target (virus) can prevent the detection of other targets present in lower concentrations that would have been detected if the high concentration target was absent. This can thus lead to false-negative results for the lower concentration target.

3.1.5. Postamplification Analysis

After PCR amplification of a target nucleic acid, additional analyses must be performed to interpret the results of the assay as outlined in Table 5.3. The simplest method is to perform gel electrophoresis using all or a portion of the PCR reaction mix. Molecular weight markers are run concurrently and the presence of a band of the expected amplicon product size (based on the genomic location targeted by the primers) is intepreted as a positive result. Although this approach is simple, the occurrence of nonspecific amplification can lead to bands that are of the expected size but are not virus-specific (Atmar et al., 1996). This approach is applied most commonly with nested

Table 5.3 Postamplification Analysis Strategies for Target Amplification

1. Gel electrophoresis
2. Restriction
3. Hybridization
   (a) Solid phase
       Slot/dot blot
       Southern blot
       Microarrays
   (b) Liquid phase
4. Sequencing

PCR assays, but caution must be used because this amplification strategy can result in nonspecific amplification. One of the following approaches may provide additional reassurance as to the specificity of the assay.

Restriction analysis combines gel electrophoresis with digestion of virus-specific amplicons using a restriction endonuclease. With this strategy, amplicons of the expected size must be generated, and the amplicons must have a specific restriction site. Digestion of the virus-specific amplicons leads to the generation of shorter fragments whose size can be predicted based on the location of the restriction site. Generation of bands of the expected sizes after restriction is interpreted as a positive result.

Hybridization (see Sec. 2.0) is the most common approach used to identify and confirm a positive PCR result. Many different hybridization formats are used, including dot/slot blots, Southern blots, and liquid hybridization. Southern blot hybridization has the advantage of providing information about the product size in addition to reactivity with a virus-specific probe. The disadvantage of Southern blot hybridization is the additional time and effort required to perform the assay. This method usually adds a day to the overall assay. In contrast, liquid hybridization assays can yield results within 1 hr after the completion of the PCR step.

Hybridization assay formats have been developed that allow multiple probes to be used. The reverse line blot arrays oligonucleotide probes on a solid matrix (e.g., nylon membrane), and these probes are hybridized to denatured amplicons. One of the primers used during the amplification process is biotinylated and hybridization of the strand containing the biotinylated primer to a virus-specific probe is detected using a streptavidin reporter, such as a peroxidase enzyme that can react with an appropriate substrate. This method has been used to not only identify virus-specific amplicons but also to further characterize norovirus strains using genotype-specific probes (Vinjé and Koopmans, 2000). This technology can be taken further with the use of DNA microarrays. Hundreds or thousands of probes are fixed to a surface (such as a silica wafer or glass slide) and hybridization of labeled products to specific probes is detected (Nolte and Caliendo, 2003). Microarrays have been used in combination with RT-PCR for the detection and genotyping of group A rotaviruses (Chizhikov et al., 2003; Lovmar et al., 2003). With a large enough number of probes, it should even be possible to deduce the sequence of virus-specific amplicons based on the hybridization patterns. However, improvements in the current technology are needed to decrease the complexity and costs of microarrays so that their potential can be fully realized (Nolte and Caliendo, 2003).

The development and increased availability of automated sequencers have led to the use of direct sequencing of amplicons as a measure of specificity. The sequence data can provide information not only confirming the specificity of the amplification but also for genotyping or classifying virus strains (Robertson et al., 1991). The information can be combined with epidemiologic data for use in surveillance and outbreak investigations (Koopmans et al., 2003). A greater quantity of amplicons is needed to generate sequence data, making this method less sensitive than the hybridization techniques described above.

3.1.6. Real-Time PCR

Automated instruments have been developed to allow the specific detection of amplified nucleic acids in a closed system. Real-time PCR can improve the efficiency of the analytic process while decreasing the risk of carryover contamination by eliminating the need for post-PCR manipulation of amplicons in confirmatory tests of specificity. Disadvantages include the capital expense of the equipment (thermal cycler and amplicon detection equipment), limited ability to perform multiplex assays, and the inability to monitor the amplicon size (Mackay et al., 2002).

There are two principal approaches for the detection of amplified products: use of DNA-binding fluorophores (fluorescing dyes) and use of specific oligoprobes. The fluorophores intercalate with double-stranded DNA and fluoresce after exposure to a specific wavelength of light. SYBR Green is the most commonly used fluorophore, but ethidium bromide and YO-PRO-1 are also used. A melting curve analysis is used to distinguish virus-specific amplicons from nonspecific primer-dimers with the former having higher dissociation temperatures. The utility of fluorophores for amplicon detection is limited by the inability of this approach to identify amplicons that result from nonspecific amplification in the initial PCR steps, especially when the target is present in low concentrations (as might be expected in contaminated foods) (Mackay et al., 2002).

Fluorescently labeled oligoprobes are the principal means for specific amplicon detection in real-time PCR assays (Fig. 5.2). The two most common methods use oligoprobes that are dual-labeled with a reporter fluorophore and a quencher fluorophore. As the name implies, the quencher fluorophore will quench the signal from the reporter fluorophore when the two are in close proximity and are exposed to a certain wavelength of light (also called fluorescence resonance energy transfer, or FRET). 5′ nuclease oligoprobes have the reporter fluorophore on the 5′ end of the oligoprobe and the quencher fluorophore on the 3′ end, and their melting temperature is gen-

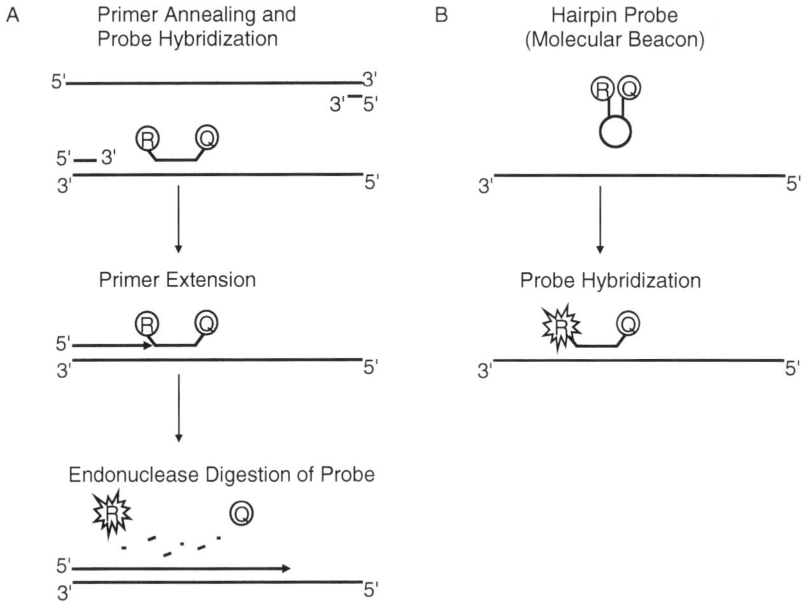

Figure 5.2 Schematic of fluorescent probe detection of PCR products in real-time PCR assays. (A) 5′ nuclease oligoprobes. A probe with a fluorescent reporter (R) and quencher (Q) hybridizes with the target amplicons strand. The quencher suppresses the signal generated by the reporter. The strand-specific primer is extended during the PCR reaction by the polymerase, and the exonuclease activity of the polymerase separates the reporter from the quencher, allowing increased fluorescent signal to be generated. (B) Hairpin oligoprobes or molecular beacons. The fluorescent reporter (R) and quencher (Q) are juxtaposed by hybridization of the 5′ and 3′ ends of the probe, allowing the quencher to suppress the signal of the reporter. Hybridization of the probe to the target physically separates the reporter and quencher, allowing an increase in the reporter's fluorescent signal.

erally ~10°C higher than those of the primers used for amplification. If virus-specific amplicons are generated, the oligoprobe will hybridize to its target sequence and the 5′-3′ exonuclease activity of the Taq polymerase will release the reporter fluorophore by hydrolysis. Fluorescence from the freed reporter fluorophore can then be detected and as the reaction proceeds, the amount of fluorescence increases proportional to the amount of target amplicons generated. Hairpin oligoprobes, or molecular beacons, are the other commonly used dual-labeled probe. The probes are designed to form a hairpin structure so that the reporter and quencher fluorophores are juxtaposed in the absence of a specific DNA target. When a target sequence with a complementary sequence is present, the probe can hybridize and assume a linear conformation. Fluorescent signal can then be detected after the spatial separation of the quencher from the reporter fluorophore (Mackay et al., 2002).

3.1.7. Application to Food-borne Viruses

PCR assays have been developed for food-borne viruses listed in Table 5.1. The utility of these assays has varied based on the availability of other diagnostic tests. PCR assays for each of these viruses are described in the following paragraphs.

The adenovirus hexon gene is the most common target of PCR assays for these viruses, although other viral genes have also served as targets (Allard et al., 1992; Xu et al., 2000). Enteric adenoviruses (group F) can be further differentiated from nonenteric adenoviruses through the use of group-specific primers, postamplification restriction analysis, and hybridization analyses (Rousell et al., 1993; Allard et al., 1994; Soares et al., 2004). To date, no real-time PCR assays for enteric adenoviruses have been developed although such assays are described for nonenteric adenoviruses (Gu et al., 2002). The sensitivity of these assays is similar to that of antigen detection assays so that the general use of PCR for the detection of adenoviruses in clinical and environmental samples is limited.

RT-PCR assays are the most sensitive means of detecting astroviruses. Nonstructural genes (putative protease region [ORF1a] and RNA-dependent, RNA polymerase region [ORF1b]) and the 3′ noncoding region are the most common regions of the viral genome targeted for amplification, although the capsid gene (ORF2) has also been targeted, especially when further characterization of strains within a serotype is desired (Jonassen et al., 1995; Noel et al., 1995; Belliot et al., 1997). Passage in cell cultures for up to 48 hr can expand the quantity of virus in a sample prior to amplification (integrated cell culture/RT-PCR), and this approach may improve sensitivity of detection as compared with that of RT-PCR alone (Mustafa et al., 1998). Although real-time RT-PCR assays for astrovirus have been described, additional studies are needed to evaluate the performance characteristics of these assays (Grimm et al., 2004; Le Cann et al., 2004).

RT-PCR assays are now the primary means of diagnosis of human calicivirus infections, supplanting the prior use of electron microscopy (Atmar and Estes, 2001). The RNA-dependent RNA polymerase and the capsid genes are the primary targets for amplification, although primers that amplify other regions of the genome have also been described (Matsui et al., 1991). Despite the evaluation of many different primer sets, no single detection assay is able to detect all strains because of the genetic diversity of these viruses (Atmar and Estes, 2001; Vinjé et al., 2003). Most laboratories will select one or a few primer sets (e.g., different pairs for genogroup I and II noroviruses and for sapoviruses) for use in initial screening of samples. If no virus is detected with the selected primers but the index of suspicion for the presence of caliciviruses remains high, the use of additional primer pairs can lead to successful virus detection (Le Guyader et al., 2004). Nested RT-PCR is also used in some laboratories (Schreier et al., 2000). Real-time RT-PCR assays that use probe detection or SYBR Green for amplification detection are also available for noroviruses, although the general utility of these assays remains to be determined (Kageyama et al., 2003; Pang et al., 2004a). The genetic diversity of noroviruses is likely to adversely affect the utility of real-time RT-PCR assays for virus quantitation because base mismatches between the primers and target sequence will decrease the efficiency of amplification. The number of strains detected may also be limited as seen with standard RT-PCR assays.

Hepatitis A virus was one of the first enteric viruses for which an RT-PCR assay was developed (Jansen et al., 1990). Many human strains can be amplified using a single primer set, although nested PCR assays have been used to increase assay sensitivity (Robertson et al., 1991; Hutin et al., 1999). Genes of the structural proteins (VP1-2A junction; VP3-VP1 junction) are most commonly targeted for amplification, although the 5′ noncoding region of the viral genome is also used (Pina et al., 2001). Real-time RT-PCR assays are available as is an integrated cell culture system, but the utility of these assays is still under investigation (Abd El Galil et al., 2002; Jiang et al., 2004). Because hepatitis A virus (HAV) is difficult to cultivate and grows very slowly in cell cultures, RT-PCR assays now are considered to be the best means of direct HAV detection.

Initial studies on hepatitis E virus (HEV) suggested little genetic heterogeneity among these strains, but only a limited number of strains had been analyzed. Use of degenerate primers and progressive decrease in the annealing temperature during PCR amplification (touchdown PCR) have allowed the amplification of additional strains of HEV and have demonstrated more genetic diversity among these viruses than previously realized (Schlauder et al., 2000; Schlauder and Mushahwar, 2001). More recently, an RT-PCR assay was designed to amplify all known strains, but thus far it has been validated with only a limited number of strains (Grimm and Fout, 2002). A real-time RT-PCR assay has also been described using SYBR Green detection of amplified products, but it has only been evaluated using a single HEV strain (Orrù et al., 2004). Additional studies are needed to determine the utility of these assays for the detection of HEV.

Rotaviruses are double-stranded RNA viruses, and the performance of RT-PCR assays is complicated by difficulties in denaturing the genome for the cDNA synthesis step. Dimethyl sulfoxide (DMSO) is often used in the reverse transcription step to help denature the double-stranded RNA. This approach has been used in assays developed for the detection of human rotavirus groups A, B, and C (Gouvea et al., 1990, 1991). These assays have not proved to be much more sensitive than the ELISA tests, although the performance characteristics of PCR can be improved when used in a nested or semi-nested format and when strain-specific oligonucleotides are used for cDNA synthesis (Gouvea et al., 1990; Buesa et al., 1996; Iturriza-Gomara et al., 1999). RT-PCR assays also are used to genotype strains, and the results correlate very well with serotyping assay results (Gouvea et al. 1990; Gentsch et al., 1992). Most of the RT-PCR assays developed and described in the 1990s target the structural genes VP4, VP6, and VP7 and amplify large segments of the genes. More recently, a real-time assay that amplifies an 87-base-pair segment of the nonstructural protein-3 (NSP-3) gene was described, which achieves a diagnostic sensitivity comparable to that seen with nested PCR assays without the risk of cross-contamination inherent in nested assays (Pang et al., 2004b). Additional studies with this assay are needed to determine how many different strains can be recognized, as only a limited number of G-types (G1, G2, and G4) have been evaluated and detected.

3.2. Transcription-Based Amplification

Transcription-based amplification systems represent another approach to detecting viruses by amplifying a portion of the genome and then detecting the amplified products. Two variations of transcription-based amplification systems have been described: nucleic acid sequence-based amplification (NASBA) and transcription-mediated amplification (TMA). NASBA is the principal method that has been developed thus far for the detection of enteric viruses, so this method will be described below while noting the methodologic differences between NASBA and TMA.

3.2.1. NASBA

NASBA is performed at a single temperature (isothermal) and uses two virus-specific primers, avian myeloblastosis virus (AMV) reverse transcriptase, RNase H, and T7 RNA polymerase in the reaction tube. The NASBA reaction leads to the generation of single-stranded RNA transcripts that are then detected by probe hybridization. In the initial step (Fig. 5.3), one of the virus-specific oligonucleotides that also contains a T7 promoter sequence at its 5′ end binds to the RNA target and primes the synthesis of a cDNA by the AMV reverse transcriptase. RNase H digests the RNA in the cDNA:RNA hybrid, and the second virus-specific oligonucleotide binds to the cDNA and primes the synthesis of a second strand of DNA using the DNA-dependent DNA polymerase activity of the reverse transcriptase. The T7 RNA polymerase recognizes the double-stranded DNA promoter region and generates RNA transcripts that can then feed back into a continuous cycle of cDNA synthesis, RNase H digestion, second-strand synthesis, and RNA transcript production. Up to a billion-fold amplification of the target RNA can be attained within two hr (Nolte and Caliendo, 2003).

3.2.2. TMA

The principal difference between NASBA and TMA is that TMA uses a reverse transcriptase with endogenous RNase H activity, whereas in NASBA the AMV reverse transcriptase lacks RNase H activity and this enzyme must be added separately.

3.2.3. Application of NASBA to Food-borne Viruses

The use of NASBA assays for the detection of enteric viruses is less well studied than the RT-PCR assays. NASBA assays have only been described for hepatitis A virus, noroviruses, astroviruses, and rotaviruses (Jean et al., 2001, 2002; Greene 2003; Tai et al., 2003; Moore et al., 2004). However, the sensitivity of these assays is as good or better than comparable RT-PCR assays (Jean et al., 2001; Greene et al., 2003; Tai et al., 2003). NASBA assays

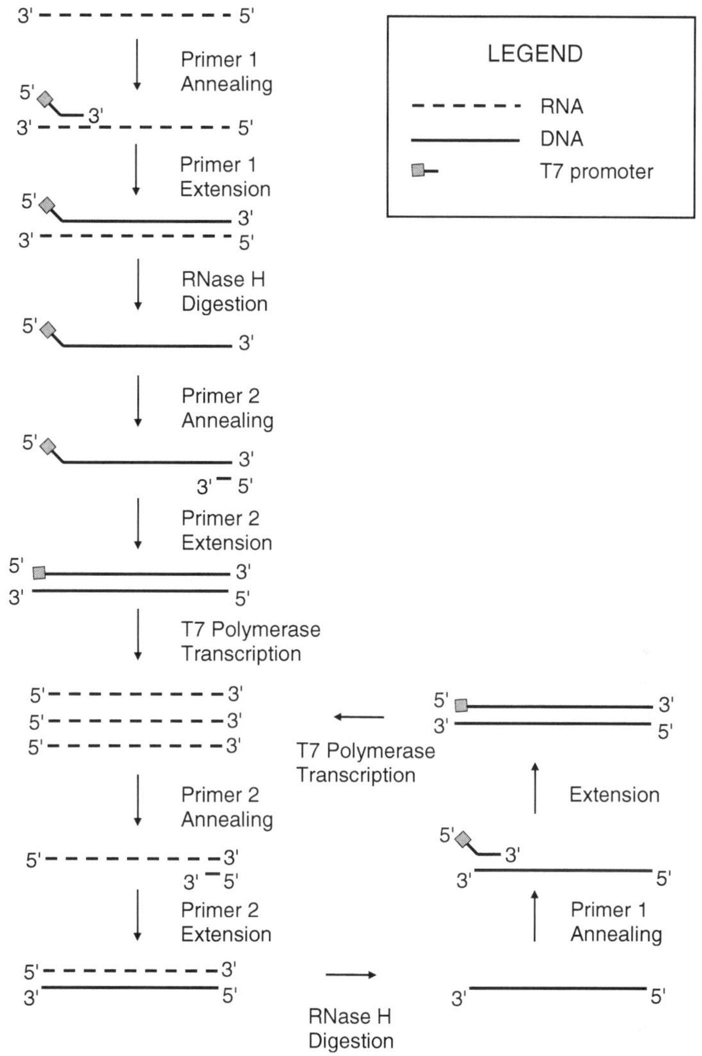

Figure 5.3 Schematic of NASBA reaction. Primer 1 anneals to the target RNA, and cDNA is made by reverse transcriptase. RNase H degrades the RNA strand, and primer 2 anneals to the cDNA and is extended by the polymerase activity of the reverse transcriptase. The T7 promoter is recognized by the T7 polymerase, which then generates numerous RNA transcripts. Primer 2 anneals to the RNA transcript, and cDNA is synthesized by the reverse transcriptase. RNase H degrades the RNA strand. Primer 1 anneals to the cDNA target, a second strand of DNA is synthesized, and T7 polymerase generates more RNA transcripts.

can also be multiplexed or coupled to RT-PCR assays to enhance virus detection (Jean et al., 2002, 2003, 2004). No TMA assays for enteric viruses have been described, and the overall utility of this approach remains to be determined. The simplicity, rapidity, and sensitivity of NASBA assays compared with RT-PCR assays suggest that further development of this approach should be pursued.

3.3. Other Signal Amplification Methods

Strand displacement amplification (SDA) and loop-mediated isothermal amplification (LAMP) are two additional target amplification methods that are available for viral diagnostics (Walker et al., 1992; Notomi et al., 2000). Both methods are based on amplification of DNA using a DNA-dependent DNA polymerase and at least four virus-specific primers. Thus, a cDNA must first be made from RNA targets. Although neither of these methods has been applied to enteric viruses or to detection of pathogens in foods, they are able to detect low numbers of nucleic acid targets as is possible with RT-PCR and NASBA (Parida et al., 2004).

3.4. Signal Amplification

Although target amplification systems are able to detect low copy numbers of a virus, issues related to carryover contamination leading to false-positive results have plagued these assays. A strategy to circumvent this problem has been the development of signal amplification systems in which neither the target nor the probe is amplified. Instead, the amount of signal generated is directly proportional to the amount of target nucleic acid. Several reporter molecules are present for each copy of target nucleic acid. This approach has the additional advantage of not being dependent on enzymatic reactions and thus is not subject to enzyme inhibitors that may be present in a sample (Nolte and Caliendo, 2003).

The principal method of signal amplification that is in use is the branched DNA (bDNA) assay, although to date no assays have been developed for any of the food-borne viruses. Oligonucleotide probes anchored to a solid substrate capture the target nucleic acid (RNA or DNA) (Urdea et al., 1991). Additional target-specific probes also bind the target. The second set of target-specific probes also binds to preamplifier molecules that will bind to bDNA amplifiers. Each bDNA amplifier then hybridizes to conjugate-labeled probes. With this approach, the signal is amplified several hundred-fold compared with a simple hybridization assay. The ease of assay performance, simple sample preparation method, and quantitative results obtained suggest that the development of this approach for the detection of food-borne viruses is warranted.

3.5. Probe Amplification

Probe amplification is a third strategy for the detection of a target nucleic acid. The only sequences contained in amplification products are those that were present in the original probe. The ligase chain reaction (LCR) is the most well developed of the probe amplification methods. In LCR, two target-specific oligonucleotide probes hybridize to adjacent sequences forming a nick. A thermostable DNA ligase closes the nick, joining the 3′ end of one oligonucleotide to the 5′ end of the other nucleotide. Two additional oligonucleotides that recognize the complementary strand of the original target as well as the ligated product generated in the initial ligation reaction are also added to the reaction and can be ligated together in subsequent cycles. Thus, logarithmic amplification can occur with each cycle of denaturation, annealing, and ligation (Wu and Wallace, 1989). The oligonucleotides are labeled with haptens so that one hapten will be captured on a solid substrate and the other will react with an antibody conjugate. Signal generated by the conjugate will be present only if the two oligonucleotides are ligated (Nolte and Caliendo, 2003). Because ligation products can serve as template, this probe amplification method is subject to carryover contamination similar to that seen with target amplification methods. No probe amplification methods have been developed for any of the food-borne viruses.

4.0. SPECIMEN PREPARATION

The goal of specimen preparation is to concentrate and purify viral nucleic acids from the sample being analyzed so that even small amounts of virus can be detected. Because many of the molecular assays described above require the use of functional enzymes (e.g., Taq polymerase, reverse transcriptase), substances that inhibit enzymatic activity can lead to false-negative results. Thus, a major objective of the specimen preparation process is the removal of PCR inhibitory substances. The steps used in sample preparation generally consist of one or more of the following (Table 5.4): (1) elution of the virus from food; (2) extraction with an organic solvent; (3) concentration of the virus; and (4) extraction of viral nucleic acids. Each of these steps is discussed below.

4.1. Elution

One of the first steps in many sample processing protocols is to separate the virus from the food matrix using an elution procedure. In foods that have been contaminated superficially, virus can be eluted by simply rinsing the food with a saline solution or glycine buffer (Schwab et al., 2000; Jean et al., 2004; Le Guyader et al., 2004b). In most other circumstances, the food is homogenized prior to the elution procedure. Virus may then be eluted directly from the food matrix using either a direct elution procedure or an adsorption-elution method.

A variety of buffers and solutions have been used successfully in direct elution protocols, including solutions of glycine and sodium chloride, borate and beef extract, saline and beef extract, and beef extract alone (Lees et al., 1994; Traore et al., 1998; Leggitt and Jaykus, 2000; Sair et al., 2002; Le Guyader et al., 2004b). Only a few studies have reported direct comparisons of dif-

Table 5.4 Major Steps in Concentration and Extraction of Viral Nucleic Acids

1. Elution
Beef extract ± saline or borate
Borate
Glycine/NaCl buffer (high pH)
Saline or phosphate-buffered saline
2. Organic solvent extraction
Choroform: butanol
Freon
3. Concentration
Antibody capture
Ligand capture
Organic flocculation
Polyethylene glycol (PEG) precipitation
Ultracentrifugation
4. Nucleic acid extraction
Guanidinium isothiocyanate (GITC)
Phenol : chloroform

ferent eluting buffers but, in these reports, glycine-based buffers have generally performed better than the other solutions tested (Traore et al., 1998; Le Guyader et al., 2004b).

Adsorption-elution also is used for virus extraction from foods. With this method, the pH and solution conductivity are lowered to adsorb the virus to solids in the homogenate followed by elution of virus using a glycine/sodium chloride buffer (pH 7.5). Adsorption-elution has primarily been applied to shellfish but has also been used successfully to detect enteric viruses in blueberries (Jaykus et al., 1996; Shieh et al., 1999; Calder et al., 2003).

4.2. Organic Solvent Extraction

Extraction with an organic solvent can be used to remove organic compounds that are insoluble or poorly soluble in water (e.g., lipids). An organic solvent can be particularly useful if homogenization of the food is part of the sample processing procedure. Trichlorotrifluoroethane (freon) has been the most common organic solvent used, but with the increased concern about environment effects of fluorocarbons, alternative solvents have been studied (Atmar et al., 1993, 1995; Le Guyader et al., 2004). Of these, chloroform: butanol (1:1, vol:vol) has performed well as an alternative (Le Guyader et al., 1996, 2000, 2004b; Dubois et al., 2002).

4.3. Virus Concentration

A variety of strategies have been used to concentrate viruses. In most instances, an elution step is used prior to the concentration step. An antibody capture assay using virus-specific antibodies attached to a fixed substrate was the first virus concentration procedure that allowed successful detection by

RT-PCR of HAV in shellfish implicated in an outbreak (Jansen et al., 1990). Since that time, additional antibody capture assays have been described using immune serum globulin and hyperimmune serum (Schwab et al., 1996; Gilpatrick et al., 2000; Kobayashi et al., 2004). Antibody is attached to a solid substrate such as paramagnetic beads, and after virus capture, repeated wash steps are used to remove substances that inhibit virus detection by molecular methods. A similar approach has been described using carbohydrate ligands for norovirus capture (Harrington et al., 2004). The potential utility of this approach is limited by the availability of broadly reactive antibodies to target viruses. For example, there is only one serotype of HAV, and antibody capture assays have worked well for this virus (Cromeans et al., 1997). This approach may not work with other enteric viruses that have multiple serotypes or antigenic types. The genetically and antigenically diverse noroviruses are an example: hyperimmune serum generated against recombinant viral protein only recognizes strains closely related to the one from which the recombinant protein was derived (Atmar and Estes, 2001). Although broadly cross-reactive monoclonal antibodies to noroviruses have been used to develop antigen ELISAs (Hale et al., 2000), the utility of these antibodies in antibody capture assays remains unproved.

Organic flocculation and polyethylene glycol (PEG) precipitation were originally used to concentrate viruses from shellfish for analysis in infectivity assays (Williams and Fout, 1992). In organic flocculation, virus is precipitated in the presence of beef extract by lowering the pH of the solution (to 3.0–4.5), and the pellet obtained after centrifugation is suspended in a small volume of sodium phosphate buffer (pH 9–9.5). No adjustments in pH are needed for the PEG precipitation. PEG-6000 was able to concentrate a variety of viruses (Lewis and Metcalf, 1988) and performed better than PEG-8000 in a direct comparison (Traore et al., 1998).

Ultracentrifugation is commonly used in research laboratories to concentrate viruses. The method has been used successfully to concentrate viruses extracted from shellfish prior to extraction of viral nucleic acids (Pina et al., 1998; Muniain-Mukija et al., 2000). The utility of the method is limited by the expense of an ultracentrifuge and the time required to pellet the virus (time of centrifugation increases as the *g*-force generated by centrifuge decreases). For these reasons, relatively few laboratories have used this approach for virus concentration from food samples.

4.4. Nucleic Acid Extraction

There are two main approaches that have been used to extract viral nucleic acids from concentrated samples: phenol : chloroform extraction and guanidinium isothiocyanate extraction. Enzymatic digestion (e.g., proteinase K) may be used as an initial step prior to phenol : choroform digestion. Several additional steps have been used to further clean-up the sample, including the use of organic solvents, suspended silica, affinity purification, and certain proprietary compounds (Jaykus, 2003). One of the more robust methods that can be adapted for nucleic acid extraction from a variety of substrates was originally described by Boom et al. (1990). The method uses chaotropic agent guanidinium thiocyanate to extract nucleic acids while inactivating nucleases. The nucleic acids are adsorbed onto a solid silica substrate following which impurities and inhibitors can be washed away before the nucleic acids are eluted and used for analysis.

When immunocapture methods are used for virus purification and concentration, the viral RNA can be released by heat denaturation into a small volume of buffer (Gilpatrick et al., 2000). Heat release is a rapid and simple method to use, but the nucleic acid must be analyzed the same day or it may be lost due to degradation by residual nucleases.

5.0. QUALITY CONTROL

Quality-control measures must be incorporated in any molecular diagnostic method. For example, the ability of PCR to detect as few as 10 copies of target sequence can lead to problems with cross-contamination between experiments or even within the same experiment and lead to false-positive results. On the other hand, failure to remove inhibitory substances during nucleic acid purification steps or the inefficient recovery of virus from a sample can lead to false-negative results. Steps (Table 5.5) to prevent or detect these problems should be routinely incorporated into any molecular detection procedure.

Table 5.5 Quality-Control Measures for Molecular Detection Assays

1. Prevention of cross-contamination
   - (a) Engineering controls
     - Separation of specimen intake, processing, set-up, and analysis areas
     - Use of dedicated equipment
     - Use of plugged pipette tips
     - Use of gloves, dedicated lab coats
     - Unidirectional workflow
     - Use of closed systems
   - (b) Experimental procedures
     - Uracil-*N*-glycolase
     - Photochemical inactivation: isopsoralen
2. Use of positive and negative controls
   - (a) Extraction controls
     - Known negative sample
     - Seeded positive sample
   - (b) Experimental control
     - Reagent control
     - Positive control
3. Inhibitor detection
   - (a) Internal standard
   - (b) Housekeeping genes

5.1. Prevention of Cross-Contamination

Cross-contamination of samples is one of the major problems encountered with the application of target amplification systems, especially with PCR assays. Millions to billions of copies of DNA amplicons per microliter are generated in each reaction. Thus, each nanoliter of reaction mix contains thousands to millions of copies of amplicons. Unrecognized microscopic contamination of equipment, clothes, and the environment can occur with daily laboratory activities and lead to false-positive results. Thus, it is important to take steps to prevent this problem and to identify the event if it occurs. The most experience in dealing with cross-contamination has been with PCR-based assays as described below.

A number of engineering controls should be used to prevent cross-contamination (Kwok and Higuchi, 1989). Ideally, the laboratory should have separate areas (rooms) for the following activities: specimen intake (cataloging and storage of samples to be analyzed), specimen processing (virus concentration and nucleic acid extraction), assay set-up, and postassay analytic assays. Workflow for an individual also should proceed in only one direction. Thus, a technician should not perform post-PCR analyses and then process new samples or set up new assays on the same day, unless she or he showers and changes clothes. Because aerosols can be generated during pipeting procedures, laboratories that perform large numbers of assays may incorporate measures to control airflow, with postanalysis areas being negative pressure in relation to other parts of the laboratory and preassay areas being positive pressure.

The work area should be cleaned with sodium hypochlorite solution to remove nucleic acids at the beginning and end of each work day and at any other time when there is concern that contamination of the work space may have occurred. Dedicated laboratory coats should be worn in only one area, gloves should be worn and changed frequently, and laboratory equipment should be dedicated to a single work area. The use of plugged pipette tips or positive displacement pipetors can help prevent contamination of equipment. Contamination of the work area with amplicons can also be prevented through the use of closed assay systems (e.g., as with real-time PCR assays). The sample to be analyzed is added to an analysis device that is then sealed and not opened again. Thus, there is no opportunity for the nucleic acids generated in the amplification reaction to contaminate the laboratory.

Additional control measures can be used to inactivate nucleic acids that may escape the engineering controls described above. Uracil-*N*-glycosylase (UNG) is a DNA enzyme that removes uracil residues from DNA molecules (Pang et al., 1992). If dUTP is used in place of dTTP in the reaction mix, the generated amplicons will contain uracil in place of thymidine. Treatment with UNG will degrade contaminating amplified DNA amplicons that contain uracil residues but not the thymidine-containing DNA (e.g., from native DNA). Because UNG has no activity against RNA molecules, it would not be effective in NASBA assay systems. UNG is added to the sample at the beginning of the reaction and is then heat inactivated during the initial heat denaturation. Potential problems that may be encountered when using UNG include the following: decreased amplification efficiency due to less efficient incorporation of uracil into DNA, need to re-optimize amplification reaction conditions when replacing dTTP with dUTP, inability to have primer annealing temperatures below 55°C (due to residual enzymatic activity that could degrade newly formed amplicons), and inefficient degradation of short amplicons (Espy et al., 1993; Nolte and Caliendo, 2003).

Another approach that can be used to control DNA carryover contamination in PCR assays is the addition of isopsoralen to the reaction mix. After amplification, and before opening the reaction vessel, the sample is exposed to long-wave ultraviolet light, which cross-links isopsoralen to the DNA and renders it resistant to further amplification. Although this approach is simple, it is inefficient at inactivating short (<100 bp) amplicons. In addition, cross-linking may interfere with hybridization and isopsoralen may inhibit amplification (Nolte and Caliendo, 2003).

5.2. Use of Positive and Negative Controls

Several controls should be incorporated into each assay. Negative controls should include an extraction control and a reagent control. The use of an extraction control (a known negative sample) can allow the identification of cross-contamination during the virus concentration and nucleic acid extraction processes. A negative reagent control allows the identification of contamination of one or more of the reagents used to perform the assay. Some investigators choose to use more than one extraction or reagent control in an assay. Although this approach is more likely to detect the low-frequency environmental contamination events (e.g., contamination of a reaction tube with an aerosol containing amplicons), it increases the costs of performing the assay.

Positive controls should also be incorporated into each assay. An extraction positive control can demonstrate the efficiency of a concentration/extraction procedure. The other positive control (purified nucleic acids added to the reaction mix) will demonstrate that all of the necessary reagents are added to the reaction mix and that all enzymes are functional. Care must be taken not to allow the positive control to cross-contaminate other samples. It is prudent to use the smallest amount of positive control necessary to consistently obtain a positive result.

5.3. Inhibitor Detection

Even after extensive sample processing, inhibitors of reverse transcription or of PCR may remain. A variety of biologic and chemical substances that are present in foods or are used during sample processing have been found to act as inhibitors, including polysaccharides, heme, phenol, and cations. These inhibitors can affect the results obtained with both qualitative and quantitative assays (Abu Al-Soud and Radstrom, 1998). Thus, it is important to minimize the presence of inhibitors that may lead to a false negative.

Several approaches are used to detect inhibitors in a sample. A housekeeping gene that is copurified with the target nucleic acid is often used as a second target to demonstrate that amplification can occur. However, this approach has not been used for foods because of the problem of identifying an appropriate housekeeping gene. More commonly, a standard control is added to the sample. The control may be nucleic acids that are amplified and detected by the virus-specific primers and are amplified in the same reaction vessel or they may be added to a separate reaction vessel and be amplified by a different set of primers. A disadvantage of adding the internal standard directly to the reaction mix is that amplification of the standard may competitively inhibit amplification of target viral nucleic acids. On the other hand, the use of different primers to amplify a standard added to a separate reaction vessel may fail to detect inhibition if the virus amplification assay is more easily inhibited than the assay for the standard. Assay-specific internal standards have been described for noroviruses, hepatitis A virus, and rotavirus (Atmar et al., 1995; Le Guyader, 2000; Parshionikar et al., 2004).

6.0. RESULT INTERPRETATION

Several factors must be considered in evaluating the results of molecular assays described above. First, all control samples must be examined. If negative control samples (extraction controls, reagent controls) are positive, then any other positive samples in the assay are suspect due to the possibility that cross-contamination has occurred. If a positive control sample turns out to be negative, it may mean inefficient processing (positive extraction control), inhibited enzymatic activity (internal standard control), or failure to add (or inadequate activity of) one of the reagents (e.g., enzymes, oligonucleotide primers, buffers).

Even if all controls work properly, it should be recognized that additional factors may affect the interpretation of the results. The sensitivity of the assay may be insufficient to detect virus that is present in the sample at a level below the limit of detection. A positive result indicates that the viral nucleic acid is present but it does not address the viability of the virus; nucleic acids can be detected using molecular assays even after the virus is no longer viable (Nuanualsuwan and Cliver, 2002). This observation has raised concern about the utility of molecular methods in the assessment of food safety (Richards, 1999). The importance of contamination of food with nonviable viral nucleic acids needs to be determined.

7.0. APPLICATION TO FOODS

Much of the work and early success in the development and application of molecular methods to foodstuffs have involved shellfish. Methods used to partially purify and concentrate cultivable enteric viruses were developed in the 1970s and 1980s (Lewis and Metcalf, 1988). These methods were modified to allow detection of these and noncultivable viruses using hybridization and RT-PCR assays (Zhou et al., 1991; Atmar et al, 1993, 1995; Lees et al., 1994, 1995). As RT-PCR methods were developed, they were applied to shellfish samples implicated in outbreaks of HAV (Jansen et al., 1990) and norovirus (Le Guyader et al., 1996; Sugieda et al., 1996; Shieh et al., 1999; Prato et al., 2004). In some cases, sequence information obtained from infected persons was used to select primers to be used for assaying the shellfish (Le Guyader et al., 1996; Prato et al., 2004). These assays have also been used to identify virus contamination in shellfish in different geographic areas and during different seasons (Le Guyader et al., 2000; Formiga-Cruz et al., 2002) and to assess the level and duration of contamination in shellfish implicated in an outbreak (Le Guyader et al., 2003).

Methods to detect viruses in other types of foods have also been developed. These include melons, lettuce, berries, hamburger meat, and sliced deli meats (Gouvea et al., 1994; Bidawid et al., 2000; Leggitt and Jaykus, 2000; Schwab et al., 2000; Jean et al., 2004; Le Guyader et al., 2004b). A limited number of studies have successfully applied these methods to detect viruses in foods associated with outbreaks of human disease. Schwab et al. (2000) detected a norovirus strain on contaminated ham associated with a cafeteria-associated outbreak, Le Guyader et al. (2004a) identified a norovirus strain in contaminated raspberries implicated in an outbreak, and Calder et al. (2003) identified hepatitis A virus in blueberries associated with a hepatitis A outbreak. These studies demonstrate that the tools to evaluate foods implicated epidemiologically during outbreak investigations are increasingly available. Additional investigation is needed before these methods can be applied for the screening of foods for contamination with food-borne viruses.

8.0. SUMMARY

Molecular assays are now available for the detection of the enteric viruses most commonly associated with the transmission of food-borne illness. Some of these assays, such as RT-PCR and NASBA, are able to detect very small quantities of virus genome and are suitable tools for the identification of viruses in foods. The application of these assays to foods has been facilitated by the development of methods for the concentration and purification of viral nucleic acids, but inhibitory substances may persist in processed samples and can lead to false-negative results. Thus, it is important to use positive and negative controls in every analysis to allow appropriate interpretation of assay results. There are now several reports of the successful identification of the same virus identified as the cause of a food-borne outbreak in the food implicated by the epidemiologic investigation. Additional studies are needed to determine the role molecular assays will have in food-safety programs.

9.0. REFERENCES

Abd El Galil, K. H., El Sokkary, M. A., Kheira, S. M., Salazar, A. M., Yates, M. V., Chen, W., and Mulchandani, A., 2004, Combined immunomagnetic separation-molecular beacon-reverse transcription-PCR assay for detection of hepatitis A virus from environmental samples. *Appl. Environ. Microbiol.* 70:4371–4374.

Abu Al-Soud, W., and Radstrom, P., 1998, Capacity of nine thermostable DNA polymerases to mediate DNA amplification in the presence of PCR-inhibiting samples. *Appl. Environ. Microbiol.* 64:3748–3753.

Allard, A., Albinsson, B., and Wadell, G., 1992, Detection of adenoviruses in stools from healthy persons and patients with diarrhea by two-step polymerase chain reaction. *J. Med. Virol.* 37:149–157.

Allard, A., Kajon, A., and Wadell, G., 1994, Simple procedure for discrimination and typing of enteric adenoviruses after detection by polymerase chain reaction. *J. Med. Virol.* 44:250–257.

Atmar, R. L., and Estes, M. K., 2001, Diagnosis of noncultivatable gastroenteritis viruses, the human caliciviruses. *Clin. Microbiol. Rev.* 14:15–37.

Atmar, R. L., Metcalf, T. G., Neill, F. H., and Estes, M. K., 1993, Detection of enteric viruses in oysters by using the polymerase chain reaction. *Appl. Environ. Microbiol.* 59:631–635.

Atmar, R. L., Neill, F. H., Romalde, J. L., Le Guyader, F., Woodley, C. M., Metcalf, T. G., and Estes, M. K., 1995, Detection of Norwalk virus and hepatitis A virus I inshellfish tissues with the PCR. *Appl. Environ. Microbiol.* 61:3014–3018.

Atmar, R. L., Neill, F. H., Woodley, C. M., Manger, R., Fout, G. S., Burkhardt, W., Leja, L., McGovern, E. R., Le Guyader, F., Metcalf, T. G., and Estes, M. K., 1996, Collaborative evaluation of a method for the detection of Norwalk virus in shellfish tissues by PCR. *Appl. Environ. Microbiol.* 62:254–248.

Belliot, G., Laveran, H., and Monroe, S. S., 1997, Detection and genetic differentiation of human astroviruses: phylogenetic grouping varies by coding region. *Arch. Virol.* 142:1323–1334.

Bidawid, S., Farber, J., and Sattar, S., 2000, Rapid concentration and detection of HAV from lettuce and strawberries. *J. Virol. Methods* 88:175–185.

Boom, R., Sol, C. J., Salimans, M. M., Jansen, C. L., Wertheim-van Dillen, P. M., and van der Noordaa, J., 1990, Rapid and simple method for purification of nucleic acids. *J. Clin. Microbiol.* 28:495–503.

Buesa, J., Colomina, J., Raga, J., Villanueva, A., and Prat, J. 1996, Evaluation of reverse transcription and polymerase chain reaction (RT/PCR) for the detection of rotaviruses: applications of the assay. *Res. Virol.* 147:353–361.

Calder, L., Simmons, G., Thornley, C., Taylor, P., Pritchard, K., Greening, G., and Bishop, J., 2003, An outbreak of hepatitis A associated with consumption of raw blueberries. *Epidemiol. Infect.* 131:745–751.

Chizhikov, V., Wagner, M., Ivshina, A., Hoshino, Y., Kapikian, A. Z., and Chumakov, K., 2002, Detection and genotyping of human group A rotaviruses by oligonucleotide microarray hybridization. *J. Clin. Microbiol.* 40:2398–2407.

Cline, J., Braman, J. C., and Hogrefe, H. H., 1996, PCR fidelity of Pfu DNA polymerase and other thermostable DNA polymerases. *Nucleic Acids Res.* 24:3546–3551.

Cromeans, T. L., Nainan, O. V., and Margolis, H. S., 1997, Detection of hepatitis A virus RNA in oyster meat. *Appl. Environ. Microbiol.* 63:2460–2463.

Dimitrov, D. H., Graham, D. Y., and Estes, M. K., 1985, Detection of rotaviruses by nucleic acid hybridization with cloned DNA of simian rotavirus SA11 genes. *J. Infect. Dis.* 152:293–300.

Dubois, E., Agier, C., Traore, O., Hennechart, C., Merle, G., Cruciere, C., and Laveran, H., 2002, Modified concentration method for the detection of enteric viruses on fruits and vegetables by reverse transcriptase-polymerase chain reaction or cell culture. *J. Food Prot.* 65:1962–1969.

Espy, M. J., Smith, T. F., and Persing, D. H., 1993, Dependence of polymerase chain reaction product inactivation protocols on amplicon length and sequence composition. *J. Clin. Microbiol.* 31:2361–2365.

Formiga-Cruz, M., Tofino-Quesada, G., Bofill-Mas, S., Lees, D. N., Henshilwood, K., Allard, A. K., Conden-Hansson, A. C., Hernroth, B. E., Vantarakis, A., Tsibouxi, A., Papapetropoulou, M., Furones, M. D., and Girones, R., 2002, Distribution of human virus contamination in shellfish from different growing areas in Greece, Spain, Sweden, and the United Kingdom. *Appl. Environ. Microbiol.* 68:5990–5998.

Gentsch, J. R., Glass, R. I., Woods, P., Gouvea, V., Gorziglia, M., Flores, J., Das, B. K., and Bhan, M. K., 1992, Identification of group A rotavirus Gene 4 types by polymerase chain reaction. *J. Clin. Microbiol.* 30:1365–1373.

Gilpatrick, S. G., Schwab, K. J., Estes, M. K., and Atmar, R. L., 2000, Development of an immunomagnetic capture reverse transcription-PCR assay for the detection of Norwalk virus. *J. Virol. Methods* 90:69–78.

Gouvea, V., Glass, R. I., Woods, P., Taniguchi, K., Clark, H. F., Forrester, B., and Fang, Z.-F., 1990, Polymerase chain reaction amplification and typing of rotavirus nucleic acid from stool specimens. *J. Clin. Microbiol.* 28:276–282.

Gouvea, V., Allen, J. R., Glass, R. I., Fang, Z.-Y., Bremont, M., Cohen, J., McCrae, M. A., Saif, L. J., Sinarachatanant, P., and Caul, E. O., 1991, Detection of group B and C rotaviruses by polymerase chain reaction. *J. Clin. Microbiol.* 29:519–523.

Gouvea, V., Santos, N., Carmo Timenetsky, M., Estes, M. K., 1994, Identification of Norwalk virus in artificially seeded shellfish and selected foods. *J. Virol. Methods* 48:177–187.

Greene, S. R., Moe, C. L., Jaykus, L.-A., Cronin, M., Grosso, L., and van Aarle, P., 2003, Evaluation of the NucliSens® basic kit assay for detection of Norwalk virus RNA in stool specimens. *J. Virol. Methods* 108:123–131.

Grimm, A. C., and Fout, G. S., 2002, Development of a molecular method to identify hepatitis E virus in water. *J. Virol. Methods* 101:175–188.

Grimm, A. C., Cashdollar, J. L., Williams, F. P., and Fout, G. S., 2004, Development of an astrovirus RT-PCR detection assay for use with conventional, real-time, and integrated cell culture/RT-PCR. *Can. J. Microbiol.* 50:269–278.

Hale, A. D., Tanaka, T. N., Kitamoto, N., Ciarlet, M., Jiang, X., Takeda, N., Brown, D. W., and Estes, M. K., 2000, Identification of an epitope common to genogroup 1 Norwalk-like viruses. *J. Clin. Microbiol.* 38:1656–1660.

Haqqi, T. M., Sarkar, G., David C.S., and Sommer, S. S., 1988, Specific amplification with PCR of a refractory segment of genomic DNA. *Nucleic Acids Res.* 16:11844.

Harrington, P. R., Vinjé, J., Moe, C. L., and Baric, R. S., 2004, Norovirus capture with histo-blood group antigens reveals novel virus-ligand interactions. *J. Virol.* 78: 3035–3045.

Hutin, Y. J. F., Pool, V., Cramer, E. H., Nainan, O. V., Weth, J., Williams, I. T., Goldstein, S. T., Gensheimer, K. F., Bell, B. P., Shapiro, C. N., Alter, M. J., and Margolis, J. S., for the National Hepatitis A Investigation Team, 1999, A multistate, foodborne outbreak of hepatitis A. *N. Engl. J. Med.* 340:595–602.

Iturriza-Gomara, M., Green, J., Brown, D. W. G., Desselberger, U., and Gray, J. J., 1999, Comparison of specific and random priming in the reverse transcriptase polymerase chain reaction for genotyping group A rotaviruses. *J. Virol. Methods* 78:93–103.

Jansen, R. W., Newbold, J. E., and Lemon, S. M., 1985, Combined immunoaffinity cDNA-RNA hybridization assay for detection of hepatitis A virus in clinical specimens. *J. Clin. Microbiol.* 22:984–989.

Jansen, R. W., Siegl, G., and Lemon, S. M., 1990, Molecular epidemiology of human hepatitis A virus defined by an antigen-capture polymerase chain reaction. *Proc. Natl. Acad. Sci. U.S.A* 87:2867–2871.

Jaykus, L. A., 2003, Challenges to developing real-time methods to detect pathogens in foods. *ASM News* 69:341–347.

Jean, J., Blais, B., Darveau, A., and Fliss, I., 2001, Detection of hepatitis A virus by the nucleic acid sequence-based amplification technique and comparison with reverse transcription-PCR. *Appl. Environ. Microbiol.* 67:5593–5600.

Jean, J., Blais, B., Darveau, A., and Fliss, I., 2002, Simultaneous detection and identification of hepatitis A virus and rotavirus by multiplex nucleic acid sequence-based amplification (NASBA) and microtiter plate hybridization system. *J. Virol. Methods* 105:123–132.

Jean, J., D'Souza, D., and Jaykus, L.-A., 2003, Transcriptional enhancement of RT-PCR for rapid and sensitive detection of noroviruses. *FEMS Microbiol. Lett.* 226:339–345.

Jean, J., D'Souza, D., and Jaykus, L.-A., 2004, Multiplex nucleic acid sequence-based amplification for simultaneous detection of several enteric viruses in model ready-to-eat foods. *Appl. Environ. Microbiol.* 70:6603–6610.

Jiang, X., Estes, M. K., and Metcalf, T. G., 1989, In situ hybridization for quantitative assay of infectious hepatitis A virus. *J. Clin. Microbiol.* 27:874–879.

Jiang, X., Wang, J., Graham, D. Y., and Estes, M. K., 1992, Detection of Norwalk virus in stool by polymerase chain reaction. *J. Clin. Microbiol.* 30:2529–2534.

Jiang, Y.-J., Liao, G.-Y., Zhao, W., Sun, M.-B., Qian, Y., Bian, C.-X., and Jiang, S.-D., 2004, Detection of infectious hepatitis A virus by integrated cell culture/strand-specific reverse transcriptase-polymerase chain reaction. *J. Appl. Microbiol.* 97: 1105–1112.

Jonassen, T. O., Monceyron, C., Lee, T. W., Kurtz, J. B., and Grinde, B., 1995, Detection of all serotypes of human astrovirus by the polymerase chain reaction. *J. Virol. Methods* 52:327–334.

Kageyama, T. S., Kojima, S., Shinohara, M., Uchida, K., Fukushi, S., Hoshino, F. B., Takeda, N., and Katayama, K., 2003, Broadly reactive and highly sensitive assay for Norwalk-like viruses based on real-time quantitative reverse transcription-PCR. *J. Clin. Microbiol.* 41:1548–1557.

Kobayashi, S., Natori, K., Takeda, N., and Sakae, K., 2004, Immunomagnetic capture rt-PCR for detection of norovirus from foods implicated in a foodborne outbreak. *Microbiol. Immunol.* 48:201–204.

Koopmans, M., Vennema, H., Heersma, H., van Strien, E., van Duynhoven, Y., Brown, D., Reacher, M., and Lopman, B., for the European Consortium on Foodborne Viruses, 2003, Early identification of common-source foodborne virus outbreaks in Europe. *Emerg. Infect. Dis.* 9:136–1142.

Kwok, S., and Higuchi, R., 1989, Avoiding false positives with PCR. *Nature* 339:237–238.

Le Cann, P., Ranarijaona, S., Monpoeho, S., Le Guyader, F., and Ferré, V., 2004, Quantification of human astroviruses in sewage using real-time RT-PCR. *Res. Microbiol.* 155:11–15.

Le Guyader, F. S., Neill, F. H., Estes M.K., Monroe, S. S., Ando, T., and Atmar, R. L., 1996, Detection and analysis of a small round-structured virus strain in oysters implicated in an outbreak of acute gastroenteritis. *Appl. Environ. Microbiol.* 62:4268–4272.

Le Guyader, F. S., Haugarreau, L., Miossec, L., Dubois, E., and Pommepuy, M., 2000, Three-year study to assess human enteric viruses in shellfish. *Appl. Environ. Microbiol.* 66:3241–3248.

Le Guyader, F. S., Neill, F. H., Dubois, E., Bon, F., Loisy, F., Kohli, E., Pommepuy, M., and Atmar, R. L., 2003, A semiquantitative approach to estimate Norwalk-like virus contamination of oysters implicated in an outbreak. *Int. J. Food Microbiol.* 87:107–12.

Le Guyader, F. S., Mittelholzer, C., Haugarreau, L., Hedlund, K.-O., Alsterlund, R., Pommepuy, M., and Svensson, L., 2004a, Detection of noroviruses in raspberries associated with a gastroenteritis outbreak. *Int. J. Food Microbiol.* 97:179–186.

Le Guyader, F. S., Schultz, A.-C, Haugarreau, L., Croci, L., Maunula, L., Duizer, E., Lodder-Verschoor, F., von Bonsdorff, C.-H., Suffredini, E., van der Poel, W. M. M., Reymundo, R., and Koopmans, M., 2004b, Round-robin comparison of methods for the detection of human enteric viruses in lettuce. *J. Food Protect.* 67:2315–2319.

Lees, D. N., Henshilwood, K., and Doré, W. J., 1994, Development of a method for detection of enteroviruses in shellfish by PCR with poliovirus as a model. *Appl. Environ. Microbiol.* 60:2999–3005.

Lees, D. N., Henshilwood, K., Green, J., Gallimore, C. I., and Brown, D. W. G., 1995, Detection of small round structured viruses in shellfish by reverse transcription-PCR. *Appl. Environ. Microbiol.* 61:4418–4424.

Leggitt, P. R., and Jaykus, L. A., 2000, Detection methods for human enteric viruses in representative foods. *J. Food Prot.* 63:1738–1744.

Lewis, G. D., and Metcalf, T. G., 1988, Polyethylene glycol precipitation for recovery of pathogenic viruses, including hepatitis A virus and human rotavirus, from oyster, water, and sediment samples. *Appl. Environ. Microbiol.* 54:1983–1988.

Lovmar, L., Fock, C., Espinoza, F., Bucardo, F., Syvänen, A.-C., and Bondeson, K., 2003, Microarrays for genotyping human group A rotavirus by multiplex capture and type-specific primer extension. *J. Clin. Microbiol.* 41:5153–5158.

Mackay, I. M., Arden, K. E., and Nitsche, A., 2002, Real-time PCR in virology. *Nucleic Acids Res.* 30:1292–1305.

Moore, C., Clark, E. M., Gallimore, C. I., Corden, S. A., Gray, J. J., and Westmoreland, D., 2004, Evaluation of a broadly reactive nucleic acid sequence based amplification assay for the detection of noroviruses in faecal material. *J. Clin. Virol.* 29:290–296.

Muniain-Mujika, I., Girones, R., Lucena, F., 2000, Viral contamination of shellfish: evaluation of methods and analysis of bacteriophages and human viruses. *J. Virol. Methods* 89:109–118.

Mustafa, H., Palombo, E. A., and Bishop, R. F., 1998, Improved sensitivity of astrovirus-specific RT-PCR following culture of stool samples in CaCO-2 cells. *J. Clin. Virol.* 11:103–107.

Myers, T. W., and Gelfand, D. H., 1991, Reverse transcription and DNA amplification by a *Thermus thermophilus* DNA polymerase. *Biochemistry* 30:7661–7666.

Noel, J. S., Lee, T. W., Kurtz, J. B., Glass, R. I., and Monroe, S. S., 1995, Typing of human astroviruses from clinical isolates by enzyme immunoassay and nucleotide sequencing. *J. Clin. Microbiol.* 33:797–801.

Nolte, F. S., and Caliendo, A. M., 2003, Molecular detection and identification of microorganisms, in: *Manual of Clinical Microbiology*, 8th ed. (P. R. Murray, E. J. Baron, J. H. Jorgensen, M. A. Pfaller, and R. H. Yolken, eds.), ASM Press, Washington, DC, pp. 234–256.

Notomi, T., Okayama, H., Masubuchi, H., Yonekawa, T., Watanabe, K., Amino, N., and Hase, T., 2000, Loop-mediated isothermal amplification of DNA. *Nucleic Acids Res.* 28:e63.

Nuanualsuwan, S., and Cliver, D. O., 2002, Pretreatment to avoid positive RT-PCR results with inactivated viruses. *J. Virol. Methods* 104:217–225.

Orrù, G., Masia, G., Orrù, G., Romanò, L., Piras, V., and Coppola, R. C., 2004, Detection and quantitation of hepatitis E virus in human faeces by real-time quantitative PCR. *J. Virol. Methods* 118:77–82.

Pang, J., Modlin, J., and Yolken, R., 1992, Use of modified nucleotides and uracil-DNA glycosylase (UNG) for the control of contamination in the PCR-based amplification of RNA. *Mol. Cell. Probes* 6:251–256.

Pang, X., Lee, B., Chui, L., Preiksaitis, J. K., and Monroe, S. S., 2004a, Evaluation and validation of real-time reverse transcription-PCR assay using the LightCycler system for detection and quantitation of norovirus. *J. Clin. Microbiol.* 42:4679–4685.

Pang, X. L., Lee, B., Boroumand, N., Leblanc, B., Preiksaitis, J. K., and Ip, C. C. Y., 2004b, Increased detection of rotavirus using a real time reverse transcription-polymerase chain reaction (RT-PCR) assay in stool specimens from children with diarrhea. *J. Med. Virol.* 72:496–501.

Parida, M., Posadas, G., Inoue, S., Hasebe, F., and Morita, K., 2004, Real-time reverse transcription loop-mediated isothermal amplification for rapid detection of West Nile virus. *J. Clin. Microbiol.* 42:257–263.

Parshionikar, S. U., Cashdollar, J., and Fout, G. S., 2004, Development of homologous viral internal controls for use in RT-PCR assays of waterborne enteric viruses. *J. Virol. Methods* 121:39–48.

Pina, S., Puig, M., Lucena, F., Jofre, J., and Girones, R., 1998, Viral pollution in the environment and in shellfish: human adenovirus detection by PCR as an index of human viruses. *Appl. Environ. Microbiol.* 64:3376–3382.

Pina, S., Buti, M., Jardí, R., Clemente-Casares, P., Jofre, J., and Girones, R., 2001, Genetic analysis of hepatitis A virus strains recovered from the environment and from patients with acute hepatitis. *J. Gen. Virol.* 82:2955–2963.

Prato, R., Lopalco, P. L., Chironna, M., Barbuti, G., Germinario, C., and Quarto, M., 2004, Norovirus gastroenteritis general outbreak associated with raw shellfish consumption in south Italy. *BMC Infect. Dis.* 4:37.

Ratcliff, R. M., Doherty, J. C., and Higgins G. D., 2002, Sensitive detection of RNA viruses associated with gastroenteritis by a hanging-drop single-tube nested reverse transcription-PCR method. *J. Clin. Microbiol.* 40:4091–4099.

Richards, G. P., 1999, Limitations of molecular biological techniques for assessing the virological safety of foods. *J. Food Protect.* 62:691–697.

Robertson, B. H., Khanna, B., Nainan, O. V., and Margolis, H. S., 1991, Epidemiologic patterns of wild-type hepatitis A virus determined by genetic variation. *J. Infect. Dis.* 163:286–292.

Rousell, J., Zajdel, M. E., Howdle, P. D., and Blair, G. E., 1993, Rapid detection of enteric adenoviruses by means of the polymerase chain reaction. *J. Infect.* 27: 271–275.

Saiki, R. K., Scharf, S. J., Faloona, F., Mullis, K. B., Horn, G. T., Erlich, H. A., and Arnheim, N., 1985, Enzymatic amplification of beta-globin genomic sequences and restriction site analysis for diagnosis of sickle cell anemia. *Science* 230:1350–1354.

Saiki, R. K., Gelfand D. H., Stoffel, S., Scharf, S. J., Higuchi, R., Mullis, K. B., Horn G. T., and Ehrlich H. A., 1988, Primer-directed enzymatic amplification of DNA with a thermostable DNA polymerase. *Science* 239:487–491.

Sair, A. I., D'Souza, D. H., Moe, C. L., and Jaykus, L. A., 2002, Improved detection of human enteric viruses in foods by RT-PCR. *J. Virol. Methods* 100:57–69.

Schlauder, G. G., and Mushahwar, I. K., 2001, Genetic heterogeneity of hepatitis E virus. *J. Med. Virol.* 65:282–292, 2001.

Schlauder, G. G., Frider, B., Sookoian, S., Castano, G. C., and Mushahwar, I. K., 2000, Identification of two novel isolates of hepatitis E virus in Argentina. *J. Infect. Dis.* 182:294–297.

Schreier, E., Doring, F., and Kunkel, U., 2000, Molecular epidemiology of outbreaks of gastroenteritis associated with small round structured viruses in Germany in 1997/98. *Arch. Virol.* 145:443–453.

Schwab, K. J., Neill, F. H., Fankhauser, R. L., Daniels, N. A., Monroe, S. S., Bergmire-Sweat, D. A., Estes, M. K., and Atmar, R. L., 2000, Development of methods to detect "Norwalk-like viruses" (NLVs) and hepatitis A virus in delicatessen foods: application to a food-borne NLV outbreak. *Appl. Environ. Microbiol.* 66:213–218.

Schwab, K. J., Neill, F. H., LeGuyader, F., Estes, M. K., and Atmar, R. L., 2001, Development of an RT-PCR-DNA enzyme immunoassay for the detection of Norwalk viruses in shellfish. *Appl. Environ. Microbiol.* 67:742–749.

Shieh, Y.-S. C., Calci, K. R., and Baric, R. S., 1999, A method to detect low levels of enteric viruses in contaminated oysters. *Appl. Environ. Microbiol.* 65:4709–4714.

Soares, C. C., Volotao, E. M., Albuquerque, M. C. M., Nozawa, C. M., Linhares, R. E. C., Volokhov, D., Chizhikov, V., Lu, X., Erdman, D., and Santos, N., 2004, Genotyping or enteric adenoviruses by using single-stranded conformation polymorphism analysis and heteroduplex mobility assay. *J. Clin. Microbiol.* 42:1723–1726.

Sugieda, M., Nakajima, K., and Nakajima, S., 1996, Outbreaks of Norwalk-like virus-associated gastroenteritis traced to shellfish: coexistence of two genotypes in one specimen. *Epidemiol. Infect.* 116:339–346.

Svensson, L., 2000, Diagnosis of foodborne viral infections in patients. *Int. J. Food Microbiol.* 59:117–126.

Tai, J. H., Ewert, M. S., Belliot, G., Glass, R. I., and Monroe, S. S., 2003, Development of a rapid method using nucleic acid sequence-based amplification for the detection of astrovirus. *J. Virol. Methods* 110:119–127.

Takiff, H. E., Seidlin, M., Krause, P. Rooney, J., Brandt, C., Rodriguez, W., Yolken, R., and Straus, S. E., 1985, Detection of enteric adenoviruses by dot-blot hybridization using a molecularly cloned viral DNA probe. *J. Med. Virol.* 16:107–118.

Traore, O., Arnal, C., Mignotte, B., Maul, A., Laveran, H., Billaudel, S., and Schwartzbrod, L., 1998, Reverse transcriptase PCR detection of astrovirus, hepatitis A virus, and poliovirus in experimentally contaminated mussels: comparison of several extraction and concentration methods. *Appl. Environ. Microbiol.* 64:3118–3122.

Urdea, M. S., Horn, T., Fultz, T. J., Anderson, M., Running, J. A., Hamren, S., Ahle, D., and Chang, C. A., 1991, Branched DNA amplification multimers for the sensitive, direct detection of human hepatitis viruses. *Nucleic Acids Symp. Ser.* 24:197–200.

Vinjé, J., and Koopmans, M. P. G., 2000, Simultaneous detection and genotyping of "Norwalk-like viruses" by oligonucleotide array in a reverse line blot hybridization format. *J. Clin. Microbiol.* 38:2595–2601.

Vinjé, J., Vennema, H., Maunula, L., von Bonsdorff, C. H., Hoehne, M., Schreier, E., Richards, A., Green, J., Brown, D., Beard, S. S., Monroe, S. S., de Bruin, E., Svensson, L., and Koopmans, M. P., 2003, International collaborative study to compare reverse transcriptase PCR assays for detection and genotyping of noroviruses. *J. Clin. Microbiol.* 41:1423–1433.

Walker, G. T., Little, M. C., Nadeau, J. G., and Shank, D. D., 1992, Isothermal in vitro amplification of DNA by a restriction enzyme/DNA polymerase system. *Proc. Natl. Acad. Sci. U.S.A.* 89:392–396.

Willcocks, M. M., Carter, M. J., Silcock, J. G., and Madeley, C. R., 1991, A dot-blot hybridization procedure for the detection of astrovirus in stool samples. *Epidemiol. Infect.* 107:405–410.

Williams, F. P. Jr., and Fout, G. S., 1992, Contamination of shellfish by stool-shed viruses: methods of detection. *Environ. Sci. Technol.* 26:689–696.

Wu, D. Y., and Wallace, R. B., 1989, The ligation amplification reaction (LAR)—amplification of specific DNA sequences using sequential rounds of template-dependent ligation. *Genomics* 4:560–569.

Xu, W., McDonough, M. C., and Erdman, D. D., 2000, Species-specific identification of human adenoviruses by a multiplex PCR assay. *J. Clin. Microbiol.* 38:4114–4120.

Zhou, Y.-J., Estes, M. K., Jiang, X., and Metcalf, T. G., 1991, Concentration and detection of hepatitis A virus and rotavirus from shellfish by hybridization tests. *Appl. Environ. Microbiol.* 57:2963–2968.

CHAPTER 6

Survival and Transport of Enteric Viruses in the Environment

Albert Bosch, Rosa M. Pintó, and F. Xavier Abad

1.0. VIRUSES IN THE ENVIRONMENT

1.1. Viruses and Environmental Virology

Environmental virology may be defined as the study of viruses that can be transmitted through various environments (water, sewage, soil, air, or surfaces) or food and persist enough in these vehicles to represent a health threat. A wide variety of different viruses, representing most of the families of animal viruses, can be present in human and animal fecal wastes and urine. Especially important are a variety of nonenveloped human and animal enteric pathogenic viruses that can enter the environment through the discharge of waste materials from infected individuals; contaminate food products and drinking and recreational waters; and be transmitted back to susceptible individuals to continue the cycle of infection (Table 6.1). It is estimated that billions of cases of gastrointestinal illness occur annually worldwide (Parashar et al., 1998; Oh et al., 2003). A good deal of these diarrheal cases are to some extent the result of fecal contamination of the environment (Cabelli et al., 1982; Koopman et al., 1982; Fattal and Shuval, 1989; Moore et al., 1994) while outbreaks of hepatitis A and E are associated with water, shellfish, and crops (Melnick, 1957; Reid and Robinson, 1987; Halliday et al., 1991; Bosch et al., 1991, 2001).

The significance to human health of many of the non-human animal viruses present in environmental samples is less well understood and remains uncertain or unknown for many of them. It is remarkable, however, that zoonotic viruses infecting humans continue to be discovered or appear to reemerge as important human pathogens. One example of an emerging disease is severe acute respiratory syndrome, or SARS, reported in November 2002 (Ksiazek et al., 2003). The primary mode of transmission of the SARS coronavirus appears to be direct mucous membrane contact with infectious respiratory droplets and/or through exposure to fomites. Several coronaviruses are known to spread by the fecal-oral route, but there is no current evidence that this mode of transmission plays a key role in the transmission of SARS, although there is a considerable shedding of the virus in stools (Tsang, 2003).

As a scientific discipline, environmental virology was born after a large hepatitis outbreak occurred in New Delhi between December 1955 and January 1956. The origin of the outbreak, which was attributed to hepatitis A at the time but now confirmed to be hepatitis E, was the contamination

Table 6.1 Human Enteric Viruses with Potential Environmental Transmission

Genus	*Popular Name*	*Disease Caused*
Enterovirus	Polio	Paralysis, meningitis, fever
	Coxsackie A, B	Herpangina, meningitis, fever, respiratory disease, hand-foot-and-mouth disease, myocarditis, heart anomalies, rash, pleurodynia, diabetes?
	Echo	Meningitis, fever, respiratory disease, rash, gastroenteritis
Hepatovirus	Hepatitis A	Hepatitis
Reovirus	Human reovirus	Unknown
Rotavirus	Human rotavirus	Gastroenteritis
Mastadenovirus	Human adenovirus	Gastroenteritis, respiratory disease, conjunctivitis
Norovirus	Norwalk-like virus	Gastroenteritis
Sapovirus	Sapporo-like virus	Gastroenteritis
Hepervirus	Hepatitis E	Hepatitis
Mamastrovirus	Human astrovirus	Gastroenteritis
Parvovirus	Human parvovirus	Gastroenteritis
Coronavirus	Human coronavirus	Gastroenteritis, respiratory disease
Torovirus	Human torovirus	Gastroenteritis

by sewage, from 1 to 6 weeks prior to the epidemic, of the Jumna River, the source of water for the treatment plant. Alum and chlorine treatment prevented bacterial infections, but 30,000 cases of hepatitis occurred among the population. As a consequence of this outbreak, studies in water and environmental virology began with efforts to detect poliovirus in water around 50 years ago. Since that time, other enteric viruses responsible for gastroenteritis and hepatitis have replaced enteroviruses as the main target for detection in the environment, although the near eradication of poliomyelitis from the globe calls for exhaustive studies on the occurrence of wild-type and vaccinal-type polioviruses in environmental samples.

1.2. Waterborne Transmission of Enteric Viruses

Figure 6.1 illustrates the possible routes of waterborne transmission of enteric viruses. Viruses can be transmitted by a variety of routes, including direct and indirect contact, vector transmission, and vehicle transmission. Viruses are shed in extremely high numbers in the feces of infected individuals; patients suffering from diarrhea or hepatitis may excrete from 10^5 to 10^{11} virus particles per gram of stool (Farthing, 1989). Furthermore, a single episode of vomit of a patient with norovirus gastroenteritis may contain around 10^7 particles (Cheesbrough et al., 1997). Ingestion of sewage-contaminated water or food is the main route of infection with human enteric viruses, although the role of inanimate surfaces serving as vehicles for virus infection must not be underestimated. Viruses with a viremic phase, such as the hepatitis viruses, may also be parenterally transmitted, although these days it is considered to be a much less frequent mode of transmission.

A poorly understood aspect in the epidemiology of several enteric viruses is the role of animal viruses in human disease. Nucleotide sequence analysis of some human enteric viruses has indicated a high degree of sequence similarity with animal strains. Notably, hepatitis E virus–related sequences have been detected in pigs (Meng et al., 1997; van der Poel et al., 2001; Banks et al., 2004) and birds (Huang et al., 2002). The threat of zoonotic infections may be either through direct transmission, suspected for hepatitis E virus (HEV; Reyes, 1993) and caliciviruses (Humphrey et al., 1984), or through incidental coinfection of a host with animal and human viruses, resulting in the mixing of genes and generation of novel variants (recombination/reassortment; Unicomb et al., 1999). Recombination has been demonstrated as a mechanism for rapid expansion of diversity for noroviruses and rotaviruses, but it is likely to be a common feature of the RNA viruses involved (Jiang et al., 1999; Unicomb et al., 1999). Viruses related to the human rotaviruses, astroviruses, noroviruses, sapoviruses, and HEV circulate in several animal species, providing a huge reservoir for virus diversity (Shirai et al., 1985; Meng et al., 1997; van der Poel et al., 2001; Huang et al., 2002).

In the water environment, the fate of microbial enteric pathogens may take several potential routes (Fig. 6.2). Mankind is exposed to waterborne

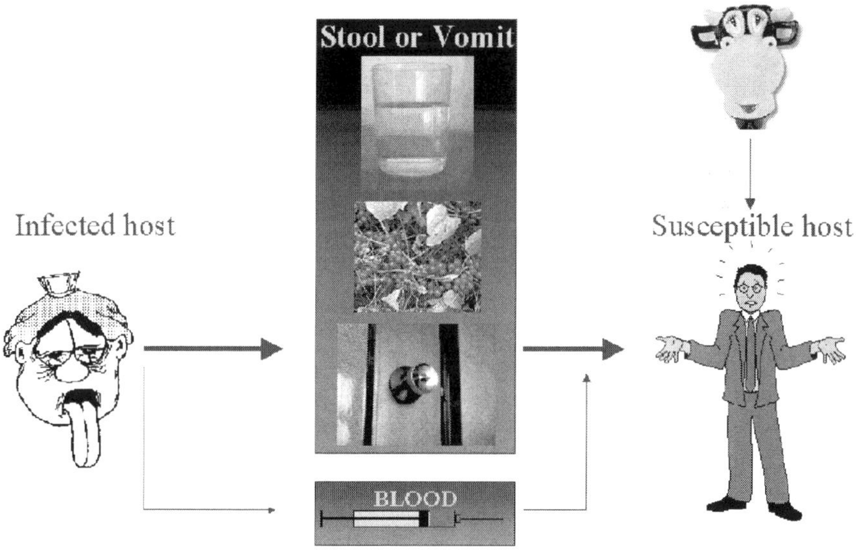

Figure 6.1 Routes of enteric virus transmission. Thick and thin arrows depict the main and minor routes of virus transmission, respectively.

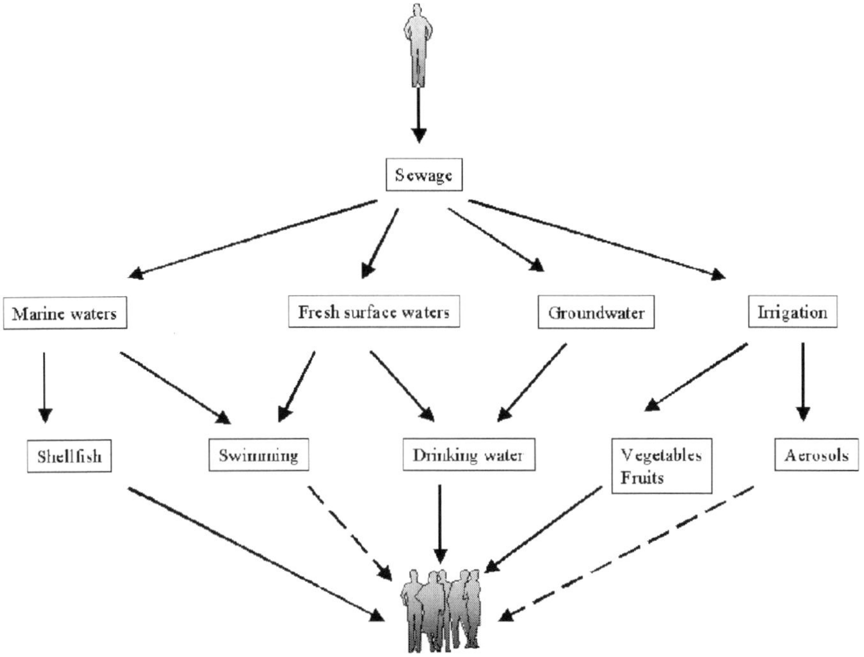

Figure 6.2 Waterborne transmission of enteric virus infections. Dashed lines depict unconfirmed transmission.

enteric virus infections through shellfish grown in contaminated waters, contaminated drinking water, food crops grown in land irrigated with wastewater and/or fertilized with sewage, and, to a lesser extent, sewage-polluted recreational waters (Tables 6.2 and 6.3).

Studies have documented the presence of enteric viruses in raw and treated drinking water (Keswick et al., 1984), and they are also frequently isolated from freshwater environments (Table 6.4). However, epidemiological proof of human infection caused by these viruses as a result of water consumption is scarce. Water-system deficiencies that caused or contributed to these outbreaks may be categorized under five major headings: (a) use of contaminated, untreated surface water; (b) use of contaminated, untreated groundwater; (c) inadequate or interrupted treatment; (d) distribution network problems; and (e) miscellaneous.

Pathogenic viruses are routinely introduced into the environment through the discharge of treated and untreated wastes, as current treatment practices are unable to provide virus-free wastewater effluents. Virus concentrations of 5,000 to 100,000 pfu/L are commonly reported in raw sewage (Rao and Melnick, 1986) and may be greatly reduced during treatment; however an average of 50 to 100 pfu/L are normally found in effluents from wastewater treatment plants (Rao and Melnick, 1986).

Table 6.2 Examples of Waterborne Viral Disease Outbreaks

Type of Water Implicated	*Virus*	*Disease*	*Reference*
Drinking water	Polio	Poliomyelitis	Mosley, 1967; Lippy and Waltrip, 1984
	Echo	Meningitis	Cliver, 1984; Amvrosieva et al., 2001
	Rotavirus	Gastroenteritis	Murphy et al., 1983; Hopkins et al., 1984; Hung et al., 1984; Craun et al., 2002; Villena et al., 2003
	Norovirus	Gastroenteritis	Kaplan et al., 1982; Blacklow and Cukor, 1982; Kukkula et al., 1999; Craun et al., 2002
	Adenovirus	Gastroenteritis	Murphy et al., 1983
	Hepatitis A	Hepatitis	Craun, 1988; Bosch et al., 1991
	Hepatitis E	Hepatitis	Khuroo, 1980; Ramalingaswami and Purcell, 1988
	Parvovirus	Gastroenteritis	Lippy and Waltrip, 1984
Recreational seawater	Rotavirus	Gastroenteritis	Fattal and Shuval, 1989
	Adenovirus		Foy et al., 1968; D'Angelo et al., 1979
	Hepatitis A	Hepatitis	Birch and Gust, 1989
Recreational freshwater	Coxsackie		Cabelli, 1983
	Enterovirus	Gastroenteritis	Lenaway et al., 1989
	Rotavirus	Gastroenteritis	Andersson and Stenström, 1987
	Norovirus	Gastroenteritis	Koopman et al., 1982
	Hepatitis A	Hepatitis	Bryan et al., 1974

Table 6.3 Examples of Large Outbreaks (Over 100 Cases) Linked to Shellfish Consumption

Year	*Country*	*Shellfish*	*No. of Cases*	*Responsible Virus*	*Reference*
1976–1977	Great Britain	Clams	800	SRSV	Appleton and Pereira, 1977
1978	Australia	Oysters	2,000	NoV	Murphy et al., 1979
1978	Australia	Oysters	150	NoV	Linco and Grohmann, 1980
1980–1981	Great Britain	Cockles	424	NoV	O'Mahony et al., 1983
1982	USA	Oysters	472	NoV	Richards, 1985
1983	Great Britain	Oysters	181	SRSV	Gill et al., 1983
1983	Malaysia	Cockles	322	HAV	Goh et al., 1984
1986	USA	Clams Oysters	813 204	NoV	Morse et al., 1986
1988	China	Clams	292, 301	HAV	Halliday et al., 1991
1999	Spain	Clams	183	HAV	Bosch et al., 2001

SRSV, small round structured viruses; NV, Norovirus; HAV, hepatitis A virus.

Table 6.4 Examples of Human Enteric Virus Isolations from Freshwater

River	*Virus Type*	*MPNCU/Liter*	*Reference*
Loire (France)	Enteroviruses and Adenoviruses	1.39	Le Bris et al., 1983
Ripoll (Spain)	Enteroviruses	15.5	Bosch et al., 1986
Besos (Spain)	Enteroviruses	16.2	Bosch et al., 1986
Tiber (Italy)	Hepatitis A virus	+[a]	Divizia et al., 1989a
Undetermined rivers (Germany)	Enteroviruses	0.5 to 56	Walter et al., 1989
Saint-Lawrence (Canada)	Culturable enteric viruses[b]	0.1 to 29	Payment et al., 2000

MPNCU, most probable number of cytopathic units of virus.
[a] Molecular detection of viral RNA.
[b] Unidentified virus grown in MA104 cells and reacting with human immunoglobulin.

Sewage sludge, a by-product of wastewater treatment, is a complex mixture of solids of biological and mineral origin that is removed from wastewater in sewage treatment plants. The sewage may undergo primary treatment (physical sedimentation or settling); secondary treatment (primary sedimentation plus high-rate biological processes, such as trickling filter/activated sludge); secondary treatment plus disinfection (chlorination, peracetic acid, UV or ozone); tertiary treatment (advanced wastewater treatment, including primary sedimentation, secondary treatment plus, for example, coagulation–sand filtration, UV, microfiltration); tertiary treatment plus disinfection; and lagooning (low-rate biological treatment). In any case, the type of treatment will determine the concentration of pathogens in a wastewater effluent and the relative risk of its disposal.

An overview of the fate of enteric viruses in coastal environments is depicted in Figure 6.3. Domestic sewage (in the form of raw sewage, treated effluent or sewage sludge) may be disposed of directly in the marine environment by coastal outfalls or by dumping from barges. In any case, viruses readily adsorb onto the abundant suspended solids present in the sewage and are discharged solid-associated into the marine environment (Fig. 6.3A). Whereas viruses associated with small-size (<3 μm) particulate material tend to float in the water column (Table 6.5), viruses adsorbed onto large/medium (>6 μm) particles readily settle down in the bottom sediment (Table 6.6). Viruses accumulate in the loose fluffy top layer of the compact bottom sediment (Fig. 6.3B) and are thereby protected from inactivation by natural or artificial processes (Rao et al., 1986; Sobsey et al., 1988). Sediments in coastal seawaters act as reservoirs from which viruses may be subsequently resuspended by several natural or artificial phenomena. Shellfish (Fig. 6.3C), being filter feeders, tend to concentrate viruses and bacteria in their edible tissues, and concentrations of these microorganisms in shellfish may be much higher than in the surrounding water. Shellfish grown in and harvested from waters receiving urban contaminants (Fig. 6.3D) have been implicated in outbreaks of viral diseases, notably viral hepatitis and gastroenteritis (Halliday et al., 1991; Le Guyader et al., 1996; Christensen et al., 1998; Bosch et al., 2001; Kingsley et al., 2002). Many of these outbreaks were related to water or shellfish meeting legal standards based on bacteriological criteria. This evidence supports the recommendation of monitoring shellfish and their overlying waters for viral contamination including the adoption of guidelines including virus standards.

The possibility nowadays to detect the presence of human enteric viruses in different types of water samples and foodstuff, in particular shellfish samples, should be a valuable tool in the prevention of waterborne and food-borne diseases. Unfortunately, in most outbreaks, virus detection is not attempted until after the outbreak and hence no prophylactic measures can

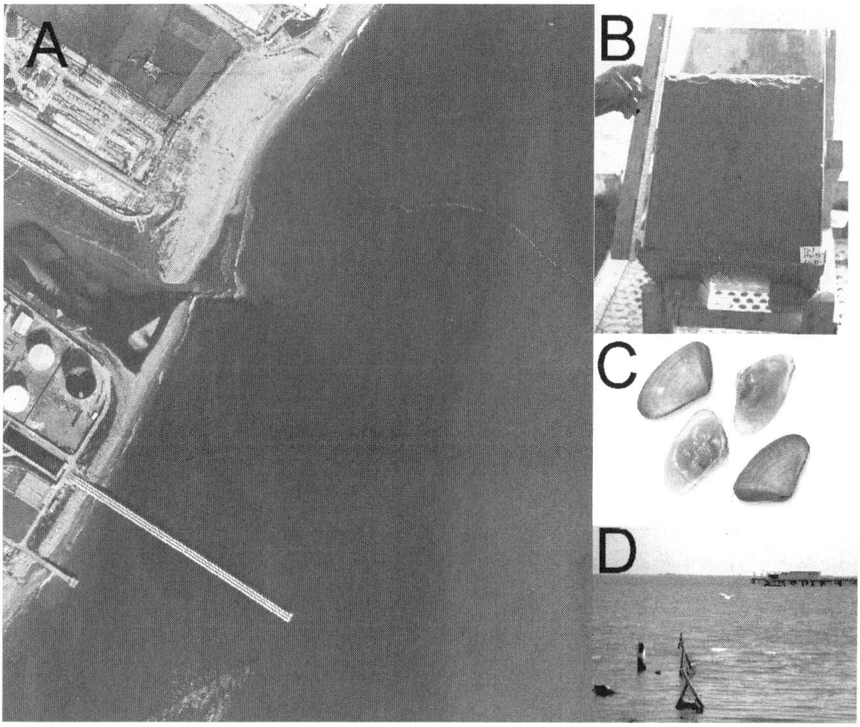

Figure 6.3 Fate of enteric viruses in coastal marine environments. (A) A heavily polluted river with abundant particulate material discharges into the sea. (B) Undisrupted marine sediment with the fluffy top layer where viruses accumulate. (C) Coquina clams and other bivalves readily adsorb pathogenic viruses within their edible tissues. (D) Shellfish grown in areas receiving urban sewage contamination is responsible for outbreaks of gastroenteritis and infectious hepatitis.

Table 6.5 Examples of Human Enteric Virus Isolations from Seawater

Site	*Virus Type*	*Virus Numbers/Liter*	*Reference*
Italy	Enteroviruses	0.4 to 16 $TCID_{50}$	De Flora et al., 1975
USA (Texas)	Enteroviruses	0.01 to 0.44 pfu	Goyal et al., 1979
USA (New York)	Poliovirus Echovirus	0 to 2.1 pfu	Vaughn et al., 1979
France	Enteroviruses Adenoviruses	0.05 to 6.5 MPNCU	Hugues et al., 1980
Spain	Enteroviruses	0.12 to 1.72 MPNCU	Finance et al., 1982
USA (Florida)	Enteroviruses	0.05 to 0.14 pfu	Schaiberger et al., 1982
Israel	Enteroviruses	1 to 6 pfu	Fattal et al., 1983
USA (Texas)	Enteroviruses	0.06 to 0.026 pfu	Rao et al., 1984
Spain	Poliovirus Echovirus	0.12 to 0.15 MPNCU	Lucena et al., 1985
USA (Texas)	Rotaviruses	0.007 to 2.6 pfu	Rao et al., 1986

$TCID_{50}$, tissue culture infectious $dose_{50}$; MPNCU, Most probable number of cytopathic units; pfu, plaque forming units.

Table 6.6 Examples of Human Enteric Virus Isolations from Marine Sediments

Site	*Virus Type*	*Virus Numbers/Kilogram*	*Reference*
Italy	Enteroviruses Reovirus	0.4 to 40 $TCID_{50}$	De Flora et al., 1975
USA (Florida)	Enteroviruses	0 to 112 pfu	Schaiberger et al., 1982
USA (Texas)	Enterovirus	39 to 398 pfu	Rao et al., 1984
USA (Texas)	Rotavirus	800 to 3800 pfu	Rao et al., 1986
Spain	Enterovirus	5 to 73 pfu	Bosch et al., 1988a
Spain	Enterovirus Rotavirus	130 to 200 pfu 57 to 140 FF	Jofre et al., 1989
Spain	Rotavirus Hepatitis A virus	0 to 560 FF +RNA	Bosch and Pintó, 1992
France	Enterovirus Rotavirus Hepatitis A virus	+RNA +RNA +RNA	Le Guyader et al., 1994

$TCID_{50}$, tissue culture infectious $dose_{50}$; pfu, plaque forming units; FF, fluorescent foci; RNA, detected by molecular hybridization.

be undertaken to decrease the severity of the outbreak. Methods for the detection of viruses in food are discussed elsewhere in this book.

The basic steps in virological analysis of water are sampling, concentration, decontamination/removal of inhibitors, and specific virus detection. Sample concentration is a particularly critical step because the viruses may be present in such low numbers that concentration of the water samples is indispensable to reduce the volume to be assayed to a few milliliters or even microliters. In relatively nonpolluted waters, the virus levels are likely to be so low that optimally hundreds, or even thousands, of liters should be sampled to increase the probability of virus detection.

A good concentration method should fulfill several criteria: it should be technically simple, fast, provide high virus recoveries, be adequate for a wide range of enteric viruses, provide a small volume of concentrate, and be inexpensive. Table 6.7 shows a broad selection of currently available and widely employed procedures; some of them require large equipment. Details on virus concentration procedures have been published elsewhere (American Public Health Association, 1998; Environmental Protection Agency, 1984). All available concentration methodologies have important limitations, and their virus concentration efficiency depends, in part, on the quality of the sampled water. Basically, all available procedures have been evaluated using samples spiked with known viruses. It is known that the recovery efficiency recorded with experimentally contaminated water dramatically decrease when the method is applied in actual field trials. Additionally, none of the existing concentration procedures has been tested with all of the medically important virus groups; normally, a few specific enteric viruses have been employed to conduct the evaluation trials. However, several virus concentration methods have been used successfully to recover naturally occurring enteric viruses in water (Finance et al., 1982; Gerba and Goyal, 1982; Goyal

Table 6.7 Procedures for the Concentration of Viruses from Water Samples

Principle	*References*
Adsorption-elution methods	
Negatively charged filters	Farrah et al., 1976
Positively charged filters	Sobsey and Jones, 1979; Gilgen et al., 1997
Glass powder	Gajardo et al., 1991; Sarrette et al., 1977; Schwartzbrod and Lucena, 1978
Glass fiber	Vilaginès et al., 1997
Precipitation methods	
Organic flocculation	Katzenelson et al., 1976
Ammonium sulfate precipitation	Bosch et al., 1988b; Shields and Farrah, 1986
Polyethylene glycol hydroextraction	Farrah et al., 1978; Lewis and Metcalf, 1988
Ultracentrifugation	Mehnert et al., 1997; Steinman, 1981
Lyophilization	Gajardo et al., 1995; Pintó et al., 2001
Ultrafiltration	Divizia et al., 1989b

and Gerba, 1983; Environmental Protection Agency, 1984; Rao et al., 1986; Lewis and Metcalf, 1988; Henshilwood et al., 1998).

Most of the procedures for concentrating and extracting viruses make use of the properties of the viral proteinaceous macromolecules. Certain protein structures confer on viruses in an aquatic environment the properties of a hydrophilic colloid of an amphoteric nature whose electric charge varies according to the pH and the ionic force of the environment. Viruses can therefore be adsorbed onto and then detach themselves from different substrates that are positively or negatively charged depending on their pH. Methods based on the adsorption of viruses from the sampled water onto a suitable solid surface from which they may subsequently be eluted into a much smaller volume are preferred for use with large-volume samples.

Different types of filters have been evaluated for the recuperation of aquatic viruses, in the form of flat membranes or cartridges. Cartridge-type filters have the advantage to allow filtration of large volumes of moderately turbid water within a relatively short time. Their chemical composition, and porosity vary enormously. A whole range of "negatively" or "positively" charged filters now exist. Their efficiency depends on the type of water being tested and the presence of interfering substances such as detergents, suspended solid matter, or organic matter, which can affect the adsorption of viruses on these filters (Sobsey and Glass, 1984; Sobsey and Hickey, 1985, Gilgen et al., 1997).

The disadvantage of the negatively charged membranes or cartridges (Farrah et al., 1976) is that the water sample must be pretreated prior to concentration. This includes acidification of water and addition of salts to the water sample to facilitate virus adsorption because electronegative filters do not adsorb viruses well under ambient water conditions (Rao and Melnick, 1986). The necessity of this pretreatment step limits the on-location use of this method to a certain extent, although automatic injection systems do exist for treating several hundred liters of water. Virus concentration with electropositive filters may be performed on location at ambient conditions and without any prior amendment of the sample, which make this procedure most suited for in-field studies, provided that the sample pH is lower than 8.5 (Sobsey and Jones, 1979). Glass powder (Sarrette et al., 1977; Schwartzbrod and Lucena, 1978; Gajardo et al., 1991) or glass fiber (Vilaginès et al., 1997) have also been satisfactorily used in different laboratories as adsorbent materials for virus concentration.

Viruses in eluate volumes too large to be conveniently and economically assayed directly for viruses, such as those obtained from processing large volumes of water through cartridge or large disk filters, can be reconcentrated by several methods. Obviously, the recovery of small quantities of viruses from natural waters is dependent not only on the efficacy of primary concentration from the original large volume but also on the reconcentration of the primary eluate to a smaller volume.

Methods such as aluminum hydroxide adsorption-precipitation (American Public Health Association, 1998), polyethylene glycol hydroextraction

(Farrah et al., 1978; Lewis and Metcalf, 1988), organic flocculation (Katzenelson et al., 1976), and ammonium sulfate precipitation (Shields and Farrah, 1986; Bosch et al., 1988b) that are impractical for processing large fluid volumes are suitable for second-step concentration procedures. Alternatively, viruses can be sedimented depending on their molecular weight using ultracentrifugation (Steinman, 1981; Mehnert et al., 1997). Freeze-drying of samples (Gajardo et al., 1995; Pintó et al., 2001) and rehydration in a smaller volume provides a procedure for both virus concentration and removal of PCR inhibitors. Ultrafiltration (Divizia et al., 1989b) can utilize size exclusion rather than adsorption and (or) elution to concentrate viruses can provide consistent recoveries of different viruses under widely varying water conditions.

Because an evaluation of the presence of viruses in sediment provides an additional insight into long-term water-quality conditions, several methods for the detection of viruses have been developed. These methods consist of virus elution from the solid materials followed by concentration of the eluted viruses. Viruses are usually eluted from the sediments by using alkaline buffers (Gerba et al., 1977; Bosch et al., 1988a; Jofre et al., 1989) or chaotropic agents (Wait and Sobsey, 1983; Lewis and Metcalf, 1988; Jofre et al.,1989). Procedures such as organic flocculation (Wait and Sobsey, 1983), ammonium sulfate precipitation (Jofre et al., 1989), polyethylene precipitation (Lewis and Metcalf, 1988), or ultrafiltration (Gerba et al., 1977) are commonly employed to concentrate viruses from the eluate.

1.3. Viruses in Soil

Diseases associated with soil have been categorized according to the origin of the etiological agent as follows (Weissman et al., 1976; Santamaría and Toranzos, 2003): (a) soil-associated diseases that are caused by opportunistic or emerging pathogens belonging to the normal soil microbiota; (b) soil-related diseases that result in intoxication from the ingestion of food contaminated with entero- or neurotoxins; (c) soil-based diseases caused by pathogens indigenous to soil; and (d) soil-borne diseases caused by enteric pathogens that get into soil by means of human or animal excreta. In this latter category are included viruses transmitted through the fecal-oral route.

The transport of viruses through soil to groundwater and then to the community has been a topic of great concern. Many epidemics of infectious diseases have been attributed to the consumption of contaminated groundwater, casting soil as a vector and source of important human disease agents. There is a concern about a possible increase in soil-borne diseases in human population, given the land disposal practices of sewage and sewage sludge. In developing countries, untreated domestic wastewater is used in agricultural irrigation, presenting a high risk to farm workers and to consumers of food products irrigated with wastewater (Strauss, 1994). In spite of the clear public health implications of the occurrence and survival of viruses in the soil compartment, studies on the fate of viruses in soil are scarce due to the complexity of the methodologies for virus extraction from soil.

Table 6.8 Factors Influencing Virus Transport in Soil

Factor	*Effects*
Flow rate	Rate of movement increases with increased flow rate of water.
Hydraulic condition	Rate of movement is greater in saturated than unsaturated soil flow.
Soil texture	Fine-grained soils retain more viruses than coarse-grained soils.
Soil solution	Greater ionic strength means greater adsorption of viruses.
pH	Higher pH leads to greater adhesion to soil.
Virus type	Adsorption varies according to the strain and type of virus.
Humic substances	Organic matter may retard virus adhesion to soil.
Cations	Adsorption increases in the presence of cations.

The most relevant factors controlling virus transport through soil are soil type, water saturation state, pH, conductivity of the percolating water, and soluble organic matter (Table 6.8). The type of soil has a great influence on the level of viral transport. Fine-textured soils tend to absorb viruses more readily than coarsely textured soils. As a general rule, sandy soils are relatively poor adsorbents of enteric viruses, whereas soils with clay content of 30% to 100% are excellent adsorbents (Sobsey et al., 1980). In consequence, viral adsorption increases with increasing clay mineral content (Gerba et al., 1981). The high adsorptive properties of a clay soil will prevent virus transport to another matrix, such as groundwater, whereas coarse soil will not.

Microbial movement in soils is also greatly dependent on the water saturation state. When the soil is saturated, all pores are filled with water, which allows faster virus transport through the soil because virus contact with the soil has been diminished. When the flow is unsaturated, the viruses are in closer contact with the soil, thus promoting virus adsorption to the soil (Santamaría and Toranzos, 2003).

Goyal and Gerba (1979) considered soil pH as the single most important factor influencing viral adsorption, although the combined effect of organic matter and clay content, and cation-exchange capacity, could surpass the sole soil pH effect. At ambient conditions, viruses are usually negatively charged, thus being attracted to and entrapped by positively charged material in soil (Sobsey et al., 1980). In neutral and alkaline soil situations, viruses will not bind to any particulate matter and will be allowed to move freely in soil. There are, however, many exceptions to these general rules.

Virus absorption to soil is also affected by cation concentration. Cations favor virus adsorption to soil by reducing their repulsive forces. Sewage wastes provide an environment that enhances virus retention to soil, while this retention would be low in distilled water. As a matter of fact, distilled water may actually lead to the elution of viruses from soils, favoring virus mobilization and transport through soil. On the other hand, soluble organic matter will compete with the virus for soil adsorption sites. Likewise, humic and fulvic acids will also compete with the virus and will reduce the level of adsorption of viruses to the soil (Sobsey and Hickey, 1985).

2.0. VIRUS PERSISTENCE IN THE ENVIRONMENT

Persistence is the term of choice to describe the capacity of a given virus to retain its infectivity in a given scenario. However, some authors unfamiliar with environmental virology claim that this term is confusing because it also describes the ability of certain viruses to produce infections in which, contrary to what applies in acute infections, a degree of equilibrium is established between the virus and the host (i.e., a cell or a whole animal). Other authors avoid the use of the term survival to describe the natural persistence of virus infectivity, based on the ambiguity of the "live" condition of viruses. Keeping in mind that a virus will be able to maintain its infectious status provided that all the virion components remain unaltered, the term *stability* may also be properly employed in this context.

One critical question in environmental virology is whether or not viruses can persist long enough, and in high enough concentrations in the environment, to cause disease in individuals who are in contact with polluted recreational water, soil, or fomites, or who consume contaminated water or seafood. Because viruses outside their hosts are inert particles, their chances of transmission from host to host are greatly dependent on the degree of their robustness, which allows them to remain infectious during the various conditions they may encounter in the environment.

Numerous physical, chemical, and biological factors influence virus persistence in the environment (Table 6.9). Some of the primary factors affecting the survival of viruses in liquid environmental matrices or media are temperature, ionic strength, chemical constituents, microbial antagonism, the sorption status of the virus, and the type of virus. Considerable differences have been observed in the survival of viruses in different types of environmental samples. Different behaviors and inactivation rates have been observed not only among viruses of different families and genera, but also among viruses of the same family, genus, and even among similar types or strains of virus (Block, 1983).

Among the chemical constituents of liquid or semisolid (feces, human night soil, biosolids, animal manures, etc.) environmental matrices, the amount and type of organic matter and specific antiviral chemicals (such as ammonia at elevated pH levels) play a role in virus stability. Of the physical factors influencing virus persistence in liquid media, temperature, sunlight, and virus association with solids are among the most important. Soil moisture, temperature, sunlight, and other soil characteristics may influence the persistence of viruses in soil. On inanimate surfaces, the most important factors that affect virus stability are the type of virus and surface, relative

Table 6.9 Factors Affecting Virus Persistence in the Environment

Factor	*Effect*
Physical	
Heat	Inactivation is directly proportional to temperature
Light	Light, especially its UV component, is germicidal
Desiccation or drying	Usually increased inactivation at lower relative humidity
Aggregation/Adsorption	Protection from inactivation
Pressure	High pressure induces inactivation
Chemical	
pH	Worst stability at extreme pH values
Salinity	Increased salt concentrations are virucidal
Ammonia	Virucidal
Inorganic ions	Some (e.g., Pt, Pd, Rh) are virucidal
Organic matter	Dissolved, colloidal, and solid organic matter protect from inactivation
Enzymes	Proteases and nucleases contribute to inactivation
Biological	
Microbial activity	Contributes to inactivation
Protozoal predation	Contributes to removal/death
Biofilms	Adsorption to biofilms protects from inactivation, while microbial activity in biofilms may be virucidal
Type of virus	Stability varies according to the strain and type of virus

humidity, moisture content, temperature, composition of the suspending medium, light exposure, and presence of antiviral chemical or biological agents. Most of these factors are also relevant for the ability of viruses to persist in aerosolized droplets, together with the moisture content and the size of the aerosol particles, and the air quality.

Some enteric virus infections follow a seasonal pattern, whereas others fail to do so. In regions with temperate climates, infections due to enteroviruses generally reach a peak in summer and early fall (Moore, 1982). On the contrary, rotavirus, norovirus, and astrovirus infections occur mainly during the cooler months (McNulty, 1978; Mounts et al., 2000; Guix et al., 2002), although seasonal and nonseasonal distributions of rotavirus in sewage have been described (Hejkal et al., 1984; Bosch et al., 1988c). On the other hand, cases of hepatitis A do not show a clear seasonal pattern (Lemon, 1985), whereas enteric adenovirus infections are reported to peak in midsummer (Wadell et al., 1989). These data suggest that temperature, and probably relative humidity, may be meaningful in the seasonal distribution of outbreaks of certain human enteric viruses (Enright, 1954), due to the influence of these factors on virus persistence.

Understanding environmental virus stability, and elucidating the factors that affect it, may shed some light on the potential public health risk associated with these environmental pollutants and at the same time provide tools to interrupt the chain of fecal-oral virus transmission. In this chapter, only studies involving the persistence of enteric viruses in the absence of any deliberately applied inactivation process are reviewed. Neither work on virus disinfection nor studies conducted with potential indicators, such as bacteriophages, are considered because they will be discussed in other chapters.

2.1. Methods to Study Environmental Virus Persistence

Most studies to determine the potential of viruses to persist in environmental settings have been performed by artificially adding a known amount of infectious virus to a given sample, determining the reduction in the infectious titer after subjecting the spiked sample to designated conditions, and applying statistical procedures to determine the significance of virus decay. Obviously, this implies the use of virus strains that may be propagated in cell cultures and enumerated through quantal infectivity assays (e.g., plaque assays), thus greatly restricting the range of viruses that are able to be included in these studies.

Molecular detection approaches such as PCR or RT-PCR are normally employed for fastidious virus analysis. However, they are unable to differentiate between infectious and noninfectious particles (Kopecka et al., 1993; American Public Health Association, 1998) and are, therefore, unsuitable for virus persistence studies, even when quantitative real-time procedures are employed. Although reports on the presence of norovirus sequences in bottled mineral water raised a lot of concern (Beuret et al., 2000, 2002), many authors have shown the lack of correlation between virus persistence and molecular detection of virus genomes. It now seems obvious that infectious particles are degraded more rapidly than virus genomes.

Most enteric viruses of public health concern consist of RNA genomes. In studies employing RT-PCR, it has been shown that poliovirus genomic RNA is not stable in nonsterilized water (Tsai et al., 1995). Although free DNA is fairly stable, it is unlikely that a free single-stranded RNA genome of noroviruses, astrovirus, poliovirus, or hepatitis A virus would remain stable without its protein coat in the environment. This presumption is less clear for the double-stranded RNA genome of rotaviruses. Nevertheless, it has been shown that altered nucleocapsids of noninfectious virions may still encapsidate a RT-PCR detectable single-stranded RNA (Gassilloud et al., 2003).

Amplification of a piece of the virus genome is not indicative of the presence of the infectious agent. It can be assumed that even when different target sequences from unrelated parts of the genome are detected by molecular amplification, there is still no indication of the presence of unaltered capsid with functional surface residues involved in receptor recognition and cell attachment.

From a strictly theoretical point of view, the use of an antigen-capture PCR assay, involving virus binding with a conformationally dependent monoclonal antibody and amplification of different unrelated genomic targets, could provide a fair estimation on the stability of a nonculturable virus. However, this approach requires a lot of experimentation before it can be considered adequate to be applied in virus persistence studies. In the meantime, infectious surrogates are usually employed to generate data on unculturable virus survival; for example, feline calicivirus has been used to mimic norovirus behavior (Thurston-Enriquez et al., 2003a, 2003b). Another promising approach to increase the likelihood of detecting intact and potentially infectious viruses in cell cultures is to pretreat the virions with proteolytic enzymes and nucleases prior to nucleic acid extraction, amplification, and detection, thereby eliminating the detection of free nucleic acids or nucleic acids associated with damaged, inactivated virions (Nuanualsuwan and Cliver, 2002).

Some health significant enteric viruses, such as rotavirus, astrovirus, and enteric adenovirus, replicate poorly in cell cultures; yet their persistence may be evaluated by integrated cell culture RT-PCR assays (Pintó et al., 1995; Reynolds et al., 1996; Abad et al., 1997; Reynolds et al., 2001). For this purpose, cells supporting the propagation of a wide variety of enteric viruses, such as CaCo-2 (colonic carcinoma) or PLC/PRF/5 cells (human liver hepatoma), are used for an *in vivo* amplification step prior to molecular amplification (Grabow et al., 1993; Pintó et al., 1994). It should be recognized, however, that most of the studies on virus persistence in the environment were performed under laboratory conditions and that data obtained from these studies may not truly represent their behavior under actual field conditions.

2.2. Virus Persistence in Environmental Waters

The survival of viruses in environmental waters has been extensively reviewed (Bitton, 1980; Kapucinski and Mitchell, 1980; Block, 1983; Bosch, 1995). As previously mentioned, the most relevant factors affecting virus survival in the water environment are temperature (Akin et al., 1971; Raphael et al., 1985; Bosch et al., 1993), virus association with solids (Gerba and Schaiberger, 1975; La Belle et al., 1980; Rao et al., 1984; Sobsey et al., 1988), exposure to UV (Bitton et al., 1979; Bitton, 1980), and the presence of microbial flora (Gunderson et al., 1968; Fujioka et al., 1980; Toranzo et al., 1983; Ward et al., 1986; Gironés et al., 1989, 1990).

The effect of temperature on viral persistence in water may be due to several mechanisms including protein denaturation, RNA damage, and influence on microbial or enzymatic activity (Dimmock, 1967; Melnick and Gerba, 1980; Deng and Cliver, 1995). Early studies pointed to damage to virion proteins as the primary target for viral inactivation at high temperatures, although damage to both protein and RNA occurs at all temperatures (Dimmock, 1967). Even though all viruses persist better at lower temperatures than at higher temperatures, some viral strains, such as hepatitis A virus and parvovirus, do exhibit higher thermal resistance than other viruses.

As mentioned earlier in this chapter, virus adsorption to particulate material increases the persistence of enteric viruses in the water environment (Gerba and Schaiberger, 1975; La Belle et al., 1980; Rao et al., 1984; Sobsey et al., 1988), although differences have been observed among study locations

(La Belle et al., 1980). The increased virus survival in the presence of sediment has important implications in the marine environment, because fecal contamination of coastal areas results in contamination of shellfish harvesting areas, accumulation of solid-associated viruses in sediments with sediments acting as virus reservoirs, and finally accumulation of viruses in shellfish. Additionally, virus uptake by molluskan bivalves is enhanced by the presence of particulate material (Landry et al., 1983).

Although self-purification processes are reported to be more pronounced in seawater than in river water (Matossian and Garabedian, 1967; Gironés et al., 1989), the effect of salinity on virus stability is variable. Thus, many studies have reported enhanced removal of virus infectivity in saline solution compared with distilled water (Dimmock, 1967; Salo and Cliver, 1976), whereas others report no significant effect of salinity on virus persistence (Lo et al., 1976; Fujioka et al., 1980). In any case, the self-purification capacity of water is finite.

Several observations demonstrate the potential involvement of native aquatic microorganisms in the inactivation of viruses, particularly in marine habitats. However, data on the successful isolation of microorganisms with virucidal properties are scarce (Fujioka et al., 1980; Girones et al., 1990; Bosch et al., 1993). Additionally, the ability of bacteria to inactivate viruses is usually lost while subculturing the microorganisms in the laboratory (Gunderson et al. 1968; Katzenelson 1978), although in a few studies, such bacteria could be subcultured for more than 1 year without losing their antiviral activity (Girones et al., 1990; Bosch et al., 1993). In some studies, the virucidal agents in the tested waters could not be separated from the microorganisms (Shuval et al., 1971; Denis et al., 1977; Fujioka et al., 1980; Ward et al., 1986; Gironés et al., 1990), whereas in others the virucidal activity could be separated from the bacteria (Matossian and Garabedian, 1967; O'Brien and Newman 1977; Toranzo et al., 1983; Bosch et al., 1993). The antiviral activity seems to be based on proteolytic bacterial enzymes that inactivate virus particles in water by cleavage of viral proteins, thus exposing the viral RNA to nuclease digestion (Toranzo et al., 1983; Gironés et al., 1990, Bosch et al., 1993).

It seems reasonable to assume that environmental factors and the compositional makeup of a given type of water may be substantially different from one geographical location to another, which implies that different data of virus persistence are produced when the same viral strain is suspended in water sampled from different sites (Bosch et al., 1993). Furthermore, it is highly likely that natural waters, particularly in the marine environment, contain a variety of potential antiviral factors, and that the antiviral action observed is generally the expression of the most dominant factor(s) present in any given water source.

2.3. Virus Persistence in Soil

As has been mentioned earlier, soil pollution with human wastes may greatly contribute to groundwater contamination. Because of the increasing emphasis placed on land application as a means of organic waste disposal, it appears relevant to evaluate the persistence of human pathogens in soil.

Viruses in moisture-saturated soils may remain infectious for long periods of time, even at ambient temperatures of 20°C, in which the soil would be microbially active. If soil moisture drops under 10%, dramatic losses in virus infectivity are observed regardless of soil temperature or the medium in which viruses are applied (Yeager and O'Brien, 1979). For example, enteric viruses survive for 15–25 days in an air-dried soil as compared with 60–90 days in soils with 10% moisture content. One of the pathways for virus removal from warm soils is through evaporation, which would account for the loss of viral pathogens from dry soils. The rate of evaporation is directly related to temperature and relative humidity. Under constant moisture of 10% or greater, the main factors controlling the inactivation of viruses appear to be not only soil temperature but also soil texture. The survival of viruses is enhanced by a combination of low soil temperature and sufficient moisture (Bitton, 1980). As temperature increases, the virus inactivation rates also increase significantly (Yeager and O'Brien, 1979; Straub et al., 1992). At 4°C and with constant moisture, viruses are able to persist for 180 days, whereas at 37°C no viruses persist after 12 days.

Certain soil characteristics also influence virus survival. For example, virus persistence has been reported to decrease as a function of increasing soil pH and resin-extractable phosphorus (Hurst et al., 1980). Increase in exchangeable aluminum, on the other hand, increased virus survival. The relative levels of clay and humic acids may also enhance virus survival (Bitton and Gerba, 1984). Viruses survive better in an adsorbed state than in suspension. Virus adsorption to clay materials through electrostatic interactions is speculated to protect viral genome against nucleases or other antagonistic factors in soil (Bitton and Gerba, 1984). Additionally, clay contributes to virus survival by retaining minimum amounts of water, even in dry soils. This water provides the moisture required for virion stability. On the other hand, poorly absorbent sandy soils can increase their viral retention in the presence of divalent cations (Mg^{2+}, Ca^{2+}) but not monovalent (Na^{+}) or trivalent (Fe^{3+}) cations (Lefler and Kott, 1974).

Clay loam soils generally afford more protection to viruses than sandy soils. However, in rapidly drying soils, virus persistence may decrease more deeply in clay soils than in sandy soils, due to the water-holding capacity of soil. Clay soils can hold more water than sandy soils, but when water is evaporating from both soils, the clay soils, because of their mineral content, will retain the remaining water more tightly than sandy soils at the same moisture content, making them less apt for biological activity (Straub et al., 1992).

The presence of indigenous microorganisms is deleterious to virus survival, although this effect is not observed at low temperature; at 1°C poliovirus remains stable through 70 days (Hurst et al., 1980). Indigenous soil aerobic microorganisms significantly reduce virus persistence, while indigenous anaerobes do not (Hurst, 1987). In a study involving a variety of soils and poliovirus, echovirus and HAV suspended in groundwater, secondary sewage effluent or primary sewage treatment effluent, HAV was usually more persistent than poliovirus and echovirus; the 99% reduction times for HAV were normally greater than 12 weeks (Sobsey et al., 1989). This indicates that HAV is an extremely stable agent, capable of persisting for more than 3 months in soil, and hence it poses a health threat.

The ultraviolet component of sunlight is destructive to viruses (Bitton, 1980). The UV has been shown to inactivate viruses at the surface of the soil but as the viruses move deeper in the soil column, it plays a minor role in inactivating viruses. Disposable diapers may contribute to soil contamination with human pathogens. A field survey of virus inactivation in diapers buried in landfills for at least 2 years showed complete inactivation (Huber et al., 1994). In laboratory conditions, HAV and poliovirus experimentally seeded in disposable diapers showed 2.5 and 4 $\log_{10}$ reduction, respectively, after 80 days at 25°C, in aerobic conditions (Gray et al., 1993).

Quantitative interpretations (Carrington et al., 1998a, 1998b) of existing data on poliovirus and cytopathic enterovirus decay rates in sludge-amended soil (Tierney et al., 1977; Hurst et al., 1978) indicated that, at the prevailing summer temperatures (19–34°C), the decimal reduction rates were between 2.7 and 3.7 days, whereas at the winter temperatures (13–26°C), it was 24 days. Carrington et al. (1998a) analyzed data reported by Straub et al. (1993) and found that decimal reduction time for poliovirus at winter temperature of 15°C and moisture levels of 15–25% was 92 days as compared with 1.2 days at summer temperatures of 27–33°C at moisture levels of 3–40%.

Most studies on virus persistence in soil have been performed in North American soil types and autochthonous climatic conditions. It has been pointed out that in other parts of the world, where mean soil temperature seldom exceeds 15°C at 10 cm depth in summer and about 5°C in winter, viral decay rates would be slow or with decimal reduction times from 24 days to more than 100 days (Carrington et al., 1998b; Rzezutka and Cook, 2004). However, the same authors suggest that cultivation of soil after sludge application would encourage virus decay by enhancing evaporation.

2.4. Virus Persistence on Fomites

Outbreaks of acute gastroenteritis and hepatitis are a matter of concern in institutions such as daycare centers, hospitals, nurseries, schools, and military quarters. Many of these outbreaks have been suspected to be caused by vehicular transmission of agents through contaminated environmental surfaces (Ryder et al., 1977; Halvorsrud and Orstavick, 1980; Rocchi et al., 1981; Sattar et al., 1986; Butz et al., 1993; Green et al., 1998). As has been mentioned earlier, stools from patients with diarrhea or hepatitis contain a very high number of the causative virus, and a single vomiting episode of an individual suffering from norovirus gastroenteritis may expel 3×10^7 virus particles, all of which are able to contaminate fomites (Cheesbrough et al., 1997; Green et al., 1998, 1994).

It has been demonstrated that human enteric viruses are able to survive on several types of materials commonly found in institutions and domestic environments long enough to represent a source for secondary transmission of disease (Hendley et al., 1973; Sattar et al., 1986, 1987; Ansari et al., 1988; Mbithi et al., 1991; Abad et al., 1994, 2001). The stability of health-significant human enteric viruses has been investigated on various nonporous (aluminum, china, glazed tile, plastic, latex, stainless steel, and polystyrene) and porous (cloth, different types of papers and cotton cloth) surfaces (Sattar et al., 1986; Abad et al., 1994, 2001). As a general conclusion, when dried on environmental fomites, hepatitis A virus and rotavirus are more resistant to inactivation than enteric adenovirus, astrovirus, and poliovirus.

The higher stability of HAV in comparison with poliovirus, both of which belong to the Picornaviridae family, is due to the inherently more stable molecular structure of HAV capsid, concordant with the special codon usage described for this virus (Sánchez et al., 2003). In fact, it appears undeniable that poliovirus, which has been extensively employed as a model to elucidate enteric virus behavior in many scenarios, may fail to provide an adequate indication of the persistence of other human enteric viruses, such as HAV, astrovirus, or rotavirus, dried on fomites (Sobsey et al., 1988; Mbithi et al., 1991; Abad et al., 1994, 2001).

The resistance to desiccation appears to be of major significance in determining the ability of a virus strain to survive on fomites. A pronounced loss in virus titer at this stage dramatically reduces the chances of subsequent virus persistence. On the contrary, viruses involved in outbreaks probably transmitted through fecally contaminated environmental surfaces (i.e., HAV, rotavirus, or astrovirus) show little decay at the desiccation step. On the contrary, HAV and HRV, which have been involved in outbreaks probably transmitted through fecally contaminated environmental surfaces, show little decay at the desiccation step (Mahl and Sadler, 1975; Keswick et al., 1983; Sattar et al., 1986; Sobsey et al., 1988; Abad et al., 1994, 2001).

In spite of the experimental data on virus persistence on environmental surfaces, it is generally very difficult to determine whether, and to what extent, fomites play a role in the spread of infectious agents. Keswick et al. (1983) have suggested that the prevalence of asymptomatic infections in daycare facilities may make contaminated surfaces in these environments a reservoir of infection for previously uninfected inmate children and their family contacts.

As mentioned previously, there is a considerable shedding of the SARS coronavirus in stools, where it remains stable at room temperature for several days (Tsang, 2003). Although epidemiological evidence suggests that the major mode of transmission for SARS coronavirus is by close personal contact with an infected individual, contact with environmental surfaces contaminated with respiratory secretions or other body fluids may also play a role in transmission (Tsang, 2003). In addition, SARS coronavirus has been detected in a variety of environmental surfaces, such as the toilet and floor in the aparment of an infected individual and the walls and rooftop of a building with multiple cases (Tsang, 2003).

Hands are frequently in contact with environmental surfaces, and the potential for transfer of virus between surfaces and hands has been studied

(Hendley et al., 1973; Ansari et al., 1988; Mbithi et al., 1992). It was ascertained in these studies that rotavirus and hepatitis A virus could retain infectivity for several hours on skin and could be transferred in an infectious state from fingertips to other surfaces and vice versa. Enteric virus transfer between hands was apparently influenced by moisture. Moisture would mediate suspension of virus particles and facilitate their movement between touching surfaces; drying would reduce this effect. Laboratory studies have shown that viruses persist better in the environment at high relative humidity and at low temperatures (Moe and Shirley, 1982; Sattar et al., 1988; Sobsey et al., 1988; Abad et al., 1994). However, data on the effect of relative humidity on enteric virus survival is contradictory. These reported differences, particularly affecting rotavirus persistence, are difficult to explain but may be due to differences in the methodologies between these studies.

Temperature substantially affects the survival of feline calicivirus, an infectious surrogate for human norovirus, which is able to persist for long periods of time dried onto glass coverslips with log reductions of 4.75 after 2 months and 3 weeks, at 4°C and room temperature, respectively (Doultree et al., 1999). The authors suggested that the effect of temperature on feline calicivirus stability may reflect the greater prevalence of norovirus infections in cooler seasons (Lopman et al., 2003).

Because the fecal-oral route is the common means of enteric virus transmission, it seems reasonable to evaluate the effect of fecal material on the persistence of virus on fecally contaminated fomites. Again, data on the protective effect of feces on viruses are contradictory; fecal matter appears to affect the survival of enteric viruses in opposite ways, depending on the type of surface and the virus strain (Keswick et al., 1983; Sobsey et al., 1988; Abad et al., 1994).

2.5. Virus Persistence in Aerosols

Aerosols are an important means of virus transmission in humans. Various authors have reported the isolation of enteric viruses from aerosols produced by sludge-treatment plants (Fannin et al., 1985; Fattal et al., 1987; Pfirrmann and Bossche, 1994; Alvarez et al., 1995). The presence of microorganisms in aerosols generated from wastewater-treatment processes or in treated wastewater for agricultural irrigation is a potential danger to human health (Teltsch et al., 1980; Alvarez et al., 1995). In hospitals, aerosolization of vomit was reported to be of major importance in the transmission of norovirus infection during outbreaks, while cleaning vomit or feces from patients did not significantly increase the risk of developing gastroenteritis (Chadwick and McCann, 1994). Members of the *Caliciviridae* family have been reported to be fairly stable in aerosols (Donaldson and Ferris, 1976). The most important factors affecting the stability of viruses in the aerosol state are temperature, pH, relative humidity, moisture content, size of the aerosol particle, composition of the suspending medium, sunlight exposure, air quality, and virus type.

The basis of virus inactivation in aerosols is poorly understood, although mechanisms for bacteriophage inactivation in aerosols have been proposed

(Trouwborst et al., 1974). At high relative humidity, surface alteration of the virion has been reported, whereas at low relative humidity virus inactivation appears to be mediated by the removal of structural water molecules. Relative humidity seems to confer a protective effect on aerosolized nonenveloped virus particles. Thus, poliovirus was more stable in aerosol at 22°C at high relative humidity than at low relative humidity (Hemes et al., 1960; Harper, 1961). Picornavirus infectious RNA may be detected at all humidity levels, suggesting that virus inactivation is caused by virion capsid damage (Akers and Hatch, 1968).

High relative humidity and low temperature enhance the persistence of bovine rotavirus in aerosols (Moe and Harper, 1983; Ijaz et al., 1985), although simian rotavirus SA11 survival in aerosols seems to be the best at intermediate relative humidity levels (Sattar et al., 1984). In any case, human, simian, and calf rotavirus strains may be detected in aerosols after as long as 10 days (Moe and Harper, 1983; Sattar et al., 1984; Ijaz et al., 1985), although discrepancies, probably due to methodological differences, are found among these studies. Aerosolized adenovirus particles also show increased persistence at high relative humidity and low temperature (Miller and Artenstein, 1967; Elazhary and Derbyshire, 1979).

Contrarily to nonenveloped viruses, viruses with an outer lipid envelope seem to be more stable at lower relative humidity (Hemmes et al., 1960). After 6 days at 20°C and 50% relative humidity, infectious human coronavirus particles could be recovered in aerosols (Ijaz et al., 1985). Virus infectivity in aerosols is also affected by solutes in the suspending media used for aerosolization. Addition of salts and proteins in the suspending media provides a protective effect against dehydration and thermal inactivation of aerosolized picornaviruses (McGeady et al., 1979; Reagan et al., 1981) and may also influence the rehydration rate during sample rehumidification prior to the infectivity assay (Benbough, 1969).

2.6. Virus Persistence in Food

Outbreaks of viral infection attributed to the consumption of contaminated soft fruit, salad vegetables, and other foods are increasingly reported (Mead et al., 1999; Lopman et al., 2003). A recent example is an outbreak of hepatitis A virus in western Pennsylvania in late 2003, which affected more than 600 people and resulted in three fatalities (MMWR, 2003). The incident involved green onions imported from Mexico and added to the restaurant's homemade salsa. Those green onions were stored in a single container for up to 5 days in the ice used for shipping them. Some of the uncooked green onions were used in the restaurant's mild salsa that was prepared in large batches and stored for up to 3 days. If the shipment ice was contaminated, prolonged exposure combined with the relatively long storage of salsa may account for why so many patrons became infected. Green onions, which are multilayered and can retain soil particles that could harbor fecal contaminants, were probably contaminated during harvesting and packing. Alternatively, HAV-contaminated water used for irrigation, processing, and storage

Table 6.10 Examples of Virus Persistence in Food

Food	*Virus*	*Temperature*	*Storage Time*	*Log_{10} Titer Reduction*	*Reference*
Lettuce	HAV	4°C	7 days	2.03	Croci et al., 2002
Carrot	HAV	4°C	7 days	≥2.44	
Lettuce	HAV	RT	6 days	≤1.00	Bidawid et al., 2001
		4°C	6 days	≤0.50	
		RT	12 days	4.00	
		4°C	12 days	≤0.50	
Cabbage	Polio	8–17°C	5 days	6.15	Ward et al., 1982
		13–22°C	2 days	5.55	
Grass[a]	Polio	4–16°C	40 hr	2.40	Badawy et al., 1990
	Rotavirus	4–16°C	40 hr	≥4.87	
	Polio	22–41°C	40 hr	≥4.39	
	Rotavirus	22–41°C	40 hr	2.99	
Creme sandwich	HAV	21°C	7 days	2.05	Sobsey et al., 1988

RT, room temperature.
[a] Bermuda hybrid grass and rye grass.

may have been the source of contamination. The high environmental persistence of HAV makes any of these scenarios possible.

Data on the potential of enteric viruses to persist between the preparation of food and its consumption are required to ascertain the risk of virus transmission through food. This information is also important for the development of treatments applied to food in order to inactivate contaminant viruses. Disinfection practices for food are reviewed in another chapter.

Examples of studies on virus persistence in food are depicted in Table 6.10. Studies have shown that viruses remain infectious for several days or weeks on vegetable crops irrigated with contaminated sewage effluent or sludge (Tierney et al., 1977; Ward and Irving, 1987). Several enteroviruses have been reported to survive during commercial and household storage for periods of up to 5 weeks on vegetables irrigated with contaminated effluent (Larkin et al., 1976; Ward and Irving, 1987).

The factors that affect virus survival in the environment, especially on fomites, are also relevant for the fate of viruses in food products. Among them, temperature has a great influence on virus stability in food as in any other suspending matrix; the higher the temperature, the more pronounced the virus decay. Natural or added constituents of food may influence the rate of virus inactivation by temperature (Cliver and Riemann, 1999). For instance, salt used in pickling sausage batter has been shown to protect viruses from thermal inactivation (Grausgruber, 1963), whereas acidity often enhances the virucidal effect of temperature (Cliver et al., 1970; Salo and Cliver, 1976). Additionally, viruses appear to resist thermal inactivation during cooking when fat levels are high (Filippi and Banwart, 1974).

Furthermore, some ingredients may have a direct virucidal effect, as has been elucidated for free, unsaturated fatty acids with enveloped viruses (Kohn et al., 1980). Naturally occurring substances in fruit juices have been reported to bear a reversible inactivating effect on enteroviruses, attributed to plant polyphenols such as tannins (Konowalchuk and Speirs, 1976, 1978; Cliver and Kostenbader, 1979).

Although the presence of fecal material and high relative humidity strongly enhances virus persistence (Konowalchuk and Speirs, 1975), the effect of modified atmosphere packaging does not appear to be significant on virus persistence (Bidawid et al., 2001). As is the case with fomites, a rapid and marked decline in virus titer on crops/vegetables is attributed to drying/desiccation (Larkin et al., 1976; Tierney et al., 1977; Ward and Irving, 1987) combined with the action of sunlight and temperature (Kott and Fishelson, 1974). Direct sunlight irradiation (particularly its UV component) by itself is able to induce a pronounced reduction in virus numbers in food (Badawy et al., 1990).

3.0. CONCLUSIONS

Further work is required to develop robust and reliable quantitative methods to recover and detect health significant viruses in environmental and food samples. These procedures should also be adequate for newly recognized emerging pathogens of concern, as well as for non-human viruses capable of zoonotically infecting humans and having greater potential to cause human infection and illness. Simple standardized diagnostic procedures for selected pathogens are needed to establish specific virological guidelines in selected food products, notably shellfish or food imports from regions with endemic infections.

Molecular characterization of agents responsible for waterborne and food-borne outbreaks will provide relevant information on the prevalence of infections among the population, which may be important in the development and/or efficacy of vaccines. The long pursued objective of the eradication of poliomyelitis will require comprehensive surveys on the occurrence of wild-type and vaccinal-type poliovirus in environmental samples that may represent potential reservoirs and vehicles of transmission.

Another important issue in environmental studies is microbial source tracking, which is imperative for the maintenance of microbiological quality and safety of water systems used for drinking, recreation, and in seafood harvesting, because contamination of these systems can represent high risks to human health and significant economic losses due to closure of beaches and shellfish harvesting areas. As mentioned earlier and discussed elsewhere in this book, bacteriophages and other microorganisms of fecal flora have been proposed as models of virus behavior. However, from the strictly structural point of view, there is no better surrogate of an actual virus pathogen to track their behavior in the environment than a noninfectious virus-like particle

(VLP) of the same virus. Recombinant VLPs of health-significant viruses as norovirus and rotavirus have been employed to investigate the influence of electrostatic interactions in the filtration of norovirus in quartz sand and rotavirus behavior under disinfection conditions, respectively (Redman et al., 1997; Caballero et al., 2004). As model systems, recombinant tracers are perfectly adequate for field studies of microbial tracking, as they may be produced in extremely high numbers (several milligram amounts). Additionally, their noninfectious nature, due to the lack of a nucleic acid, makes them suitable for use in scenarios where the use of actual pathogenic viruses is not prudent; for example, drinking-water treatment plants, shellfish growing waters, or selected food products.

4.0. REFERENCES

Abad, F. X., Pinto, R. M., and Bosch, A., 1994, Survival of enteric viruses on environmental fomites. *Appl. Environ. Microbiol.* 60:3704–3710.

Abad, F. X., Pintó, R. M., Villena, C., Gajardo, R., and Bosch, A., 1997, Astrovirus survival in drinking water. *Appl. Environ. Microbiol.* 63:3119–3122.

Abad, F. X., Villena, C., Guix, S., Caballero, S., Pinto, R. M., and Bosch, A., 2001, Potential role of fomites in the vehicular transmission of human astroviruses. *Appl. Environ. Microbiol.* 67:3904–3907.

Akers, T. G., and Hatch, M. T., 1968, Survival of a picornavirus and its infectious ribonucleic acid after aerosolization. *Appl. Microbiol.* 16:1811–1813.

Akin, E. W., Benton, W. H., and Hill, J. W. Jr., 1971, Enteric viruses in ground and surface waters: a review of their occurrence and survival, in: *Virus and Water Quality: Occurrence and Control* (V. Griffin and J. Snoeyink, eds.), University of Illinois Press, Urbana, IL, pp. 59–74.

Alvarez, A., Buttner, M. P., and Stetzenbach, L., 1995, PCR for bioaerosol monitoring: sensitivity and environmental interference. *Appl. Environ. Microbiol.* 61: 3639–3644.

American Public Health Association, 1998, *Standard Methods for the Examination of Water and Wastewater*, 20th ed., American Public Health Association, American Water Works Association, Water Pollution Control Federation, Washington, DC.

Amvrosieva, T. V., Titov, L. P., Mulders, M., Hovi, T., Dyakonova, O. V., Votyakov, V. I., Kvacheva, Z. B., Eremin, V. F., Sharko, R. M., Orlova, S. V., Kazinets, O. N., and Bogush, Z. F., 2001, Viral water contamination as the cause of aseptic meningitis outbreak in Belarus. *Cent. Eur. J. Publ. Health* 9:154–157.

Andersson, Y., and Stenström, G., 1987, Waterborne outbreaks in Sweden: causes and etiology. *Water Sci. Technol.* 19:575–580.

Ansari, S. A., Sattar, S. A., Springthorpe, V. S., Wells, G. A., and Tostowaryk, W., 1988, Rotavirus survival on human hands and transfer of infectious virus to animate and nonporous inanimate surfaces. *J. Clin. Microbiol.* 26:1513–1518.

Appleton, H., and Pereira, M. S., 1977, A possible virus aetiology in outbreaks of food poisoning from cockles. *Lancet* 1:780–781.

Badawy, A. S., Rose, J. B., and Gerba, C. P., 1990, Comparative survival of enteric viruses and coliphage on sewage irrigated grass. *J. Environ. Sci. Health* 25: 937–952.

Banks, M., Bendall, R., Grierson, S., Heath, G., Mitchell, J., and Dalton, H., 2004, Human and porcine hepatitis E virus strains, United Kingdom. *Emerg. Infect. Dis.* 10:953–955.

Benbough, J. E., 1969, The effect of relative humidity on the survival of airborne Semliki forest virus. *J. Gen. Virol.* 4:473–437.

Beuret, C., Kohler, D., and Lüthi, T. M., 2000, Norwalk-like virus sequences detected by reverse transcription-polymerase chain reaction in mineral waters imported into or bottled in Switzerland. *J. Food Prot.* 63:1576–1582.

Beuret, C., Kohler, D., Baumgartner, A., and Lüthi, T. M., 2002, Norwalk-like virus sequences in mineral waters: one-year monitoring of three brands. *Appl. Environ. Microbiol.* 68:1925–1931.

Bidawid, S., Farber, J. M., and Sattar, S. A., 2001, Survival of hepatitis A virus on modified atmosphere-packaged (MAP) lettuce. *Food Microbiol.* 18:95–102.

Birch, C., and Gust, I., 1989, Sewage pollution of marine waters: the risks of viral infection. *Med. J. Aust.* 151:609–610.

Bitton, G., 1980, *Introduction to Environmental Virology*, John Wiley & Sons, New York.

Bitton, G., and Gerba, C. P., 1984, Groundwater pollution microbiology: the emerging issues, in *Groundwater Pollution Microbiology* (G. Bitton and C.P. Gerba, eds.), John Wiley & Sons, New York. pp. 1–8.

Bitton, G, Fraxedas, R., and Gifford, G. E., 1979, Effect of solar radiation on poliovirus: preliminary experiments. *Water Res.* 13:225–228.

Blacklow, N. R., and Cukor, G., 1982, Norwalk virus: a major cause of epidemic gastroenteritis. *Am. J. Public. Health.* 72:1321–1323.

Block, J. C., 1983, Viruses in environmental waters, in: *Viral Pollution of the Environment* (G. Berg, ed.), CRC Press, Inc., Boca Raton, FL, pp. 117–145.

Bosch, A., 1995, The survival of enteric viruses in the water environment. *Microbiologia SEM* 11:393–396.

Bosch, A., and Pintó, R. M., 1992, Human enteric viruses in the environment, in: *Environmental Protection, Vol. 3* (A.Z. Keller and H.C. Wilson, eds.), University of Bradford Press, Bradford, UK, pp 63–71.

Bosch, A., Lucena, F., Girones, R., and Jofre, J., 1986, Survey of viral pollution in Besos River. *J. Water Pollut. Control Fed.* 58:87–91.

Bosch, A., Lucena, F., Girones, R., and Jofre, J., 1988a, Occurrence of enterovirus on marine sediment along the coast of Barcelona (Spain). *Can. J. Microbiol.* 34:921–924.

Bosch, A., Pintó, R. M., Blanch, A. R., and Jofre, J., 1988b, Detection of human rotavirus in sewage through two concentration procedures. *Water Res.* 22:343–348.

Bosch, A., Pintó, R. M., and Jofre, J., 1988c, Non-seasonal distribution of rotavirus in Barcelona raw sewage. *Zbl. Bakt. Hyg. B* 186:273–277.

Bosch, A., Lucena, F., Diez, J. M., Gajardo, R., Blasi, M., and Jofre, J., 1991, Waterborne viruses associated with a hepatitis outbreak. *J. Am. Water Works Assoc.* 83:80–83.

Bosch, A., Gray, M., Diez, J. M., Gajardo, R., Abad, F. X., Pintó, R. M., and Sobsey, M. D., 1993, The survival of human enteric viruses in seawater. *MAP Tech. Rep. Ser.* 76:1–7.

Bosch, A., Sánchez, G., Le Guyader, F., Vanaclocha, H., Haugarreau, L., and Pintó, R. M., 2001, Human enteric viruses in coquina clams associated with a large hepatitis A outbreak. *Water Sci. Tech.* 43:61–66.

Bryan, J. A., Lehmann, J. D., Setiady, T. F., and Hatch, M. H., 1974, An outbreak of hepatitis A associated with recreational lake water. *Am. J. Epidemiol.* 99:145–154.
Butz, A. M., Fosarelli, P., Dick, J., Cusack, T., and Yolken, R., 1993, Prevalence of rotavirus on high-risk fomites in day-care facilities. *Pediatrics* 92:202–205.
Caballero, S., Abad, F. X., Loisy, F., Le Guyader, F. S., Cohen, J., Pintó, R. M., and Bosch, A., 2004, Rotavirus virus-like particles as surrogates in environmental persistence and inactivation studies. *Appl. Environ. Microbiol.* 70:3904–3909.
Cabelli, V., 1983, Public health and water quality significance of viral diseases transmitted by drinking water and recreational water. *Water Sci. Technol.* 15:1–15.
Cabelli, V. J., Dufour, A. P., McCabe, L. J., and Levin, M. A., 1982, Swimming-associated gastroenteritis and water quality. *Am. J. Epidemiol.* 115:606–616.
Carrington, E. G., Davis, R. D., and Pike, E. B., 1998a, Review of the scientific evidence relating to the controls on the agricultural use of sewage sludge. Part 1—the evidence underlying the 1989 Department of the Environment code of practice for agricultural use of sludge and the sludge (use in agriculture) regulations. WRC report no. DETR 4415/3. WRC Medmenham, Marlow, UK, p. 38.
Carrington, E. G., Davis, R. D., and Pike, E. B., 1998b, Review of the scientific evidence relating to the controls on the agricultural use of sewage sludge. Part 1—evidence since 1989 relevant to controls on the agricultural use of sewage sludge. WRC report no. DETR 4415/3. WRC Medmenham, Marlow, UK, pp. 41–42.
Caul, E. O., 1994, Small round structured viruses: airborne transmission and hospital control. *Lancet* 343:1240–1242.
Chadwick, P. R., and McCann, R., 1994, Transmission of a small round structured virus by vomiting during a hospital outbreak of gastroenteritis. *J. Hosp. Infect.* 26: 251–259.
Cheesbrough, J. S., Barkess-Jones, L., and Brown, D. W., 1997, Possible prolonged environmental survival of small round structured viruses. *J. Hosp. Infect.* 35:325–326.
Christensen, B. F., Lees, D., Wood, K. H., Bjergskov, T., and Green, J., 1998, Human enteric viruses in oysters causing a large outbreak of human food borne infection in 1996/97. *J. Shellfish Res.* 17:1633–1635.
Cliver, D. O., 1984, Significance of water and the environment in the transmission of virus disease. *Monogr. Virol.* 15:30–42.
Cliver, D. O., and Kostenbader, K. D., Jr., 1979, Antiviral effectiveness of grape juice. *J. Food Prot.* 42:100–104.
Cliver, D. O., and Riemann, H. P., 1999, *Foodborne Diseases*, 2nd ed., Academic Press, New York.
Cliver, D. O., Kostenbader, K. D., Jr., and Vallenas, M. R., 1970, Stability of viruses in low moisture foods. *J. Milk Food Tech.* 33:484–491.
Craun, G. F., 1988, Surface water supplies and health. *J. Am. Water Works Assoc.* 80:40–52.
Craun, G. F., Nwachuku, N., Calderon, R. L., and Craun, M. F., 2002, Outbreaks in drinking-water systems, 1991–1998. *J. Environ. Health* 65:16–25.
Croci, L., De Medici, D., Scaslfaro, C., Fiore, A., and Toti, L., 2002, The survival of hepatitis A virus in fresh produce. *Int. J. Food Microbiol.* 73:29–34.
D'Angelo, L. H., Hierholzer, J. C., Keenlyside, R. A., Anderson, L. J., and Mortone, W. J., 1979, Pharyngoconjunctival fever caused by adenovirus type 4: report of a swimming pool related outbreak with recovery of virus from pool water. *J. Infect. Dis.* 140:42–47.
De Flora, S., De Renzi, G., and Badolati, G., 1975, Detection of animal viruses in coastal seawater and sediments. *Appl. Environ. Microbiol.* 30:472–475.

Deng, M. Y., and Cliver, D. O., 1995, Persistence of inoculated hepatitis A virus in mixed human and animal wastes. *Appl. Environ. Microbiol.* 61:87–91.

Denis, F. A., Dupwis, T., Denis, N. A., and Brisou, J. L., 1977, Survie dans l'eau de mer de 20 souches de virus a ADN et ARN. *J. Fr. Hydrol.* 8:25–36.

Dimmock, N. L., 1967, Differences between the thermal inactivation of picornaviruses at high and low temperatures. *Virology* 31:338–353.

Divizia, M., De Filippis, P., Di Napoli, A., Venuti, A., Peres, B., and Pana, A., 1989a, Isolation of wild type hepatitis A virus from the environment. *Water Res.* 23:1155–1160.

Divizia, M., Santi, A. L., and Pana, A., 1989b, Ultrafiltration: an efficient second step for hepatitis A virus and poliovirus concentration. *J. Virol. Methods* 23:55–62.

Donaldson, A. I., and Ferris, N. P., 1976, The survival of some airborne animal viruses in relation to relative humidity. *Vet. Microbiol.* 1:413–420.

Doultree, J. C., Druce, J. D., Birch, C. J., Bowden, D. S., and Marshall, J. A., 1999, Inactivation of feline calicivirus, a Norwalk virus surrogate. *J. Hosp. Infect.* 41:51–57.

Elazhary, M. A., and Derbyshire, J. B., 1979, Effect of temperature, relative humidity and medium on the aerosol stability of infectious bovine rhinotracheitis virus. *Can. J. Comp. Med.* 43:158–167.

Enright, J. R., 1954, The epidemiology of paralytic poliomyelitis in Hawaii. *Hawaii Med. J.* 13:350–354.

Environmental Protection Agency, 1984, *USEPA Manual Methods for Virology*, US Environmental Protection Agency, Research and Development, 600/4-84-013. USEPA, Cincinnati, OH.

Fannin, K. F., Vana, S. T., and Jakubowski, W., 1985, Effect of activated sludge wastewater treatment plant on ambient air densities of aerosols containing bacteria and viruses. *Appl. Environ. Microbiol.* 49:1191–1196.

Farrah, S. R., Gerba, C. P., Wallis, C., and Melnick, J. L., 1976, Concentration of viruses from large volumes of tap water using pleated membrane filters. *Appl. Environ. Microbiol.* 31:221–226.

Farrah, S. R., Goyal, S. M., Gerba, C. P., Conklin, R. H., and Smith, E. M., 1978, Comparison between adsortion of poliovirus and rotavirus by aluminum hydroxide and activated sludge flocs. *Appl. Environ. Microbiol.* 35:360–363.

Farthing, M. J. G., 1989, *Viruses and the Gut*, Smith Kline & French, Garden City, UK.

Fattal, B., and Shuval, H. I., 1989, Epidemiological research on the relationship between microbial quality of coastal seawater and rotavirus induced gastroenteritis among bathers on the Mediterranean Israeli beaches. Research project no. ICP-CEH-039-ISR-16(D). WHO, Athens, pp. 1–25.

Fattal, B., Vasl, R. J., Katzenelson, E., and Shuval, H. I., 1983, Survival of bacterial indicator organisms and enteric viruses in the Mediterranean coastal waters of Tel-Aviv. *Water Res.* 17:397–402.

Fattal, B., Margalith, M., Shuval, H. I., Wax, Y., and Morag, A., 1987, Viral antibodies in agricultural populations exposed to aerosols from wastewater irrigation during a viral disease outbreak. *Am. J. Epidemiol.* 125:899–906.

Filippi, J. A., and Banwart, G. J., 1974, Effect of the fat content of ground beef on the heat inactivation of poliovirus. *J. Food Sci.* 39:865–868.

Finance, C., Brigaud, M., Lucena, F., Aymard, M., Bosch, A., and Schwartzbrod, L., 1982, Viral pollution of seawater at Barcelona. *Zbl. Bakteriol. Mikrobiol. Hyg.* B176:530–536.

Foy, H. M., Cooney, M. K., and Halten, J. B., 1968, Adenovirus type 3 epidemic associated with intermittent chlorination of a swimming pool. *Arch. Environ. Health* 17:795–802.
Fujioka, R. S., Loh, P. C., and Lau, L. S., 1980, Survival of human enteroviruses in the Hawaiian Ocean environment: evidence for virus inactivating microorganisms. *Appl. Environ. Microbiol.* 39:1105–1110.
Gajardo, R., Díez, J. M., Jofre, J., and Bosch, A., 1991, Adsorption-elution with negatively and positively-charged glass powder for the concentration of hepatitis A virus from water. *J. Virol. Methods* 31:345–352.
Gajardo, R., Bouchriti, N., Pintó, R. M., and Bosch, A., 1995, Genotyping of rotaviruses isolated from sewage. *Appl. Environ. Microbiol.* 61:3460–3462.
Gassilloud, B., Schwartzbrod, L., and Gantzer, B., 2003, Presence of viral genomes in mineral water: a sufficient condition to assume infectious risk? *Appl. Environ. Microbiol.* 69:3965–3969.
Gerba, C. P., and Schaiberger, G. E., 1975, Effect of particulates on virus survival in seawater, *J. Water Pollut. Cont. Fed.* 47:93–103.
Gerba, C. P., and Goyal, S. M., eds., 1982, *Methods in Environmental Virology*. Marcel Decker, New York.
Gerba, C. P., Smith, E. M., and Melnick, J. L., 1977, Development of a quantitative method for detecting enteroviruses in estuarine sediments. *Appl. Environ. Microbiol.* 34:158–163.
Gerba, C. P., Goyal, S. M., Cech, I., and Bogdan, G. F., 1981, Quantitative assessment of the adsorptive behavior of viruses to soils. *Environ. Sci. Technol.* 15:940–944.
Gilgen, M., Germann, D., Lüthy, J., and Hübner, P., 1997, Three-step isolation method for sensitive detection of enterovirus, rotavirus, hepatitis A virus, and small round structured viruses in water samples. *Int. J. Food Microbiol.* 37:189–199.
Gill, O. N., Cubitt, W. D., McWiggan, D. A., Watney, B. M., and Bartlett, C. L. R., 1983, Epidemic of gastroenteritis caused by oysters contaminated with small round structured viruses. *Br. Med. J.* 287:1532–1534.
Gironés, R., Jofre, J., and Bosch, A., 1989, Natural inactivation of enteric viruses in seawater. *J. Environ. Qual.* 18:34–39.
Gironés, R., Jofre, J., and Bosch, A., 1990, Isolation of marine bacteria with antiviral properties. *Can. J. Microbiol.* 35:1015–1021.
Goh, K. T., Chan, L., Ding, J. L., and Oon C. J., 1984, An epidemic of cockles associated hepatitis A in Singapore. *WHO Bull.* 62:893–897.
Goyal, S. M., and Gerba, C. P., 1979, Comparative adsorption of human enteroviruses, simian rotavirus, and selected bacteriophages to soils. *Appl. Environ. Microbiol.* 38:241–247.
Goyal, S. M., and Gerba, C. P., 1983, Viradel method for detection of rotavirus from seawater. *J. Virol. Methods* 7:279–285.
Goyal, S. M., Gerba, C. P., and Melnick, J. L., 1979, Human enteroviruses in oysters and their overlaying waters. *Appl. Environ. Microbiol.* 37:572–581.
Grabow, W. O. K., Puttergill, D. L., and Bosch, A., 1993, Detection of adenovirus types 40 and 41 by means of the PLC/PRF/5 human liver cell line. *Water Sci. Tech.* 27:321–327.
Grausgruber, W., 1963, Investigations of the inactivation of infectious swine paralysis virus in scalded sausages. *Wiener Tierärztliche Monatsschrift* 50:678–685.
Gray, M., De León, R., Tepper, B. E., and Sobsey, M. D., 1993, Survival of hepatitis A virus (HAV), poliovirus and F-specific coliphages in disposable and landfill leachates. *Water Sci. Technol.* 27:429–432.

Green, J., Wright, P. A., Gallimore, C. I., Mitchell, O., Morgan-Capner, P., and Brown, D. W. G., 1998, The role of environmental contamination with small round structured viruses in a hospital outbreak investigated by reverse-transcriptase polymerase chain reaction assay. *J. Hosp. Infect.* 39:39–45.

Guix, S., Caballero, S., Villena, C., Bartolomé, R., Latorre, C., Rabella, N., Simó, M., Bosch, A., and Pintó, R. M., 2002, Molecular epidemiology of astrovirus infection in Barcelona, Spain. *J. Clin. Microbiol.* 40:133–139.

Gunderson, K., Brandberg, A., Magnusson, S., and Lycke, E., 1968, Characterization of a marine bacterium associated with virus inactivating capacity. *Acta Path. Microbiol. Scand.* 71:281–286.

Halliday, M. L., Kang, L. -Y., Zhou, T. -Z., Hu, M. -D., Pan, Q. -C., Fu, T. -Y., Huang, Y. S., and Hu, S. L., 1991, An epidemic of hepatitis A attributable to the ingestion of raw clams in Shanghai, China. *J. Infect. Dis.* 164:852–859.

Halvorsrud, J., and Orstavick, I., 1980. An epidemic of rotavirus-associated gastroenteritis in a nursing home for the elderly. *Scand. J. Infect. Dis.* 12:161–164.

Harper, G. J., 1961, Airborne microorganisms: survival tests with four viruses. *J. Hyg. (Cambridge)* 59:479–486.

Hejkal, T. W., Smith, E. M., and Gerba, C. P., 1984, Seasonal occurrence of rotavirus in sewage. *Appl. Environ. Microbiol.* 47:588–590.

Hemmes, J. H., Winkler, K. C., and Kool, S. M., 1960, Virus survival as a seasonal factor in influenza and poliomyelitis. *Nature* 188:430–438.

Hendley, J. O., Wenzel, R. P., and Gwaltney Jr, J. M., 1973, Transmission of rhinovirus colds by self-inoculation. *N. Engl. J. Med.* 288:1361–1364.

Henshilwood, K., Green, J., and Lees, D. N., 1998, Monitoring the marine environment for small round structured viruses (SRSVS): a new approach to combating the transmission of these viruses by molluscan shellfish. *Water Sci. Tech.* 38:51–56.

Hopkins, R. S., Gaspard, G. B., Williams, F. P. Jr, Karlin, R. J., Cukor, G., and Blacklow, N. R., 1984, A community waterborne gastroenteritis outbreak: evidence for rotavirus as the agent. *Am. J. Public Health* 74:263–265.

Huang, F. F., Haqshenas, G., Shivaprasad, H. L., Guenette, D. K., Woolcock, P. R., Larsen, C. T., Pierson, F. W., Elvinger, F., Toth, T. E., and Meng, X. J., 2002, Heterogeneity and seroprevalence of a newly identified avian hepatitis E virus from chickens in the United States. *J. Clin. Microbiol.* 40:4197–4202.

Huber, M. S., Gerba, C. P., Abbaszedagan, M., Robinson, J. A., and Bradford, S. M., 1994, Study of persistence of enteric viruses in land filled disposable diapers. *Environ. Sci. Technol.* 28:1767–1772.

Hugues, B., Cini, A., Plissier, M., and Lefebre, J. R., 1980, Recherche des virus dans le milieu marin à partir d'échantillons de volumes différents. *Eau. Ouetec.* 13:199–203.

Humphrey, T. J., Cruickshank, J. G., and Cubitt, W. D., 1984, An outbreak of calicivirus associated gastroenteritis in an elderly persons home: a possible zoonosis? *J. Hyg. (London)* 93:293–299.

Hung, T., Wang, C. H., Fang, Z., Chou, Z., Chang, X., Liong, X., Chen, G., Yao, H., Chaon T., Ye, W., Den, S., and Chang, W., 1984, Waterborne outbreak of rotavirus diarrhoea in adults in China caused by a novel rotavirus. *Lancet* 2:1139–1142.

Hurst, C. J., 1987, Influence of aerobic microorganisms upon virus survival in soil. *Can. J. Microbiol.* 34:696–699.

Hurst, C. J., Farrah, S. R., Gerba, C. P., and Melnick, J. L., 1978, Development of quantitative methods for the detection of enteroviruses in sewage sludges during activation and following land disposal. *Appl. Environ. Microbiol.* 36:81–89.

Hurst, C. J., Gerba, C. P., and Cech, I., 1980, Effects of environmental variables and soil characteristics on virus survival in soil. *Appl. Environ. Microbiol.* 40:1067–1079.

Ijaz, M. K., Brunner, A. H., Sattar, S. A., Nair, R. C., and Johnson-Lussenburg, C. M., 1985a, Survival characteristics of airborne human coronavirus 229E. *J. Gen. Virol.* 66:2743–2748.

Ijaz, M. K., Sattar, S. A., Johnson-Lussenburg, C. M., and Springthorpe, V. S., 1985b, Comparison of the airborne survival of calf rotavirus and poliovirus type 1 (Sabin) aerosolized as a mixture, *Appl. Environ. Microbiol.* 49:289–293.

Jiang, X., Espul, C., Zhong, W. M., Cuello, H., and Matson, D. O., 1999, Characterization of a novel human calicivirus that may be a naturally occurring recombinant. *Arch. Virol.* 144:2377–2387.

Jofre, J., Blasi, M., Bosch, A., and Lucena, F., 1989, Occurrence of bacteriophages infecting *Bacteroides fragilis* and other viruses in polluted marine sediments. *Water Sci. Tech.* 21:15–19.

Kaplan, J. E., Godman, R. A., Schonberger, L. B., Lippy, E. C., and Gary, W., 1982, Gastroenteritis due to Norwalk virus: an outbreak associated with a municipal water system. *J. Infect. Dis.* 146:190–197.

Kapuscinski, R. B., and Mitchell, R., 1980, Processes controlling virus inactivation in coastal waters. *Water Res.* 14:363–371.

Katzenelson, E., 1978, Survival of viruses, in: *Indicators of Viruses in Water and Food* (G. Berg, eds.), Ann Arbor Sci., Ann Arbor, MI, pp. 39–50.

Katzenelson, E., Fattal, B., and Hostovesky, T., 1976, Organic flocculation: an efficient second-step concentration method for the detection of viruses in tap water. *Appl. Environ. Microbiol.* 32:638–639.

Keswick, B. H., Gerba, C. P., DuPont, H. L., and Rose J. B., 1984, Detection of enteroviruses in treated drinking water. *Appl. Environ. Microbiol.* 47:1290–1294.

Keswick, B. H., Pickering, L. K., DuPont, H. L., and Woodward, W. E., 1983, Survival and detection of rotaviruses on environmental surfaces in day care centers. *Appl. Environ. Microbiol.* 46:813–816.

Khuroo, M. S., 1980, Study of an epidemic of non A, non B hepatitis—possibility of another human hepatitis virus distinct from post-transfusion non A, non B type. *Am. J. Med.* 68:818–824.

Kingsley, D. H., Meade, G. K., and Richards, G. P., 2002, Detection of both hepatitis A virus and Norwalk-like virus in imported clams associated with food-borne illness. *Appl. Environ. Microbiol.* 68:3914–3918.

Kohn, A., Gitelman, J., and Inbar, M., 1980, Unsaturated free fatty acids inactivate animal enveloped viruses. *Arch. Virol.* 66:301–307.

Konowalchuk, J., and Speirs, J. I., 1975, Survival of enteric viruses on fresh vegetables. *J. Milk Food Technol.* 38:469–472.

Konowalchuk, J., and Speirs, J. I., 1976, Virus inactivation by grapes and wines. *Appl. Environ. Microbiol.* 32:757–763.

Konowalchuk, J., and Speirs, J. I., 1978, Antiviral effect of commercial juices and beverages. *Appl. Environ. Microbiol.* 35:1219–1220.

Koopman, J. S., Eckert, E. A., Greenberg, H. B., Strohm, B. C., Isaacson, R. E., and Monto, A. S., 1982, Norwalk virus enteric illness acquired by swimming exposure. *Am. J. Epidemiol.* 115:173–177.

Kopecka, H., Dubrou, S., Prévot, J., Maréchal, J., and López-Pila, J. M., 1993, Detection of naturally occurring enteroviruses in waters by reverse transcription, polymerase chain reaction and hybridization. *Appl. Environ. Microbiol.* 59:1213–1219.

Kott, H., and Fishelson, L., 1974, Survival of enteroviruses on vegetables irrigated with chlorinated oxidation pond effluents. *Isr. J. Technol.* 12:290–297.

Ksiazek, T. G., Erdman, D., Goldsmith, C. S., Zaki, S. R., Peret, T., Emery, S., Tong, S., Urbani, C., Comer, J. A., Lim, W., Rollin, P. E., Dowell, S. F., Ling, A.-E., Humphrey, C. D., Shieh, W.-J., Guarner, J., Paddock, C. D., Rota, P., Fields, B., DeRisi, J., Yang, J.-Y., Cox, N., Hughes, J. M., LeDuc, J. W., Bellini, W. J., Anderson, L. J., and the SARS Working Group, 2003, A novel coronavirus associated with severe acute respiratory syndrome. *N. Engl. J. Med.* 348: 1953–1966.

Kukkula, M., Maunula, L., Silvennoinen, E., and von Bosndorff, C. H., 1999. Outbreak of viral gastroenteritis due to drinking water contaminated by Norwalk-like viruses. *J. Infect. Dis.* 180:1771–1776.

La Belle, R. L., Gerba, C. P., Goyal, S. M., Melnick, J. L., Lech, I., and Bogdan, G. F., 1980, Relationships between environmental factors, bacterial indicators and the occurrence of enteric viruses in estuarine sediments. *Appl. Environ. Microbiol.* 39:588–596.

Landry, E. F., Vaughn, J. M., Vicale, T. J., and Mann, R., 1983, Accumulation of sediment-associated viruses in shellfish. *Appl. Environ. Microbiol.* 45:238–247.

Larkin, E. P., Tierney, J. T., and Sullivan, R., 1976, Persistence of virus on sewage-irrigated vegetables. *J. Environ. Eng. Div.* 1:29–35.

Le Bris, J. M., Billaudel, S., Bertrand, P., Loukou, G., and et Courtieu, A. L., 1983, Recherche des virus et des salmonelles dans la Loire par une méthode d'adsorption-élution sur filtres en microfibre de verre. *Tech. Sci. Munic. L'Eau.* 6:303–306.

Lefler, E., and Kott, Y., 1974, Enteric virus behavior in sand dunes. *Isr. J. Technol.* 12:298–304.

Le Guyader, F., Dubois, E., Menard, D., and Pommepuy, M., 1994, Detection of hepatitis A virus, rotavirus, and enterovirus in naturally contaminated shellfish and sediment by reverse transcription-seminested PCR. *Appl. Environ. Microbiol.* 60:3665–3671.

Le Guyader, F., Neill, F. H., Estes, M. K., Monroe, S. S., Ando, T., and Atmar, R. L., 1996, Detection and analysis of a SRSV strain in oysters implicated in an outbreak. *Appl. Environ. Microbiol.* 62:4268–4272.

Lemon, S. M., 1985, Type A viral hepatitis—new developments in an old disease. *N. Engl. J. Med.* 313:1059–1067.

Lenaway, D. D., Brockmann, R., Dolan, G. J., and Cruz-Uribe F., 1989, An outbreak of an enterovirus-like illness at a community wading pool: implications for public health inspection programs. *Am. J. Public Health* 79:889–890.

Lewis, G. D., and Metcalf, T. G., 1988, Polyethylene glycol precipitation for recovery of pathogenic viruses, including hepatits A virus and human rotavirus, from oysters, water and sediment samples. *Appl. Environ. Microbiol.* 54:1983–1988.

Linco, S. J., and Grohmann, G. S., 1980, The Darwin outbreak of oyster associated viral gastroenteritis. *Med. J. Aust.* 1:211–213.

Lippy, E. C., and Waltrip, S. C., 1984, Waterborne disease outbreaks 1946–1980: a thirty-five-year perspective. *J. Am. Water Works Assoc.* 76:60–67.

Lo, S., Gilbert, J., and Hetrick, F., 1976, Stability of human enteroviruses in estuarine and marine waters. *Appl. Environ. Microbiol.* 32:245–248.

Lopman, B. A., Reacher, M. H., Van Duijnhoven, Y., Hanon, F. X., Brown, D., and Koopmans, M., 2003, Viral gastroenteritis outbreaks in Europe, 1995–2000. *Emerg. Infect. Dis.* 9:90–96.

Lucena, F., Bosch, A., Jofre, J., and Schwartzbrod, L., 1985, Identification of viruses isolated from sewage, river water and coastal seawater in Barcelona. *Water Res.* 19:1237–1239.

Mahl, M. C., and Sadler, C., 1975, Virus survival on inanimate surfaces. *Can. J. Microbiol.* 21:819–823.

Matossian, A. M., and Garabedian, G. A., 1967, Virucidal action of seawater. *Am. J. Epidemiol.* 85:1–8.

Mbithi, J. N., Springthorpe, V. S., and Sattar, S. A., 1991, Effect of relative humidity and air temperature on survival of hepatitis A virus on environmental surfaces. *Appl. Environ. Microbiol.* 59:3463–3469.

Mbithi, J. N., Springthorpe, V. S., Boulet, J. R., and Sattar, S. A., 1992, Survival of hepatitis A virus on human hands and its transfer on contact with animate and inanimate surfaces. *J. Clin. Microbiol.* 30:757–763.

McGeady, M. L., Siak, J. S., and Crowell, R. L., 1979, Survival of coxsackie virus B3 under diverse environmental conditions. *Appl. Environ. Microbiol.* 37:972–977.

McNulty, M. S., 1978, Rotaviruses. *J. Gen. Virol.* 40:1–18.

Mead, P. S., Slutsker, L., and Dietz, V., 1999, Food-related illness and death in the United States. *Emerg. Inf. Dis.* 5:607–625.

Mehnert, D. U., Stewien, K. E., Hársi, C. M., Queiroz, A. P. S., Candeias, J. M. G., and Candeias, J. A. N., 1997, Detection of rotavirus in sewage and creek water: efficiency of the concentration method. *Mem. Inst. Oswaldo Cruz* 92:97–100.

Melnick, J. L., 1957, A water-borne urban epidemics of hepatitis, in: *Hepatitis Frontiers* (G. A. LoGrippo, F. W. Hartman, J. G. Mateer, and J. Barron, eds.), Little Brown, Boston, pp. 211–225.

Melnick, J. L., and Gerba, C. P., 1980, The ecology of enteroviruses in natural waters. *Crit. Rev. Environ. Control.* 10:65–93.

Meng, X. J., Purcell, R. H., Halbur, P. G., Lehman, J. R., Webb, D. M., Tsareva, T. S., Haynes, J. S., Thacker, B. J., and Emerson S. U., 1997, A novel virus in swine is closely related to the human hepatitis E virus. *Proc. Natl. Acad. Sci. U. S. A.* 94:9860–9865.

Miller, W. S., and Artenstein, M. S., 1967, Aerosol stability of three acute respiratory disease viruses. *Proc. Soc. Exp. Biol. Med.* 125:222–227.

MMWR (Morbidity and Mortality Weekly Report), 2002, Surveillance for waterborne-disease outbreaks—United States, 1999–2000. *MMWR* 51:1–52.

MMWR (Morbidity and Mortality Weekly Report), 2003, Hepatitis A outbreak associated with green onions at a restaurant—Monaca, Pennsylvania, 2003. *MMWR* 52:1155–1157.

Moe, K., and Harper, G. J., 1983, The effect of relative humidity and temperature on the survival of bovine rotavirus in aerosol. *Arch. Virol.* 76:211–216.

Moe, K., and Shirley, J. A., 1982, The effects of relative humidity and temperature on the survival of human rotavirus in feces. *Arch. Virol.* 72:179–186.

Moore, M., 1982, Enteroviral disease in the United States, 1970–1979. *J. Infect. Dis.* 146:103–108.

Moore, A. C., Herwaldt, B. L., Craun, G. F., Calderon, R. L., Highsmith, A. K., and Juranek, D. D., 1994, Waterborne disease in the United States, 1991 and 1992. *J. Am. Water Works Assoc.* 86:87–99.

Morse, D. L., Guzewich, J. J., Hanrahan, J. P., Stricof, R., Shayegani, M., Deibel, R., Grabau, J. C., Nowak, N. A., Herrmann, J. E., Cukor, G., and Blacklow, N. R., 1986, Widespread outbreaks of clam- and oyster-associated gastroenteritis: role of Norwalk virus. *New Engl. J. Med.* 314:678–681.

Mosley, J. W., 1967, Transmission of viruses by drinking water, in: *Transmission of Viruses by the Water Route* (G. Berg, eds.), John Wiley & Sons, New York, pp. 5–23.

Mounts, A. W., Ando, T. A., Koopmans, M., Bresee, J. S., Inouye, S., Noel, J., and Glass, R. I., 2000, Cold weather seasonality of gastroenteritis associated with Norwalk-like viruses. *J. Infect. Dis.* 181:284–287.

Murphy, A. M., Grohmann, G. S., Christopher, R. J., Lopez, W. A., Davey, G. R., and Millsom, R. H., 1979, An Australia-wide outbreak of gastroenteritis from oysters caused by Norwalk virus. *Med. J. Aust.* 2:329–333.

Murphy, A. M., Grohmann, G. S., and Sexton, F. H., 1983, Infectious gastroenteritis in Norfolk Island and recovery of viruses from drinking water. *J. Hyg.* 91: 139–146.

Muscillo, M., La Rosa, G., Carducci, A., Cantiani, L., and Marianelli, C., 1997, Molecular analysis of poliovirus 3 isolated from an aerosol generated by a waste water treatment plant. *Water Res.* 12:3125–3131.

Nuanualsuwan, S., and Cliver, D. O., 2002, Pretreatment to avoid positive RT-PCR results with inactivated viruses. *J. Virol. Methods* 104:217–225.

O'Brien, R. T., and Newman, J. S., 1977, Inactivation of poliovirus and coxsackie viruses in surface water. *Appl. Environ. Microbiol.* 33:334–340.

Oh, D.-Y., Gaedicke, G., and Schreir, J. M., 2003, Viral agents of acute gastroenteritis in German children: prevalence and molecular diversity. *J. Med. Virol.* 71:82–93.

O'Mahony, M., Gooch, C. D., Smyth, D. A., Thrussell, A. J., Bartlett, C. L. R., and Noah, N. D., 1983, Epidemic hepatitis A from cockles. *Lancet* I:518–520.

Parashar, U. D., Holman, R. C., Clarke, M. J., Bresee, J. S., and Glass R. I., 1998, Hospitalizations associated with rotavirus diarrhea in the United States, 1993 through 1995: surveillance based on the new ICD-9-CM rotavirus-specific diagnostic code. *J. Infect. Dis.* 177:13–17.

Payment, P., Berte, A., Prevost, M., Ménard, B., and Barbeau, B., 2000, Occurrence of pathogenic microorganisms in the Saint-Lawrence river (Canada) and comparison of health risks for populations using it as their source of drinking water. *Can. J. Microbiol.* 46:565–576.

Pfirrmann, A., and Bossche, G. V., 1994, Occurrence and isolation of airborne human enteroviruses from waste disposal and utilization plants. *Zbl. Hyg.* 196:38–51.

Pintó, R. M., Diez, J. M., and Bosch, A., 1994, Use of the colonic carcinoma cell line CaCo-2 for in vivo amplification and detection of enteric viruses. *J. Med. Virol.* 44:310–315.

Pintó, R. M., Gajardo, R., Abad, F. X., and Bosch, A., 1995, Detection of fastidious infectious enteric viruses in water. *Environ. Sci. Tech.* 29:2636–2638.

Pintó, R. M., Villena, C., Le Guyader, F., Guix, S., Caballero, S., Pommepuy, M., and Bosch, A., 2001, Astrovirus detection in wastewater samples. *Water Sci. Tech.* 12:73–76.

Ramalingaswami, V., and Purcell, R. H., 1988, Waterborne non-A, non-B hepatitis. *Lancet.* 1:571–573.

Rao, V. C., and Melnick, J. L., 1986, Environmental virology, in: *Aspects of Microbiology 13* (J. A. Cole, C. J. Knowles, and D. Schlessinger, eds.), American Society for Microbiology, Washington, DC.

Rao, V. C., Seidel, K. N., Goyal, S. M., Metcalf, T. C., and Melnick, J. L., 1984, Isolation of enteroviruses from water, suspended solids and sediments from Galveston bay; survival of poliovirus and rotavirus adsorbed to sediments. *Appl. Environ. Microbiol.* 48:404–409.

Rao, V. C., Metcalf, T. G., and Melnick, J. L., 1986, Development of a method for concentration of rotavirus and its application to recovery of rotaviruses from estuarine waters. *Appl. Environ. Microbiol.* 52:484–488.

Raphael, R. A., Sattar, S. A., and Springthorpe, V. S., 1985, Long term survival of human rotavirus in raw and treated river water. *Can. J. Microbiol.* 31:124–128.

Reagan, K. J., McGeady, M. L., and Crowell, R. L., Persistence of human rhinovirus infectivity under diverse environmental conditions. *Appl. Environ. Microbiol.* 41:618–627.

Redman, J. A., Grant, S. B., Olson, T. M., Hardy, M. E., and Estes, M. K., 1997, Filtration of recombinant Norwalk Virus particles and bacteriophage MS2 in quartz sand: importance of electrostatic interactions. *Environ. Sci. Tech.* 31:3378–3383.

Reid, T. M. S., and Robinson, H. G., 1987, Frozen raspberries and hepatitis A. *Epidemiol. Infect.* 98:109–112.

Reyes, G. R., 1993, Hepatitis E virus (HEV): molecular biology and emerging epidemiology. *Prog. Liver Dis.* 11:203–213.

Reynolds, K. A., Gerba, C. P., and Pepper, I. L., 1996, Detection of infectious enteroviruses by an integrated cell culture-PCR procedure. *Appl. Environ. Microbiol.* 62:1424–1427.

Reynolds, K. A., Gerba, C. P., Abbaszadegan, M., and Pepper, I. L., 2001, ICC/PCR detection of enteroviruses and hepatitis A virus in environmental samples. *Can. J. Microbiol.* 47:153–157.

Richards, G. P., 1985, Outbreaks of shellfish-associated enteric illness in the United-States: requisite for development of viral guidelines. *J. Food Prot.* 48:815–823.

Rocchi, G., Vella, S., Resta, S., Cochi, S., Donelli, G., Tangucci, F., Manichella, D., Varveri, A., and Inglese R., 1981, Outbreak of rotavirus gastroenteritis among premature infants. *Br. Med. J.* 283:886.

Ryder, R. W., McGowan, J. E., Hatch, M. H., and Palmer, E. L., 1977, Reovirus-like agent as a cause of nosocomial diarrhoea in infants. *J. Pediatr.* 90:698–702.

Rzutka, A., and Cook, N., 2004, Survival of human enteric viruses in the environment and food. *FEMS Microbiol. Rev.* (in press).

Salo, R. J., and Cliver, D. O., 1976, Effect of acid pH, salt and temperature on the infectivity and physical integrity of enteroviruses. *Arch. Virol.* 52:269–282.

Sánchez, G., Bosch, A., and Pintó, R. M., 2003, Genome variability and capsid structural constraints of hepatitis A virus. *J. Virol.* 77:452–459.

Santamaría, J., and Toranzos, G. A., 2003, Enteric pathogens and soil: a short review. *Int. Microbiol.* 6:5–9.

Sarrette, B. A., Danglot, C. D., and Vilagines, R., 1977, A new and simple method for recuperation of enterovirus from water. *Water Res.* 11:355–358.

Sattar, S. A., Ijaz, M. K., Johnson-Lussenburg, C. M., and Springthorpe, V. S., 1984, Effect of relative humidity on the airborne survival of rotavirus SA11. *Appl. Environ. Microbiol.* 47:879–881.

Sattar, S. A., Lloyd-Evans, N., Springthorpe, V. S., and Nair, R. C., 1986, Institutional outbreaks of rotavirus diarrhoea: potential role of fomites and environmental surfaces as vehicles for virus transmission. *J. Hyg. Camb.* 96:277–289.

Sattar, S. A., Karim, Y. G., Springthorpe, V. S., and Johnson-Lussenburg, C. M., 1987, Survival of human rhinovirus type 14 dried onto nonporous inanimate surfaces: effect of relative humidity and suspending medium. *Can. J. Microbiol.* 33:802–806.

Sattar, S. A., Dimock, K. D., Ansari, S. A., and Springthorpe, V. S., 1988, Spread of acute hemorrhagic conjunctivitis due to enterovirus 70: effect of air temperature and relative humidity on virus survival on fomites. *J. Med. Virol.* 25:289–296.

Schaiberger, G. E., Edmond, T. D., and Gerba, C. P., 1982, Distribution of enteroviruses in sediments contiguous with a deep marine sewage outfall. *Water Res.* 16:1425–1428.

Schwartzbrod, L., and Lucena, F., 1978, Concentration des enterovirus dans les eaux par adsorption sur poudre de verre: proposition d'un apareillage simplifié. *Microbia.* 4:55–68.

Shields, P. A., and Farrah, S. R., 1986, Concentration of viruses in beef extract by flocculation with ammonium sulphate. *Appl. Environ. Microbiol.* 51:211–213.

Shirai, J., Shimizu, M., and Fukusho, A., 1985, Coronavirus-, calicivirus-, and astrovirus-like particles associated with acute porcine gastroenteritis. *Nippon Juigaku Zasshi* 47:1023–1026.

Shuval, H. I., Thompson, A., Fattal, B., Cymbalista, S., and Weiner, Y., 1971, Natural virus inactivation processes in seawater. *J. San. Eng. Div. Am. Soc. Civ. Eng.* 5:587–600.

Sobsey, M. D., and Jones, B. L., 1979, Concentration of poliovirus from tap water using positively charged microporous filters. *Appl. Environ. Microbiol.* 37:588–595.

Sobsey, M. D., and Glass, J. S., 1984, Influence of water quality on enteric virus concentration by microporous filter methods. *Appl. Environ. Microbiol.* 47:956–960.

Sobsey, M. D., and Hickey, A. R., 1985, Effect of humic and fulvic acid on poliovirus concentration from water by microporous filtration. *Appl. Environ. Microbiol.* 49:259–264.

Sobsey, M. D., Dean, C. H., Knuckles, M. E., and Wagner, R. A., 1980, Interactions and survival of enteric viruses in soil materials. *Appl. Environ. Microbiol.* 40:92–101.

Sobsey, M. D., Shields, P. A., Hauchman, F. S., Davis, A. L., Rullman, V. A., and Bosch, A., 1988, Survival and persistence of hepatitis A virus in environmental samples, in: *Viral Hepatitis and Lliver Disease* (A. Zuckerman, eds.), Alan R. Liss, New York, pp. 121–124.

Sobsey, M. D., Shields, P. A., Hauchman, F. H., Hazard, R. L., and Caton, III, L. W., 1989, Survival and transport of hepatitis A virus in soils, groundwater and wastewater. *Water Sci. Technol.* 10:97–106.

Steinman, J., 1981, Detection of rotavirus in sewage. *Appl. Environ. Microbiol.* 41:1043–1045.

Straub, T. M., Pepper, I. L., and Gerba, C. P., 1992, Persistence of viruses in desert soils amended with anaerobically digested sewage sludge. *Appl. Environ. Microbiol.* 58:636–641.

Straub, T. M., Pepper, I. L., and Gerba, C. P., 1993, Virus survival in sewage sludge amended desert soil. *Water Sci. Tech.* 27:421–424.

Strauss, M., 1994, Health implications of excreta and wastewater use—Hubei environmental sanitation study, 2nd workshop, Hubei, Wuhan.

Teltsch, B., Kedmi, S., Bonnet, L., Borenzstajn-Rotem, Y., and Katzenelson, E., 1980, Isolation and identification of pathogenic microorganisms at wastewater-irrigated fields: ratios in air and wastewater. *Appl. Environ. Microbiol.* 39:1183–1190.

Thurston-Enriquez, J. A., Haas, C. N., Jacangelo, J., and Gerba, C. P., 2003a, Chlorine inactivation of adenovirus type 40 and feline calicivirus. *Appl. Environ. Microbiol.* 69:3979–3985.

Thurston-Enriquez, J. A., Haas, C. N., Jacangelo, J., Riley, K., and Gerba, C. P., 2003b, Inactivation of feline calicivirus and adenovirus type 40 by UV radiation. *Appl. Environ. Microbiol.* 69:577–582.

Tierney, J. T., Sullivan, R., and Larkin, E. P., 1977, Persistence of poliovirus 1 in soil and on vegetables grown in soil previously flooded with inoculated sewage sludge or effluent. *Appl. Environ. Microbiol.* 33:109–113

Toranzo, A. E., Barja, J. L., and Hetrick, F. M., 1983, Mechanism of poliovirus inactivation by cell-free filtrates of marine bacteria. *Can. J. Microbiol.* 29:1481–1486.

Trouwborst, T., Kuyper, S., de Jong, J. C., and Plantinga, A. D., 1974, Inactivation of some bacterial and animal viruses by exposure to liquid-air interfaces. *J. Gen. Virol.* 24:155–165.

Tsai, Y.-L., Tran, B., and Palmer, C. J., 1995, Analysis of viral RNA persistence in seawater by reverse transcriptase-PCR. *Appl. Environ. Microbiol.* 61:363–366.

Tsang, T., 2003, Environmental issues. WHO Global Conference on Severe Acute Respiratory Syndrome (SARS), Kuala Lumpur, Malaysia, 17–18 June 2003.

Unicomb, L. E., Podder, G., Gentsch, J. R., Woods P. A., Hasan, K. Z., Faruque, A. S., Albert, M. J., and Glass, R. I., 1999, Evidence of high-frequency genomic reassortment of group A rotavirus strains in Bangladesh: emergence of type G9 in 1995. *J. Clin. Microbiol.* 37:1885–1891.

van der Poel, W. H. M., Verschoor, F., van der Heide, R., Herrera, M. I., Vivo, A., Kooreman M., and de Roda Husman, A. M., 2001, Hepatitis E virus sequences in swine related to sequences in humans, the Netherlands. *Emerg. Infect. Dis.* 6:970–976.

Vaughn, J. M., Landry, E. F., Thomas, M. Z., Vicale, T. J., and Penello, W. F., 1979, Survey of human enterovirus occurrence in fresh and marine surface waters on Long-Island. *Appl. Environ. Microbiol.* 38:290–296.

Vilaginès, P., Sarrette, B., Champsaur, H., Hugues, B., Dubrou, S., Joret, J.-C., Laveran, H., Lesne, J., Paquin, J. L., Delattre, J. M., Oger, C., Alame, J., Grateloup, I., Perrollett, H., Serceau, R., Sinègre, F., and Vilaginès, R., 1997, Round robin investigation of glass wool method for poliovirus recovery from drinking water and sea water. *Water Sci. Tech.* 35:445–449.

Villena, C., Gabrielli, R., Pintó, R. M., Guix, S., Donia, D., Buonomo, E., Palombi, L., Cenko, F., Bino, S., Bosch, A., and Divizia, M., 2003, A large infantile gastroenteritis outbreak in Albania caused by multiple emerging rotavirus genotypes. *Epidemiol. Infect.* 131:1105–1110.

Wadell, G., Allard, A. Svensson, L., and Uhnoo, I., 1989, Enteric adenoviruses, in: *Viruses and the Gut* (M. J. G. Farthing, eds.), Smith, Kline and French, Welwyn Garden City, UK., pp. 70–78.

Wait, D. A., and Sobsey, M. D., 1983, Method for recovery of enteric viruses from estuarine sediments with caotropic agents. *Appl. Environ. Microbiol.* 46:379–385.

Walter, R., Macht, W., Durkop, J., Becht, R., Hornig, U., and Schulze, P., 1989, Virus levels in river waters. *Water Res.* 21:133–138.

Ward, B. K., and Irving, L. G., 1987, Virus survival on vegetables spray-irrigated with wastewater. *Water Res.* 21:57–63.

Ward, B. K., Chenoweth, C. M., and Irving, L. G., 1982, Recovery of viruses from vegetable surfaces. *Appl. Environ. Microbiol.* 44:1389–1394.

Ward, R. L., Knowlton, D. R., and Winston, P. E., 1986, Mechanism of inactivation of enteric viruses in fresh water. *Appl. Environ. Microbiol.* 52:450–459.

Weissman, J. B., Craun, G. F., Lawrence, D. N., Pollard, R. A., Saslaw, M. S., and Gangarosa, E. J., 1976, An epidemic of gastroenteritis traced to a contaminated public water supply. *Am. J. Epidemiol.* 103:391–398.

Yeager, J. G., and O'Brien, R. T., 1979, Enterovirus inactivation in soil. *Appl. Environ. Microbiol.* 38:694–701.

CHAPTER 7

Bacterial Indicators of Viruses

Samuel R. Farrah

1.0. INTRODUCTION

Viruses that are transmitted by the fecal-oral route can cause disease in humans after the consumption of contaminated food or water. The diseases caused by many of these viruses are usually mild and self-limiting such as gastroenteritis caused by noroviruses (Nishida et al., 2003; Parshionikar et al., 2003). However, some viruses, such as hepatitis A virus, may cause more serious diseases that require hospitalization or that may be fatal (Halliday et al., 1991; Tang et al., 1991; Niu et al., 1992; Hutin et al., 1999). Prevention of contamination of food and water by pathogenic microorganisms is one method by which their transmission can be reduced/eliminated. However, water and wastewater treatment procedures do not always eliminate infectious viruses. In addition, food handlers who have mild or asymptomatic infection may be responsible for contaminating food at several stages during production and processing.

Detection of infectious viruses directly in food or water before they are released to consumers would be another way to prevent the transmission of these viruses. This is usually not done, mainly because methods are not available that can detect infectious units in foods with a certain degree of accuracy. Some viruses do not grow easily in cell cultures and may require inoculation of humans or other animals for detecting them as infectious viruses. Molecular methods such as RT-PCR (reverse transcription–polymerase chain reaction) are increasingly being used to detect these viruses (Le Guyader et al., 1993; Dore et al., 2000). However, only a limited number of laboratories are equipped to routinely conduct such tests. Also, the detection of viral genome in foods may or may not indicate the presence of infectious viruses. In fact, viral genomes have been reported to survive much longer in mineral water than infectious viruses (Gassilloud et al., 2003). One way to avoid this problem is to use integrated cell culture-PCR to determine if the genomes detected using molecular methods are infectious (Blackmer et al., 2000). Because of the above problems associated with the detection of pathogenic viruses in food and water, attempts have been made to find a suitable indicator that, when present, would indicate that viral contamination of food or water has occurred.

2.0. DESIRABLE CHARACTERISTICS OF INDICATORS

Because enteric viruses are usually transmitted by the fecal-oral route, components of feces have been used to detect the presence of fecal pollution in food and water. It was hoped that simple tests for components of sewage or feces could be used to indicate the presence of pathogenic microorganisms including viruses. Both chemical and microbiological indicators have been developed. Chemicals that may indicate the presence of fecal pollution include coprostanol (Dutka et al., 1974). However, tests for chemicals are of limited value because they require specialized equipment and are limited in their sensitivity. Because microbiological tests can detect one or a few living microorganisms, they are generally more sensitive than the chemical tests. Also, tests for bacteria usually require minimal equipment and are not too complicated to perform. Therefore, bacteria have been used as indicators of viruses in most studies.

For use as viral indicators, bacteria should possess certain characteristics as discussed by Berg and Metcalf (1978). The most important requirements are that the presence of bacteria should correlate well with the presence of enteric viruses in a given environment; viruses should not be found when the indicator is absent or is present in low numbers; the viruses should be frequently found when the number of indicator bacteria is high; and the indicators should not be pathogenic themselves, thus simplifying the procedures for their culture and identification and reducing the hazards to laboratory workers from accidental contamination. Indicator bacteria should survive in a given environment for approximately as long as viruses because if they are inactivated more rapidly than viruses, they may not be detected when viruses are still present. Conversely, if they survive much longer than viruses, they may indicate a threat long after the viruses have been inactivated. In summary, if the survival of bacteria in an environment differs greatly from the survival of viruses, their usefulness as indicators is diminished.

The procedures for detecting the indicators should be relatively simple. The need for special equipment and complex procedures would reduce the number of laboratories that are capable of doing the analyses. The availability of well-equipped laboratories in many different locations will reduce the time between sample collection and analysis. Also, simple and inexpensive procedures that do not rely on special operator skills are easier to standardize, and the results of standardized procedures are easier to compare from laboratory to laboratory.

3.0. BACTERIA USED AS INDICATORS FOR VIRUSES

It was recognized early in the study of microbiology that certain bacteria, such as *Escherichia coli*, were present in the intestines and feces of humans and other animals (Escherich, 1885). Initially, tests designed to detect *E. coli* were based on its ability to ferment glucose. These tests were later modified to detect the fermentation of lactose. However, these tests were not specific for *E. coli* alone and detected a range of microorganisms. These microorganisms were collectively termed *coliform bacteria*. The latter group includes *Aerobacter aerogenes*, *Klebsiella pneumoniae*, *Enterobacter* spp., and *Serratia* spp. (Duncan and Razzell, 1972; Newton et al., 1977; Stiles and Ng, 1981), and all of these bacteria may not be associated with fecal pollution. Later, different media and incubation at higher temperatures were used to make the tests more specific for microorganisms associated with feces. The microorganisms detected in these tests were termed *fecal coliforms* or *thermotolerant coliforms* (Eijkman, 1904). Although *E. coli* is a significant component of the fecal coliform population, other thermotolerant bacteria, such as *Klebsiella pneumoniae*, may also be detected by the fecal coliform test (Bagley and Seidler, 1977; Hussong at al., 1981).

More recent modifications of the test used to detect *E. coli* and related microorganisms include the use of media that contain compounds that produce a colored or fluorescent compound when they are hydrolyzed by bacterial enzymes (Manafi et al., 1991). As efforts were being made to make these tests more specific for detecting *E. coli*, it was discovered that *E. coli* can also be found in pristine areas in tropical climates (Hazen and Toranzos, 1990). The ability of *E. coli* to exist and possibly grow in some tropical environments that are not fecally polluted should be considered when using it as an indicator organism in these environs (Hazen and Toranzos, 1990). Some of the developments in the use of *E. coli* and related bacteria as indicators of fecal pollution are given in Table 7.1.

Because the correlation between the detection of microorganisms using the above-mentioned tests and viruses has not been satisfactory in many cases, other microorganisms have been considered as indicators for viral pol-

Table 7.1 Chronology of Events Leading to the Use of *Escherichia coli* and Related Bacteria as Indicators of Fecal Pollution

Event	*Reference*
Identification of *E. coli* as an inhabitant of the intestinal tracts of warm-blooded animals	Escherich (1885)
The use of lactose fermentation at elevated temperatures (44.5–55.5°C) to test for fecal (thermotolerant) coliforms	Eijkman (1904)
The development and use of membrane filter (MF) procedures to detect coliforms and *E. coli*	Clark et al. (1951); Toranzos and McFeters (1997)
The finding of *E. coli* in pristine areas not influenced by known fecal contamination	Hazen and Toranzos (1990)
The use of chromogenic and fluorogenic compounds to detect coliforms and *E. coli*	Manafi et al. (1991)

Table 7.2 Bacteria That Have Been Considered for Use as Indicators of Fecal Pollution

Bacteria	*Comments*
Coliforms	A group of microorganisms that can ferment lactose; testing for this group may also detect bacteria that are not of fecal origin.
Fecal (thermotolerant) coliforms	This group of microorganisms is more specific for *Escherichia coli* than the coliform group; however, non–fecal coliforms may sometimes give positive results.
Fecal streptococci, enterococci	Some bacteria formerly classified as *Streptococcus* spp. have been reclassified as *Enterococcus* spp. (i.e., *Enterococcus faecalis*); some studies indicate that these bacteria may be better indicators for viral contamination of marine waters than *E. coli* and other related bacteria.
Bacteroides spp.	They are found in large numbers in the intestinal tracts of humans and other animals; because they are anaerobic bacteria, they do not survive long in aerobic environments. Both *E. coli* and *S. faecalis* were found to survive longer that *Bacteroides* spp. in water (Fiksdal et al., 1985)
Clostridium perfringens	This organism is found in human intestines and sewage and can be used to monitor water sources for fecal contamination; because the spores may survive for long periods, it may indicate historical pollution.

lution (Table 7.2). These microorganisms include fecal streptococci, anaerobic bacteria present in human intestines (*Bacteroides*), and spore-forming bacteria of the genus *Clostridium.* Besides *E. coli* and related bacteria, the enterococci are probably the most commonly used indicator bacteria. *Bacteroides* spp. are present in high numbers in the human intestine, and these bacteria have been detected in sewage and natural waters (Allsop and Stickler, 1984). However, because their numbers decline more rapidly in water that those of *E. coli* or *S. faecalis*, they have not found use as an indicator. Previously, several enteric streptococci were classified in the genus *Streptococcus.* They have now been reclassified as members of the genus *Enterococcus.* The two species most frequently found in humans are *Enterococcus faecalis* and *E. faecium.* The enterococci are distinguished by their ability to grow in 6.5% NaCl and at high temperature (45°C). Enterococci have been found to be more reliable than coliforms in determining health risks in marine waters (Cabelli et al., 1982).

Clostridium perfringens has also been used as an indicator of pollution (Fujioka and Shizumura, 1985). One problem is that this organism produces spores that may survive for long periods in natural environments. Therefore, it may indicate the presence of pollution that occurred long before the sampling.

4.0. METHODS FOR DETECTING INDICATOR BACTERIA

Total and fecal coliforms can be enumerated in standardized tests that use a series of tubes with specific bacteriological media inoculated with serial 10-fold dilutions of the initial sample (Toranzos and McFeters, 1997; Clesceri et al., 1998). The advantage of these tests is that they can be used with both liquid and solid samples. Their main disadvantage is that the most probable number (MPN) of microorganisms calculated using these tests has a relatively high degree of uncertainty. More accurate determination of the actual number of bacteria in liquid samples can be obtained using membrane filtration (MF) procedures. Larger volumes of water can be tested with this test than the MPN or tube tests, and the results obtained reflect more accurately the numbers of bacteria present in a sample (Clark et al., 1951).

Fluorogenic and chromogenic substrates have been incorporated into tests for coliform bacteria and *E. coli*, making the tests easier to perform (Manafi et al., 1991). The enzyme responsible for hydrolysis of lactose, ∃-D-galactosidase (∃-GAL), can also hydrolyze chromogenic substrates such as *o*-nitrophenyl-∃-D-galactopyranoside. The hydrolyzed product can easily be detected by change in color. A positive test is thought to indicate the presence of coliform bacteria. An enzyme that is found in a majority of *E. coli* isolates is ∃-D-glucuronidase (GUD). This enzyme can hydrolyze compounds such as 4-methylumbelliferyl-∃-D-glucuronide (MUG). The hydrolysis of MUG produces a product that fluoresces when irradiated with UV light at a wavelength of 365 nm. These materials are now included in media used for the detection of coliforms and *E. coli* in water (Feng and Hartman, 1982; Brenner et al., 1993). It should be realized, however, that not all *E. coli* strains are able to hydrolyze MUG (Chang et al., 1989).

Indicator bacteria may be injured by a variety of physical and chemical means including exposure to sublethal levels of disinfectants, UV light, high or low temperatures, freezing, copper, and starvation (Speck et al., 1975; Hackney et al., 1979; McFeters et al., 1982; Singh and McFeters, 1986; Kang and Siragusa, 1999). The injured bacteria may not be able to grow on some media used for detecting indicator bacteria and therefore escape detection. However, many injured bacteria can be detected by using procedures designed to give them an opportunity for repair (McFeters et al., 1982).

5.0. CORRELATION BETWEEN INDICATOR BACTERIA AND PATHOGENS IN WATER AND FOOD

Indicator bacteria have been found to be useful in determining the possible presence of pathogens in many cases. Hood et al. (1983) studied the relationship between indicator bacteria (fecal coliforms and *E. coli*) and bacterial pathogens (*Salmonella* spp.) in shellfish. Although *Salmonella* spp. were detected in some samples, none was present when the fecal coliform level in oysters was less than the level recommended by the National Shellfish Sanitation Program (230 fecal coliforms per 100g). The authors found that low levels of fecal coliforms was a good indication for the absence of *Salmonella* spp. However, the reverse was not true; high levels of fecal coliforms did not always indicate the presence of *Salmonella* spp. This is understandable because the indicator would likely be present in all cases of fecal pollution, but the presence of the pathogen would be variable and related to the level of infection within a given population. Natvig et al. (2002) compared the survival of *Salmonella enterica* serovar typhimurium and *E. coli* in soil contaminated with manure. The number of *E. coli* in the soil contaminated with bovine manure was always higher than the level of *S. enterica* serovar typhimurium. The authors concluded that *E. coli* was useful as an indicator for *S. enterica* serovar typhimurium under these conditions.

E. coli and *Salmonella* spp. are similar in their physiological characteristics and are likely to originate from the same source. Therefore, a correlation between their numbers and survival in natural environments would be expected. The lack of correlation between pathogens that are normally found in estuarine waters (*Vibrio* spp.) and fecal indicators is not surprising. Thus, Koh et al. (1994) found no correlation between *Vibrio* spp. and several indicator organisms (*E. coli*, enterococci, and total and fecal coliforms) in water from Apalachicola Bay.

Overall, a high correlation between indicator bacteria and viruses in water and food (especially shellfish) has not been found, although the presence or absence of indicator bacteria has been related to the presence of viruses in some cases. Sobsey et al. (1980) studied the levels of enteric viruses in oysters taken from areas closed to shellfish harvesting and those approved for this purpose. Enteric viruses were found in 12% of the oysters samples taken from areas closed to shellfish harvesting but not in samples taken from approved areas. Also, viruses were detected in samples that contained greater than the recommended 230 fecal coliforms per 100g of oyster meat.

Kingsley et al. (2002) examined imported clams that had been implicated in an outbreak of Norwalk-like gastroenteritis. These authors were able to detect both hepatitis A and Norwalk-like virus genomes in samples of clams. In addition, the clams contained high levels of fecal coliforms (93,000/100g of clam meat). Because this level was approximately 300 times the recommended level, the finding of virus genomes and the implication of the clams in disease transmission is not surprising.

The level of indicator bacteria in water or food has not been found to be correlated with the number of viruses in several studies. Gerba et al. (1979) examined waters that were approved for recreational use and for shellfish harvesting. The number of indicator bacteria (total and fecal coliforms) and enteroviruses in both the sediment and in the overlaying water was determined over a 3-year period. Enteroviruses were detected in more than 40% of samples from recreational water that met accepted standards for total and fecal coliforms. Enteroviruses were also detected in 35% of the samples taken from areas approved for harvesting shellfish. Goyal et al. (1979) determined the level of indicator bacteria and human enteroviruses in oysters and in the water overlaying the oyster harvesting areas. These authors isolated viruses from waters that met the bacteriological standards current at the time of the study. They did not find a statistical relationship between the number of viruses and indicator bacteria (total and fecal coliforms) in the oysters.

Ellender et al. (1980) examined oysters and water overlaying the oyster beds for fecal coliforms and enteroviruses. These authors selected two different Mississippi reef areas for the study, one of which was closed and the other open for shellfish harvesting. Viruses were isolated from oysters taken from both open and closed areas. However, 146 viruses were isolated from oysters taken from the closed area and only 12 viruses were isolated from oysters from the approved area. The number of fecal coliforms in water was not correlated with the number of viruses in oysters.

In a similar study, Wait et al. (1983) examined hard shell clams from beds that were open or closed for shellfish harvesting. Although the levels of total and fecal coliforms were higher in water from the closed area, enteric viruses were isolated from oysters taken from both sites. No statistically significant difference was found between the occurrence of viruses in clams from the open and closed sites despite a clear difference in the levels of indicator bacteria in the water from the two sites.

Molecular methods were used by Le Guyader et al. (1993, 1994) to study viruses in shellfish. Genomic probes were used to detect hepatitis A and enteroviruses in cockles and mussels (Le Guyader et al., 1993). No statistically significant difference was found between the presence of viral genomes and fecal coliform counts. Using RT-PCR, Le Guyader et al. (1994) detected enterovirus, rotavirus, and hepatitis A genomes in 22%, 20%, and 14% of cockles, respectively. Again, no relationship was found between viral and bacterial contamination.

Dore et al. (2000) examined oysters from four sites for the presence of *E. coli* and Norwalk-like virus (NLV). All of the samples met the standard of <230 *E. coli* per gram of shellfish meat. NLV was detected in samples from only the most polluted site, as determined by the number of *E. coli* in oysters taken directly from the site and in market-ready oysters that had been taken from the site and treated by depuration. The level of *E. coli* could be used to predict the absence of NLV at the three least polluted sites but not at the most polluted site. Skraber et al. (2004) compared coliforms and coliphages for their ability to predict viral contamination of the Mosells River. They did not detect infectious enteroviruses but did detect the genomes of enterovirus and norovirus, the presence of which was correlated with the levels of bacteriophages but not those of fecal coliforms.

It is clear from these and other studies (Table 7.3) that the level of indicator bacteria may predict the presence of human enteric viruses in some but not all cases. Viruses are more likely to be found at environmental sites and in shellfish meat that are highly contaminated with indicator bacteria, although they may also be found in the presence of low levels of indicator bacteria that meet the accepted standards.

Table 7.3 Correlation Between the Presence of Indicator Bacteria and Viruses in Water and Shellfish

Bacterial Indicator	*Virus*	*Results*	*Reference*
Total coliforms, fecal coliforms	Enteroviruses (as infectious viruses)	Viruses were isolated from estuarine water and oysters even though the water met acceptable standards for indicator bacteria.	Goyal et al. (1979)
Total coliforms, fecal coliforms	Enteroviruses (as infectious viruses)	Viruses were detected in 35% of estuarine water samples that met accepted standards for shellfish harvesting.	Gerba et al. (1979)
Total coliforms, fecal coliforms	Enteric viruses (as infectious viruses)	Enteric viruses were found in 12% of oysters taken from closed areas but were not found in oysters taken from open areas.	Sobsey et al. (1980)
Fecal coliforms	Enteroviruses (as infectious viruses)	The level of fecal coliform bacteria in water did not reflect the level of viruses in water.	Ellender et al. (1980)
Total coliforms, fecal coliforms	Enteric viruses (as infectious viruses)	Enteric viruses were isolated from clams taken from both closed and open areas.	Wait et al. (1983)
Fecal coliforms	Enteroviruses, hepatitis A, and rotaviruses (as detected by RT-PCR)	No relationship between virus and indicator bacteria in shellfish.	e Guyader et al. (1993, 1994)
Fecal coliforms, fecal streptococci	Enterovirus, rotaviruses, and hepatitis A (as detected by RT-PCR)	Viral contamination of river water was correlated with bacteriophages but not with indicator bacteria.	Baggi et al. (2001)
Fecal coliforms	Enteroviruses (as infectious viruses and by RT-PCR); noroviruses (as detected by RT-PCR)	Infectious enteroviruses were not detected in river water; coliform concentrations were not related to the presence of viral genomes.	Skraber et al. (2004)

6.0. DIFFERENTIAL SURVIVAL OF BACTERIA AND VIRUSES

Because of large differences in their size and composition, it is not surprising to note that the length of viral and bacterial survival is different under different environmental conditions. Wastewater solids that were undergoing aerobic treatment were treated with certain physical and chemical methods that reduced the activity of protozoan predators. This caused a decrease in the adsorption of bacteria to wastewater solids leading to a reduction in their rates of inactivation (Farrah et al., 1985). In contrast, the same treatments did little to change the inactivation rates of several viruses or to change the ability of these viruses to adsorb to wastewater solids. It was concluded that viruses and bacteria were inactivated by different mechanisms during aerobic treatment of wastewater solids.

Baggi et al. (2001) examined the levels of bacteria and viruses during wastewater treatment and after the discharge of the effluent to the river. The wastewater treatment plants reduced the levels of fecal coliforms and fecal streptococci in raw sewage by an average of 2.3 $\log_{10}$. In contrast, the levels of three bacteriophages were reduced by only 0.7 $\log_{10}$. This is likely one reason that contamination of the river with viruses (enteroviruses, rotaviruses, and hepatitis A) was correlated with bacteriophages but not with fecal indicator bacteria.

The observed lack of correlation between indicator bacteria and enteric viruses in shellfish may at least partly be explained by two factors: (1) selective accumulation and (2) differential survival within shellfish. Burkhardt and Calci (2000) studied the accumulation of indicator bacteria and a bacteriophage (F^+) by oysters over a 1-year period. The most significant finding of this study was the marked change in accumulation of bacteriophages between November and January. Over this period, the bioaccumulation of bacteriophages increased by a factor of 99-fold. In contrast, accumulation of *E. coli*, fecal coliforms, and *Clostridium perfringens* did not change appreciably. This selective accumulation may account for seasonal differences in viral diseases associated with the consumption of oysters and for the lack of correlation between bacterial indicators and viruses in oysters.

Often, shellfish are not sold to consumers immediately after harvest but are exposed to clean estuarine water treated with UV light. After such treatment, the levels of indicator bacteria in oyster meat may be reduced. However, several studies have demonstrated that depuration is better at reducing the levels of *E. coli* and other indicator bacteria but not of viruses. Power and Collins (1989) compared the reductions in *E. coli*, poliovirus, and bacteriophages during depuration by mussels. They found that *E. coli* was eliminated from the mussels at a rate faster rate than that of poliovirus or bacteriophage. This led them to conclude that *E. coli* was an inappropriate indicator for demonstrating virus elimination during depuration.

A significant difference in the elimination of indicator bacteria (*E. coli*, fecal streptococci, and *Clostridium* spores) and bacteriophages by mussels during depuration was also observed by Mesquita et al. (1991). Elimination rates (T_{90}) for the indicator bacteria were in the range of 50 hr while for phages it was 500 hr. In another study, T_{90} values for *E. coli* and bacteriophages during depuration of oysters were 6.5 and greater than 40 hr, respectively (Dore and Lees, 1995). Schwab et al. (1998) found that depuration of oysters and clams reduced the level of *E. coli* by 95% but reduced the titer of Norwalk virus by only 7%.

Formiga-Cruz et al. (2002, 2003) examined indicator bacteria, bacteriophages, and human viruses in shellfish in several European countries. They found that human viruses were related to all bacterial indicators and bacteriophages in heavily contaminated waters and that the current depuration treatments were effective in reducing *E. coli* in shellfish but had little effect on viruses. In another study, oysters that had undergone prolonged depuration (>72 hr), which was sufficient to greatly reduce the levels of coliforms, were implicated in an outbreak of gastrointestinal illness (Heller et al., 1986). Differences in the removal of viruses and bacteria by treatment plants, differences in their accumulation by shellfish, and differences in their elimination during depuration likely contribute to the frequently observed lack of correlation between bacterial indicators and viruses in water and shellfish.

7.0. SOURCE TRACKING

Because *E. coli* and other indicator bacteria are found in the intestines of warm-blooded animals, their presence in a sample may or may not be related to the presence of human pathogens. The lack of correlation between bacterial indicators and human viruses may in part be related to the fact that bacterial indicators in a sample may be of nonhuman origin. Identifying the source of microbial pollution is also important for controlling pollution of an area. Therefore, methods have been developed to determine the source of bacterial indicators (Table 7.4).

The fecal coliform/fecal streptococci ratio was based on the observation that human feces contained relatively more fecal coliforms than animals and

Table 7.4 Methods to Determine the Source of Bacterial Indicators

Procedure	*Comments*	*Reference*
Fecal coliform/ fecal streptococcus ratio	Relies on standard bacteriological tests; the ratio may change with time because of differences in survival rates of bacteria.	Geldreich and Kenner (1969)
Pulsed-field gel electrophoresis (PFGE)	Can detect small genetic differences; highly useful for epidemiologic studies but may be too sensitive for source-tracking studies.	Johnson et al. (1995); Parveen et al. (2001)
Multiple antibiotic resistance (MAR)	Relatively rapid and does not require special equipment; requires a database and results may be valid only for limited geographical areas.	Hagedorn et al. (1999); Kruperman (1983); Wiggins et al. (1999)
Ribotyping	A labor-intensive method that can yield reproducible results; can be used to determine the source of indicator bacteria.	Parveen et al. (1999); Carson et al. (2001)

animals had relatively more fecal streptococci. A fecal coliform/fecal streptococcus ratio of >4.0 was considered characteristic of human pollution and a ration of <0.7 was thought to indicate animal pollution (Geldrich and Kenner, 1969). The test for multiple antibiotic resistance (MAR) relies on the fact that humans and animals are exposed to different types of antibiotics and that their intestinal bacteria show different patterns of resistance to antibiotics (Kruperman, 1983; Hagedorn et al. 1999; Wiggins et al., 1999). The use of this method requires a database to be produced for a specific area, and antibiotic-resistance patterns may change rapidly because of exchange of plasmids between bacteria.

Pulsed-field gel electrophoresis (PFGE) can produce specific genomic patterns of different microorganisms. Although this technique has been used in epidemiological studies to identify the source of bacterial pathogens (Johnson et al., 1995), it may be too sensitive for source tracking studies. Parveen et al. (2001) suggested that PFGE detected small differences in genomes, which may not be associated with specific indicator characteristics such as host range. Ribotyping is a fingerprinting technique that identifies conserved sequences of rRNA. Although this technique is labor intensive, it has been used successfully in source-tracking studies (Carson et al., 2001). Using this technique, Parveen et al. (1999) were able to correctly identify 97% and 67% of *E. coli* isolates from animals and humans, respectively. Although Scott et al. (2003) could not identify the animal source of *E. coli* isolates, they concluded that ribotyping could be used in differentiating human and nonhuman isolates. Some of the methods used for microbial source tracking have been reviewed by Scott et al. (2002).

It is possible that knowing the source of bacterial indicators may make them better indicators for viruses. Current procedures detect indicators that could be from many different sources (e.g., human and non-human). In contrast, most of the viruses are mainly human pathogens. By comparing indicators of human origin with viruses of human origin, it may be possible to obtain better correlations between viruses and indicator bacteria.

8.0. SUMMARY

The association between human enteric viruses and disease is well established. However, determining the presence of all of the many types of viruses that are pathogenic to humans in food and water is not practical at this time. Because enteric bacteria are usual inhabitants of the human intestinal tract, they have been used as indicators of fecal pollution and the possible presence of enteric viruses. Several different types of bacteria have been considered for use as indicators. Currently, most tests for indicator microorganisms rely on the detection of lactose-fermenting bacteria (coliforms, fecal coliforms, *E. coli*). Food and water samples with relatively high levels of these bacteria have frequently been found to contain bacterial and viral pathogens. However, viral pathogens have also been found in food and water samples with no or acceptable levels of indicator bacteria. It may be necessary to supplement tests for bacterial indicators with tests for other indicators, such as bacteriophages (see Chapter 8). Also, it may be desirable to determine the source of indicators, at least to the extent of determining if they are from human or non-human sources. This may lead to a better correlation between the presence of human indicator bacteria and human enteric viruses.

9.0. REFERENCES

Allsop, K., and Stickler, D. J., 1984, The enumeration of *Bacteroides fragilis* organisms from sewage and natural waters. *J. Appl. Bact.* 56:15–24.

Baggi, F., Demarta, A., and Peduzzi, R., 2001, Persistence of viral pathogens and bacteriophages during sewage treatment: lack of correlation with indicator bacteria. *Res. Microbiol.* 152:743–751.

Bagley, S. T., and Seidler, R. J., 1977, Significance of fecal coliform-positive *Klebsiella*. *Appl. Environ. Microbiol.* 33:1141–1148.

Berg, G., and Metcalf, T. G., 1978, Indicators of viruses, in: *Indicators of Viruses in Water and Food* (G. Berg, ed.), Ann Arbor Science Publishers, Ann Arbor, MI, pp. 267–296.

Blackmer, F., Reynolds, K. A., Gerba, C. P., and Pepper, I. L., 2000, Use of integrated cell culture-PCR to evaluate the effectiveness of poliovirus inactivation by chlorine. *Appl. Environ. Microbiol.* 66:2267–2268.

Brenner, K. P., Rankin, C. C., Robal, Y. R., Stelma, G. N., Jr., Scarpino, P. V., and Dufour, A. P., 1993, New medium for the simultaneous detection of total coliforms and *Escherichia coli* in water. *Appl. Environ. Microbiol.* 59:3534–3544.

Burkhardt, W., III, and Calci, K. R., 2000, Selective accumulation may account for shellfish-associated viral illness. *Appl. Environ. Microbiol.* 66:1375–1378.

Cabelli, V. J., Dufour, A. P., McCabe, J. L., and Levin, M. A., 1982, Swimming-associated gastroenteritis and water quality. *Am. J. Epidemiol.* 115:606–616.

Carson, C. A., Shear, B. L., Ellersieck, M. R., and Asfaw, A., 2001, Identification of fecal *Escherichia coli* from humans and animals by ribotyping. *Appl. Environ. Microbiol.* 67:1503–1507.

Chang, G. W., Brill, J., and Lum, R., 1989, Proportion of beta-D-glucuronidase-negative *Escherichia coli* in human fecal samples. *Appl. Environ. Microbiol.* 55:335–339.

Clark, H. F., Geldreich, E. E., Jeter, H. L., and Kabler, P. W., 1951. The membrane filter in sanitary bacteriology. *Public Health Rep.* 66:951–956.

Clesceri, L. S., Greenberg, A. E., and Eaton, A. D., eds., 1998, *Standard Methods for the Examination of Water and Wastewater.* American Public Health Association, Washington, DC, p. 9-1–78.

Dore, W. J., and Lees, D. N., 1995, Behavior of *Escherichia coli* and male-specific bacteriophage in environmentally contaminated bivalve mollusks before and after depuration. *Appl. Environ. Microbiol.* 61:2830–2834.

Dore, W. J., Wood, K. H., and Lees, D. N., 2000, Evaluation of F-specific RNA bacteriophage as a candidate human enteric virus indicator for bivalve molluscan shellfish. *Appl. Environ. Microbiol.* 66:1280–1285.

Duncan, D. W., and Razzell, W. E., 1972, *Klebsiella* biotypes among coliforms isolated from forest environments and farm produce. *Appl. Microbiol.* 24:933–938.

Dutka, B. J., Chau, A. S. Y., and Coburn, J., 1974, Relationship between bacterial indicators of water pollution and fecal sterols. *Water Res.* 8:1047–1055.

Eijkman, C., 1904, Die garungsprobe bei 46° als hilfsmittel bei der trinkwassereruntersuchung. *Zentr. Bakteriol. Parasitenk.* Abt. I. Orig. 37:742.

Ellender, R. D., Mapp, J. B., Middlebrooks, B. L., Cook, D. W., and Cake, E. W., 1980, Natural enterovirus and fecal coliform contamination of gulf coast oysters. *J. Food. Prot.* 43:105–110.

Escherich, T., 1885, Die darmbakterien des neugeborenem und sauglings. *Fortshr. Med.* 3:515–522, 547–554.

Farrah, S. R., Scheuerman, P. R., Eubanks, R. D., and Bitton, G., 1985, Bacteria and viruses in aerobically digested sludge: influence of physical and chemical treatments on survival and association with flocs under laboratory conditions. *Water Sci. Tech.* 17:165–174.

Feng, P. C. S., and Hartman, P. A., 1982, Fluorogenic assays for immediate confirmation of *Escherichia coli*. *Appl. Environ. Microbiol.* 43:1320–1329.

Fiksdal, L., Maki, J. S., LaCroix, S. J., and Staley, J. T., 1985, Survival and detection of *Bacteroides* spp., prospective indicator bacteria. *Appl. Environ. Microbiol.* 49:148–150.

Formiga-Cruz, M., Tofino-Quesada, G., Bofill-Mas, S., Lees, D. N., Henshilwood, K., Allard, A. K., Conden-Hansson, A.-C., Hernroth, B. E., Vantarakis, A., Tsibouxi, A., Papapetropoulou, M., Furones, M. D., and Girones, R., 2002, Distribution of human virus contamination in shellfish from different growing areas in Greece, Spain, Sweden, and the United Kingdom. *Appl. Environ. Microbiol.* 68:5990–5998.

Formiga-Cruz, M., Allard, A. K., Conden-Hansson, A.-C., Henshilwood, K., Hernroth, B. E., Jofre, J., Lees, D. N., Lucena, F., Papapetropoulou, M., Rangdale, R. E., Tsibouxi, A., Vantarakis, A., and Girones, R., 2003, Evaluation of potential indicators of viral contamination in shellfish and their applicability to diverse geographical areas. *Appl. Environ. Microbiol.* 69:1556–1563.

Fujioka, R. S., and Shizumura, L. K., 1985, *Clostridium perfringenes*, a reliable indicator of stream water quality. *J. Water Pollut. Control Fed.* 57:986–992.

Gassilloud, B., Schwartzbrod, L., and Gantzer, C., 2003, Presence of viral genomes in mineral water: a sufficient condition to assume infectious risk? *Appl. Environ. Microbiol.* 69:3965–3969.

Geldreich, E. E., and Kenner, B. A., 1969, Concepts of fecal streptococci in stream pollution. *J. Water Pollut. Control Fed.* 41:R336–R352.

Gerba, C. P., Goyal, S. M., LaBelle, R. L., Cech, I., and Bodgan, G. F., 1979, *Am. J. Public Health.* 69:1116–1119.

Goyal, S. M., Gerba, C. P., and Melnick, J. L., 1979, Human enteroviruses in oysters and their overlaying waters. *Appl. Environ. Microbiol.* 37:572–581.

Hackney, C. R., Ray, B., and Speck, M. L., 1979, Repair detection procedure for enumeration of fecal coliforms and enterococci from seafoods and marine environments. *Appl. Environ. Microbiol.* 37:947–953.

Hagedorn, C. S., Robinson, S. L., Filtz, J. R., Grubbs, S. M., Angier, T. A., and Reneau, Jr., R. B., 1999, Using antibiotic resistance patterns in the fecal streptococci to determine sources of fecal pollution in a rural Virginia watershed. *Appl. Environ. Microbiol.* 65:5522–5531.

Halliday, M. L., Kang, L.-Y., Zhou, T.-K., Hu, M.-D., Pan, Q.-C., Fu, T.-Y., Huang, Y.-S., and Hu, S.-L., 1991, An epidemic of hepatitis A attributable to the ingestion of raw clams in Shanghai, China. *J. Infect. Dis.* 164:852–859.

Hazen, T. C., and Toranzos, G. A., 1990, Tropical source water, in: *Drinking Water Microbiology* (G. A. McFeters, ed.), Springer-Verlag, New York, pp. 32–54.

Heller, D., Gill, O. N., Raynham, E., Kirkland, T., Zadick, P. M., and Stanwell-Smith, R., 1986, An outbreak of gastrointestinal illness associated with consumption of raw depurated oysters. *Br. Med. J.* 292:1726–1727.

Hood, M. A., Ness, G. E., and Blake, N. J., 1983, Relationship among fecal coliforms, *Escherichia coli*, and *Salmonella* spp. in shellfish. *Appl. Environ. Microbiol.* 45: 122–126

Hussong, D., Damare, J. M., Weiner, R. M., and Colwell, R. R., 1981, Bacteria associated with false-positive most-probable-number coliform test results for shellfish and estuaries. *Appl. Environ. Microbiol.* 41:35–45.

Hutin, Y. J. F., Pool, V., Cramer, E. H., Nainan, O. V., Weth, J., Williams, I. T., Goldstein, S. T., Gensheimer, K. F., Bell, B. P., Shapiro, C. N., Alter, M. J., and Margolis, H. S., 1999, *N. Engl. J. Med.* 340:595–601.

Johnson, J. M., Weagant, S. D., Jinneman, K. C., and Bryant, J. L., 1995, Use of pulsed field gel electrophoresis for epidemiological study of *Escherichia coli* O157:H7 during a food-borne outbreak. *Appl. Environ. Microbiol.* 61:2806–2808.

Kang, D. H., and Siragusa, G. R., 1999, Agar underlay method for recovery of sublethally heat-injured bacteria. *Appl. Environ. Microbiol.* 65:5334–5337.

Kingsley, D. H., Meade, G. K., and Richards, G. P., 2002, Detection of both hepatitis A virus and Norwalk-like virus in imported clams associated with food-borne illness. *Appl. Environ. Microbiol.* 68:3914–3918.

Koh, E. G., Huyn, J. H., and LaRock, P. A., 1994, Pertinence of indicator organisms and sampling variables to *Vibrio* concentrations. *Appl. Environ. Microbiol.* 60: 3897–3900.

Kruperman, P. H., 1983, Multiple antibiotic resistance indexing of *Escherichia coli* to identify high-risk sources of fecal contamination of foods. *Appl. Environ. Microbiol.* 46:165–170.

Le Guyader, F., Apaire-Marchais, V., Brillet, J., and Billaudel, S., 1993, Use of genomic probes to detect hepatitis A virus and enterovirus RNAs in wild shellfish and relationship of virus contamination to bacterial contamination. *Appl. Environ. Microbiol.* 59:3963–3968.

Le Guyader, F., Dubois, E., Menard, D., and Pommepuy, M., 1994, Detection of hepatitis A virus, rotavirus, and enterovirus in naturally contaminated shellfish and sediment by reverse transcription-seminested PCR. *Appl. Environ. Microbiol.* 60:3665–3671.

Manafi, M., Kneifel, W., and Bascomb, S., 1991, Fluorogenic and chromogenic substrates used in bacterial diagnostics. *Microbiol. Rev.* 55:335–348.

McFeters, G. A., Cameron, S. C., and LeChevallier, M. W., 1982, Influence of diluents, media, and membrane filters on detection of injured waterborne coliform bacteria. *Appl. Environ. Microbiol.* 43:97–103.

Mesquita, de, M. M. F., Evison, L. M., and West, P. A., 1991, Removal of faecal indicator bacteria and bacteriophages from the common mussel (*Mytilus edulis*) under artifical depuration conditions. *J. Appl. Bact.* 70:495–501.

Natvig, E. E., Ingham, S. C., Ingham, B. H., Cooperband, L. R., and Roper, T. R., 2002, *Salmonella enterica* serovar Typhimurium and *Escherichia coli* contamination of roots and leaf vegetables grown in soils with incorporated bovine manure. *Appl. Environ. Microbiol.* 68:2737–2744.

Newton, K. G., Harrison, J. C. L., and Smith, K. M., 1977, Coliforms from hides and meats. *Appl. Environ. Microbiol.* 33:199–200.

Nishida, T., Kimura, H., Saitoh, M., Shinohara, M., Kato, M., Fukuda, S., Munemura, Y., Mikami, T., Kawamoto, A., Akiyama, M., Kato, Y., Nishi, K., Kozawa, K., and Nisho, O., 2003, Detection, quantitation, and phylogenetic analysis of noroviruses in Japanese oysters. *Appl. Environ. Microbiol.* 69:5782–5786.

Niu, M. T., Polish, L. B., Robertson, B. H., Khanna, B. K., Woodruff, B. A., Shapiro, C. N., Miller, M. A., Smith, J. D., Gedrose, J. K., Alter, M. J., and Margolis, H. S., 1992, Multistate outbreak of hepatitis A associated with frozen strawberries. *J. Infect. Dis.* 166:518–524.

Parshionikar, S. U., William-True, S., Fout, G. S., Robbins, D. E., Seys, S. A., Cassady, J. D., and Harris, R., 2003, Waterborne outbreak of gastroenteritis associated with a norovirus. *Appl. Environ. Microbiol.* 69:5263–5268.

Parveen, S., Portier, K. M., Robinson, K., Edminston, L., and Tamplin, M. L., 1999, Discriminant analysis of ribotype profiles of *Escherichia coli* for differentiating human and nonhuman sources of fecal pollution. *Appl. Environ. Microbiol.* 65:3142–3147.

Parveen, S., Hodge, N., Stall, R. E., Farrah, S. R., and Tamplin, M. L., 2001, Phenotypic and genotypic characterization of human and nonhuman *Escherichia coli. Water Res.* 35:379–386.

Power, U. F., and Collins, J. K., 1989, Differential depuration of poliovirus, *Escherichia coli*, and a coliphage by the common mussel, *Mytilus edulis. Appl. Environ. Microbiol.* 55:1386–1390.

Schwab, K. J., Neill, F. H., Estes, K. K., Metcalf, T. G., and Atmar, R. L., 1998, Distribution of Norwalk virus within shellfish following bioaccumulation and subsequent depuration by detection using RT-PCR. *J. Food Prot.* 61:1674–1680.

Scott, T. M., Rose, J. B., Jenkins, T. M., Farrah, S. R., and Lukasik, J., 2002, Microbial source tracking: current methodology and future directions. *Appl. Environ. Microbiol.* 68:5796–5803.

Scott, T. M., Parveen, S., Portier, K. M., Rose, J. B., Tamplin, M. L., Farrah, S. R., Koo, A., and Lukasik, J., 2003, Geographical variation in ribotype profiles of *Escherichia coli* isolates from humans, swine, poultry, beef, and dairy cattle in Florida. *Appl. Environ. Microbiol.* 69:1089–1092.

Singh, A., and McFeters, G. A., 1986, Recovery, growth, and production of heat-stable enterotoxin by *Escherichia coli* after copper-induced injury. *Appl. Environ. Microbiol.* 51:738–742.

Skraber, S., Gassilloud, B., and Gantzer, C., 2004, Comparison of coliforms and coliphages as tools for assessment of viral contamination of river water. *Appl. Environ. Microbiol.* 70:3644–3649.

Sobsey, M. D., Hackney, C. R., Carrick, R. J., Ray, B., and Speck, M. L., 1980, Occurrence of enteric bacteria and viruses in oysters. *J. Food Prot.* 43:111–113.

Speck, M. L., Ray, B., and Readm Jr., R. B., 1975, Repair and enumeration of injured coliforms by a plating procedure. *Appl. Microbiol.* 29:549–550.

Stiles, M. E., and Ng, L-K., 1981, Biochemical characteristics and identification of *Enterobacteriaceae* isolate from meats. *Appl. Environ. Microbiol.* 41:639–645.

Tang, T. W., Wang, J. X., Xu, Z. Y., Guo, Y. F., Qian, W. H., and Xu, J. X., 1991, A serologically confirmed, case-control study, of a large outbreak of hepatitis A in China, associated with consumption of clams. *Epidemiol. Infect.* 107:651–657.

Toranzos, G. A., and McFeters, G. A., 1997, Detection of indicator microorganisms in environmental freshwaters and drinking water, in: *Manual of Environmental Microbiology* (C. J. Hurst, G. R. Knudsen, M. J. McInerney, L. D. Stezenbach, and M. V. Walter, eds.), ASM Press, Washington, DC, pp. 184–194.

Wait, D. A., Hackney, C. R., Carrick, R. J., Lovelace, G., and Sobsey, M. D., 1983, Enteric bacterial and viral pathogens and indicator bacteria in hard shell clams. *J. Food Prot.* 46:493–496.

Wiggins, B. A., Andrews, R. W., Conway, R. A., Corr, C. L., Dobratz, E. J., Dougherty, D. P., Eppard, J. R., Knupp, S. R., Limjoco, M. C., Mettenburg, J. M., Rinehardt, J. M., Sonsino, J., Torrijos, R. L., and Zimmerman, M. E., 1999, Use of antibiotic resistance analysis to identify nonpoint sources of fecal pollution. *Appl. Environ. Microbiol.* 65:3483–3486.

CHAPTER 8

Bacteriophages as Fecal Indicator Organisms

Suresh D. Pillai

1.0. INTRODUCTION

Bacteriophages, also known as phages, are viruses that infect bacterial cells. They were first described by Frederick Twort in 1915 and then in 1917 by Felix d'Herelle. D'Herelle named them bacteriophages because of their ability to lyse bacteria on the surface of agar plates; the word *phage* is derived from the Greek for "eating" (Flint et al., 2000). A variety of bacteriophages that infect different bacterial cells have been isolated. In fact, bacteriophages exist for all known bacterial species (Joklik, 1988). Based on their structural and genetic diversity, phages have been classified into different families as shown in Table 8.1. The characteristics of some well-known bacteriophages are shown in Table 8.2.

Bacteriophages are widely distributed in the environment and have been found in groundwater (Pillai and Nwachuku, 2000; Borchardt et al., 2003), river water (Hot et al., 2003; Skraber et al., 2004), irrigation waters (Ceballos et al., 2003; Mena and Pillai, 2003), wastewaters (Ackermann and Nguyen, 1983; Ottoson and Stenstrom, 2003; Nelson et al., 2004), oceans (Paul et al., 1997; Jiang et al., 2001; Jiang and Chu, 2004), and bioaerosols (Dowd et al., 1997; Espinosa and Pillai, 2002). They have also been found in shellfish (Humphrey and Martin, 1993; Croci et al., 2000) and on the surfaces of vegetables and herbs such as carrots and parsley (Endley et al., 2003a, 2003b).

This chapter focuses on the applicability of bacteriophages as indicators of fecal pollution or contamination. Most of the available information on the use of bacteriophages as indicator organisms pertains to wastewater and drinking water microbiology and hence most of the examples cited in this chapter are from these disciplines. Only recently have reports started appearing in the food microbiology literature pertaining to bacteriophages as indicators of fecal contamination on foods (Hsu et al., 2002; Endley et al., 2003; Munian-Mujika et al., 2003; Allwood et al., 2004).

2.0. INDICATOR ORGANISMS

Ashbolt et al. (2001) have suggested that there could be different classes of indicator organisms depending on their ultimate application as process indicators, as fecal indicators, or as index or model organisms. Process indicators are organisms that are used to demonstrate the efficiency of a particular

Table 8.1 Classification of Bacteriophages

Group or Family	*Genus*	*Type Member*	*Morphology*	*Envelope*	*Type of Nucleic Acid*
Corticoviridae	Corticovirus	PM2	Isometric	No	Supercoiled dsDNA
Cystoviridae	Cystovirus	ϕ6	Isometric	Yes	3 segments of dsRNA
Inoviridae	Inovirus Plectrovirus	fd Acholeplasma phage	Rod	No	Circular ssDNA
Leviviridae	Levivirus Allolevirus	MS2 Qβ	Icosahedral	No	Linear positive strand RNA
Lipothrixviridae	Lipothrixvirus	Thermoproteus phage 1	Rod	Yes	Linear dsDNA
Microviridae	Microvirus Spirovirus	ϕX174 Spiroplasma phages, MAC-1	Icosahedral	No	Circular ssDNA
Myoviridae		T4	Tailed	No	Circular dsDNA
Plasmaviridae	Plasmavirus	Acholeplasma phage	Pleiomorphic	Yes	Circular dsDNA
Podoviridae		Coliphage T7	Tailed	No	Linear dsDNA
Siphoviridae	Lambda phage group	Coliphage lambda	Tailed	No	Linear dsDNA
Sulpholobus shibatae virus		SSV-1	Lemon-shaped	No	Circular dsDNA
Tectiviridae	Tectivirus	PRD1	Icosahedral	No	Linear dsDNA

ss, single stranded; ds, double stranded.

Table 8.2 Characteristics of Selected Bacteriophages[a]

Phage	*Common Host*	*Head Size (nm)*	*Tail Length/ Structure (nm)*	*Structure*	*Nucleic Acid Type*	*Nucleic Acid Structure*	*Mol. Weight* $(\times 10^6)$
T1	*E. coli*	50	150/simple tail	Icosahedral	dsDNA	Linear	27
T2, T4, T6	*E. coli*	85 × 110	110/complex tail with fibers	Prolate icosahedral	dsDNA	Linear	110
T3, T7	*E. coli*	60	15/short tail	Icosahedral	dsDNA	Linear	25
T5	*E. coli*	65	170/short tail	Icosahedral	dsDNA	Linear	80
λ	*E. coli*	64	140/simple tail	Icosahedral	dsDNA	Linear	32
P22	*S. typhimurium*	61	20/complex tail	Icosahedral	dsDNA	Linear	29
SPO1	*B. subtilis*	90	200/complex tail	Icosahedral	dsDNA	Linear	85
PM2	*Pseudomonas*	60	None	Icosahedral, lipid-envelope	dsDNA	Circular	6
φX174	*E. coli*	27	None	Icosahedral	ssDNA	Circular	1.7
M13	*E. coli*	7 × 900	None	Filamentous	ssDNA	Circular	2.1
MS2, Qβ	*E. coli*	24	None	Icosahedral	ssRNA	Linear	1.2
φ6	*Pseudomonas phaseolica*	65	None	Icosahedral, lipid containing envelope	dsRNA	3 linear segments	9.5

dsDNA, double-stranded DNA; ssDNA, single-stranded DNA; ssRNA, single-stranded RNA; dsRNA, double-stranded RNA.
[a] Adapted from Joklik, 1988.

Table 8.3 Ideal Characteristics of a Pathogen Indicator Organism[a]

1. Should be present when pathogens are present and absent when pathogens are absent.
2. The persistence and growth characteristics of the indicator should be similar to the pathogens.
3. The indicator organism should not multiply in the environment.
4. The ratio between the indicator organism and the pathogen should be constant.
5. The indicator organism should be present in greater concentrations than the pathogens in contaminated samples.
6. The indicator organism should be resistant or more resistant to adverse environmental factors, disinfection, and other treatment processes as pathogens.
7. The indicator organism should be non-pathogenic and easy to quantify.
8. The tests for the indicator organism should be easy and applicable to all types of samples.

[a] Modified from Goyal, 1983.

man-made or natural process; fecal indicators are organisms that are used to infer the possible presence of fecal contaminants in a milieu; and index or model organisms are indicators for the presence of a particular pathogen. Goyal (1983) recommended that indicator organisms chosen as fecal contamination indicators or pathogen indicators should satisfy certain specific criteria (Table 8.3). More recently, the IAWPRC Study Group on Health Related Water Microbiology (1991) suggested that an ideal indicator organism should meet five specific criteria as listed in Table 8.4.

Over the past years, different organisms have been proposed as indicators of fecal contamination including fecal coliforms, *Escherichia coli*, enterococci, bacteriophages, and so forth (Berg et al., 1978; Kibbey et al., 1978; Fiksdal et al., 1985; Jin et al., 2004). However, studies have repeatedly shown that bacterial indicators are not true representatives of all possible fecal contaminants, especially of the enteric viruses (Berg et al., 1978; Gerba, 1987; Wait and Sobsey, 2001; Duran et al., 2003) and that bacteriophages may be a better indicator of such pollution (Havelaar et al., 1986; Gerba, 1987;

Table 8.4 Ideal Characteristics of a Fecal Contamination Indicator Organism

1. The indicator organism should occur consistently and abundantly in fecal material, preferably exclusively in human wastes.
2. The indicator organism should not multiply in the environment or foods.
3. The indicator organism should have an ecology in the environment and foods similar to that of pathogens.
4. The indicator organism should respond to environmental stresses similar have resistance to environmental stress similar to the pathogens.
5. The laboratory analysis for the detection of the indicator organism should be simple and relatively inexpensive.

Havelaar, 1993; Hsu et al., 2002; Cole et al., 2003; Endley et al., 2003a; Hot et al., 2003).

The use of coliphages as indicators of fecal pollution is based on the assumption that their presence in water samples denotes the presence of bacteria capable of supporting their replication. The advantages of coliphages as indicators of enteric viruses (therefore fecal pollution) include their relative similarities in size, transport, survival or persistence patterns, and densities in sewage or septic samples. In addition, coliphages are found in relatively high numbers in the environment, do not readily multiply in the environment, and can be assayed at a fraction of the cost of a typical pathogenic enteric virus assay. Their relatively long persistence in the environment and resistance to common disinfectants make them conservative indicators of fecal contamination. The use of phages that infect *Bacteroides fragilis* has also been considered as good fecal pollution indicator. *Bacteroides fragilis* is an obligate anaerobe found in high concentrations in human feces, and the presence of phages that infect these bacteria is considered to be indicative of fecal pollution (ISO, 1999; Puig et al., 1999; Gantzer et al., 2002; Duran et al., 2003; Lucena et al., 2003).

3.0. COLIPHAGES

The use of coliphages as indicators of not only fecal contamination but also of enteric bacteria and viruses was suggested almost 20 years ago by Gerba (1987). Coliphages can be broadly categorized into somatic and male-specific (or F^+) coliphages. The former are phages that infect *E. coli* by attaching directly to the cell wall, whereas the latter infect the *E. coli* host by attaching to a specific bacterial appendage termed sex-pilus or F-pilus. After attachment, the viral nucleic acid is injected into the host bacteria through these appendages. Muniesa et al. (2003) reported that only a negligible number of naturally occurring bacteria can serve as potential hosts for somatic coliphages.

Coliphage numbers in humans, cows, and pigs range from approximately 10^1 to 10^7 pfu per gram of feces (Dhillon et al., 1976; Osawa et al., 1981; Havelaar et al., 1986) and they are almost always present in raw sewage at 10^4 to 10^6 pfu/ml. Coliphages do not generally multiply in the environment because of the need for a live host for multiplication. However, there is a theoretical possibility that somatic coliphages can multiply in certain environments where *E. coli* can also multiply (e.g., in raw sewage). This is considered to be a drawback for the use of coliphages as indicators (Muniesa et al., 2003). However, to date there have been no reports of coliphage multiplication in natural environments probably because the temperatures required for efficient phage infection and replication (i.e., >30°C) are rare in the open environment. In addition, optimum phage and bacterial densities and bacterial physiological conditions needed for phage replication are rarely found in the natural water environment (Muniesa and Joffre, 2004).

3.1. Somatic Coliphages

Somatic coliphages have been found in sewage-contaminated tropical waters but not in pristine waters (Toranzos et al., 1988). They have also been detected in storm-water runoff (Davies et al., 2003), graywater samples (Ottoson and Stenstrom, 2003), and in bioaerosols around wastewater treatment plants and animal-rearing operations (Espinosa and Pillai, 2002). Suan et al. (1988) found somatic coliphages to be highly correlated with fecal coliforms in tropical waters in Asia but, in Chile, Castillo et al. (1988) found low correlation between somatic coliphages and fecal coliforms and a weak correlation between somatic coliphages and total coliforms. In fact, the coliphage-to-coliform ratio has been found to vary widely in sewage, secondary effluent, and river water and is influenced by environmental temperature (Bell, 1976). In one study, Carducci et al. (1999) did not find statistical relationship between the presence of coliphages and other organisms in bioaerosols collected around a wastewater treatment facility. However, the bioaerosol sampling employed in that study was not designed for effective coliphage capture. Skraber et al. (2004) reported that somatic coliphages were less sensitive to environmental stresses than the thermotolerant coliforms and thus were much more reliable indicators of fecal contamination. Torella et al. (2003) have reported that somatic coliphages of *Salmonella* spp. and *E. coli* were more resistant than fecal coliforms to freezing and cold temperatures (4°C).

The degradation of viral genomes of somatic coliphages has been found to be much more similar to that of infectious viruses, suggesting that somatic coliphages can be used as reliable indicators of fecal or pathogen contamination even when molecular methods are used to detect them (Skraber et al., 2004). Of the 68 surface-water samples positive for somatic coliphages (10^3–10^4 pfu/L) in France, only 2 were positive for enteroviruses by virus isolation in cell cultures and 60 were positive for enteroviruses by molecular methods (Hot et al., 2003).

In a survey of shellfish in Spain, Muniain-Mujika et al. (2003) reported somatic coliphages in 90% of the shellfish samples (n = 60) collected over a 6-month period. Of the 36 shellfish (mussels) samples harvested from 3 sampling sites in the Adriatic Sea, Croci et al. (2000) detected somatic coliphages in 78% of the samples with concentrations ranging from 0.4 MPN/g to 110 MPN/g. Significantly, none of the samples were positive for male-specific coliphages, only 4 samples were positive for enteroviruses, and 13 samples were positive for hepatitis A virus. Lucena et al. (1994) have reported that somatic coliphages and Bacteroides phages appear to have the lowest decay rates compared with others in shellfish growing areas around Spain. Hsu et al. (2002) detected somatic coliphages in 88% (n = 8) of market samples of poultry as compared with male-specific coliphages that were detected in 63% of the samples. Ceballos et al. (2003) detected somatic coliphages (10^3 to 10^5 pfu/100 ml) in a river that was being used as an irrigation water source in Brazil.

3.2. Male-Specific Coliphages

Male-specific (F^+) coliphages are coliphages that infect *E. coli* via the bacterial sex-pilus, the genes for which are located on the F-plasmid. The F^+ coliphages can be RNA-containing (FRNA phages) or DNA-containing coliphages (FDNA phages). The specificity of male-specific coliphages to infect only the *E. coli* cells that produce the F-pilus is critical because the F-pilus is produced only at temperatures near 37°C or higher. Thus, the potential for male-specific coliphages to replicate in the environment, where temperatures are rarely around 37°C, is negligible. However, the potential for these coliphages to multiply in environments where the ambient temperatures may reach 37°C during certain periods of the year needs to be explored.

The FRNA phages are a relatively homogenous group of small, icosahedral coliphages belonging to group E (Leviviridae). Because FRNA phages are morphologically similar to many of the enteroviruses, have similar persistence and transport characteristics in water, and are equally resistant or more resistant to chlorination than enteroviruses, they have attracted considerable attention as an indicator of enteric viral contamination (Shah and McCamish, 1972; Duran et al., 2003; Shin and Sobsey, 2003). Although humans do not excrete large numbers of FRNA phages in their feces, they are found in significantly large numbers in human sewage (Osawa et al., 1981; Furuse et al., 1983; Havelaar et al., 1986), leading some to believe that FRNA phages may multiply in raw sewage. The use of male-specific coliphages as fecal indicators circumvents the technical complexities and costs involved in screening for enteric viral pathogens yet provides some assurance about the absence of viral fecal contaminants (Havelaar, 1993a, 1993b; Hsu et al., 1995; Allwood et al., 2003). Allwood et al. (2003) have recently suggested that the presence of male-specific coliphages may be a strong indicator for the presence of noroviruses in water samples. They based their conclusions on the survival patterns of *E. coli*, male-specific coliphages and feline calicivirus in dechlorinated water stored for 28 days at 4°C, 25°C, and 37°C.

Male-specific coliphages have also been isolated from foods and food production/processing facilities. Espinosa and Pillai (2002) reported on the detection of male-specific coliphages from bioaerosols within poultry (broiler) houses; both FRNA and FDNA phages were detected in samples collected within the buildings and on window ledges just outside the buildings. Hsu et al. (2002) detected FRNA phages in 5 of 8 (63%) "market-ready" samples of poultry meat products. Further, they were able to monitor the presence of male-specific coliphages and other indicator organisms through the evisceration, washing, and chilling processes and showed that FRNA phages were reduced by more than 1 $\log_{10}$pfu. Endley et al. (2003) found male-specific coliphages on cilantro and parsley in the absence of *E. coli*, indicating that it may be useful to monitor male-specific coliphages in addition to *E. coli* when screening for fecal contamination of herbs. Of the 18 retail samples each of cilantro and parsley, 9 cilantro samples (50%) and 7 parsley samples (39%) were positive for male-specific coliphages.

In another study, Endley et al. (2003a) demonstrated the value of male-specific coliphages as an additional fecal contamination indicator when screening vegetables such as carrots. In this study, FDNA phages were detected in 25% of the carrot samples as compared with *E. coli* and *Salmonella*, which were present in only 8% and 4% of the samples, respectively. One of the salient features of this study was the observation that the occurrence of the fecal indicator organisms was random and that the contaminated sample may at times be positive for only one or two of the indicator organisms (Table 8.5).

Croci et al. (2000) observed that neither *E. coli* nor male-specific coliphages were reliable indicators for the presence of enteric viruses in mussels from the Adriatic sea. They report that out of 36 mussel samples that were analyzed, only 3 samples (8%) were positive for male-specific coliphages. Though these 3 samples were also positive for HAV, only 1 of these 3 samples was positive for enteroviruses. Muniain-Mujika et al. (2003) studied the comparative presence of pathogenic viruses and indicator organisms in shellfish. Out of 60 shellfish samples that were collected in 3 "human impacted" areas, 47% were positive for human adenoviruses, 19% were positive for enteroviruses, and 24% for HAV. The FRNA phages were present in 17 out of 60 (28%) shellfish samples. Enteroviruses, HAV, and human adenoviruses were repeatedly detected in samples that were negative for *E. coli*. Though only four of the FRNA positive samples (6.7%) were positive for HAV, enteroviruses, and human adenoviruses, the data strongly suggest that *E. coli* occurrence is not correlated with the occurrence of viral indicators or pathogens.

Humphrey and Martin (1993) have reported that somatic coliphages (rather than male-specific coliphages) have value as a fecal contamination indicator of virus removal during relaying of Pacific oysters. Their conclusions were based on the die-off of male-specific coliphages in oyster tissues.

Dore and Lees (1995) have reported on the persistence of male-specific coliphages in the digestive glands of environmentally contaminated bivalve molluscs even after depuration, indicating that these coliphages can proba-

Table 8.5 Microbial Indicator Organisms and Pathogens on Carrots Obtained from Different Locations[a]

	Number of Positive Samples		
Organism	*Field*	*Truck*	*Packing Shed*
Male-specific coliphages	1/25 (4%)	4/25 (16%)	14/25 (56%)
E. coli	0/25 (0%)	4/25 (16%)	2/25 (8%)
Salmonella	1/25 (4%)	2/25 (8%)	0/25 (0%)

[a] Modified from Endley et al., 2003a.

bly serve as a conservative indicator to verify that all traces of fecal contamination have been removed by depuration. Allwood et al. (2004) reported a stronger correlation between the survival of feline calicivirus and the FRNA phage MS2 than between *E. coli* and any other viral pathogen on the surfaces of leafy salad vegetables. These results also support the notion that FRNA phages can serve as a conservative indicator when evaluating pathogen intervention strategies.

3.3. Bacteroides Phages

Bacteroides fragilis is an obligate anaerobic bacterium found in high concentrations in human feces, and hence the presence of phages that infect these bacteria is considered to be indicative of human fecal contamination (Chung and Sobsey, 1993; Grabow et al., 1995; Jofre et al., 1995; ISO, 1999; Lucena et al., 2003). It should be realized, however, that the numbers of Bacteroides phages will vary depending on the host strain used and on the geographical region from where samples originated (Cornax et al., 1990; Araujo et al., 1997; Puig et al., 1999). Using *B. fragilis* strain RYC20, Muniain-Mujika et al. (2003) reported high correlation between Bacteroides phages and human enteric viruses in shellfish. Lucena et al. (1994) report that the Bacteroides phages have one of the lowest decay rates among indicator organisms in shellfish and that the fate of Bacteroides phages released into the marine environment mimics that of human viruses more than any other indicator organisms. Bacteroides phages are more resistant to conventional drinking water treatment processes than even male-specific coliphages and clostridia (Jofre et al., 1995) and they are also resistant to thermal treatment processes that are commonly used in sewage and sludge treatment (Mocé-Llivina et al., 2003), indicating that these phages may be the most conservative indicators.

4.0. DETECTION OF BACTERIOPHAGES

Due to the increased interest in using coliphages and Bacteroides phages as contamination indicators, the methods to detect them have been constantly improving. One of the key issues that confront bacteriophage detection in food and water is the appropriate sample volume for analysis. Because enteric viruses are normally recovered using very large sample volumes, the current trend in bacteriophage analysis, at least in the research laboratories, is to employ large sample volumes as well. For the recovery, detection, and enumeration of phages, a variety of different methods have been reported based on sample volume, sample processing method, and the host bacteria used (Pillai and Nwachuku, 2002).

4.1. Membrane Filtration Method

Sobsey et al. (1990) have reported on a bacteriophage detection method based on membrane filtration. In the original method, the host bacterium

Salmonella typhimurium WG49 was employed to detect F-specific coliphages. However, because the F^+ plasmid has been found to be unstable within the host bacterium, the authors acknowledged that there could be interferences (false positive) from somatic coliphages. This led to the identification of *E. coli* ATCC 15597 as the host strain for detecting male-specific coliphages. The protocol is based on adding $MgCl_2$ to the water sample followed by filtration of a defined volume (usually 1,000 ml) of the sample through a 0.45-µm pore-size filter. The phages attached to the filter are eluted in a high pH buffer after which the eluate is neutralized and plated on the appropriate host strain. The use of 0.03% tetrazolium dye aids in the detection and enumeration of the phage-induced plaques due to the contrast the dye provides.

4.2. U.S. EPA Information Collection Rule (ICR) Method

The U.S. EPA published a standardized procedure for the enumeration of somatic and male-specific coliphages for use in the Information Collection Rule (USEPA, 1996). In this procedure, large volumes of water sample (usually >1,000 L) are passed through a positively charged 1MDS filter after which the adsorbed phage particles are desorbed using a high-pH buffer. There are obviously constraints related to the volume of the sample that can be passed through the filter depending on the amount of suspended solids in the sample. Lingering concerns about the stability of phages to high-pH conditions employed in this protocol have forced researchers to explore alternate sample processing strategies.

4.3. ISO Methods

The ISO (International Organization for Standardization) method for enumeration of F-specific RNA phages is the official standard for ISO, which stipulates that the samples have to be collected, transported, and stored according to specific guidelines. This method also recommends the use of *S. typhimurium* WG49, *E. coli* K-12 Hfr, or *E. coli* HS (pFamp) as the host and includes a preconcentration step for samples that may harbor low numbers of bacteriophages. This protocol is designed for all types of water samples and can be adapted for use with food samples provided careful thought is given to sample processing and purification. The salient feature of the protocol is that the method suggests confirmatory steps when sampling new sources, when there is an unexplained overabundance of F-specific phages, or when there is an indication that somatic phages are being isolated (Pillai and Nwachuku, 2002). The basic protocol consists of using semi-soft TYG (tryptone yeast extract glucose) agar amended with a calcium-glucose solution to which 1 ml of the undiluted or diluted sample is added, mixed, and poured over a bottom-agar plate. The confirmatory tests involve the use of RNase (40 µg/ml) amended TYGA media. The ISO method can also be adjusted for use with samples containing high bacterial background using nalidixic acid–resistant *E. coli* strain CN-13 (ATCC 70078), also known as WG5. For the detection of somatic coliphages in samples with low background bacterial counts, the use of *E. coli* strain C (ATCC 13706) as the host strain has been recommended. Another highlight of the ISO protocol is the built-in confirmatory steps in phage detection when sampling new sources, when there is an unexplained overabundance of F-specific coliphages, or when there is an indication that somatic phages are being isolated. The confirmation steps include the use of RNAse for selectivity. The ISO method for the enumeration of bacteriophages infecting *B. fragilis* uses *B. fragilis* RYC 2056 (ATCC 70078) as a host. The primary advantage of this host is that, although the bacterium is an obligate anaerobe, *it does not require* anaerobic handling conditions. Only the incubation has to be carried out under anaerobic conditions.

4.4. U.S. EPA Methods 1601 and 1602

The two U.S. EPA methods 1601 and 1602 are performance-based methods designed and optimized for qualitative and quantitative detection of somatic and male-specific bacteriophages. These protocols have been extensively tested in round-robin laboratory and field tests and are being considered to be included in the pending EPA Ground Water Rule. Method 1601 is a two-step enrichment procedure for the qualitative detection of male-specific and somatic coliphages (USEPA, 2000a) and has been used to detect coliphages in carrots, cilantro, parsley, and bioaerosols (Endley et al., 2003a, 2003b; Espinosa and Pillai, 2002). The method can be used with either 100-ml or 1,000-ml sample volumes. The use of large sample volumes, as mentioned earlier, increases the probability of detecting low levels of fecally associated phages. Somatic coliphages are detected using *E. coli* CN-13 (a nalidixic acid–resistant mutant of *E. coli* ATCC 700609) as the host strain, while male-specific coliphages are detected using *E. coli* F-amp (an ampicillin-streptomycin–resistant mutant of *E. coli*). The principle underlying this method is the addition of host bacterium, $MgCl_2$, and concentrated broth medium into the sample followed by overnight incubation at 37°C. After this incubation, aliquots of the "enriched" culture are spotted on plates containing pre-prepared lawns of host bacteria. The plates are then incubated overnight and the resulting plaques are counted. It should be realized, however, that plaque counts cannot be used for quantitative purposes because they have originated from enriched samples.

The U.S. EPA Method 1602 is a quantitative detection protocol (USEPA, 2000b). This method also uses *E. coli* CN-13 and *E. coli* F-amp as hosts for detecting somatic and male-specific coliphages, respectively. However, this method is capable of handling only 100-ml sample volumes. The method involves the addition of high titer host bacterium, double-strength agar medium and $MgCl_2$ to the 100-ml sample, after which the entire contents are poured into 5 to 10 Petri dishes. After overnight incubation, the plaques are counted and tallied across different plates and the results reported as pfu/100 ml.

Recent improvements to these protocols include the use of confirmatory steps for plaque visualization by "picking" plaques from the original isolation plates and respotting them on fresh spot plates (Sobsey et al., 2004). The enrichment method has been found to be extremely valuable for detecting low levels of phages in large sample volumes (Sobsey et al., 2004).

4.5. Colorimetric Method

Colorimetric, presence/absence methods for coliphage detection have also been reported (Ijzerman et al., 1993). The method is based on the induction of β-galactosidase by *E. coli*. As a result of coliphage infection/replication, the bacterial cells are lysed, and β-galactosidase hydrolyses the yellow chromogenic substrate that develops into a distinct red color in coliphage-positive samples. Coliphage-negative samples will remain yellow.

5.0. BACTERIOPHAGES FOR TRACKING SOURCES OF CONTAMINATION

In addition to detecting the presence of fecal contamination, it is also extremely important to identify the sources of fecal contamination. Only if sources are identified would it be possible to develop remediation approaches to limit the exposure of the environment to fecal contaminants. A number of studies over the recent past have attempted to come up with tools to detect sources of fecal contamination. Indicators such as *E. coli*, enterococci, and bacteriophages and molecular methods such as pulsed-field gel electrophoresis (PFGE), ribotyping, and BOX-PCR have been proposed for source-tracking purposes (Lu et al., 2004; Meays et al., 2004). The PFGE protocol involves a specialized electrophoretic separation of the total genome after restriction digestion with specific enzyme(s). The BOX-PCR protocol involves the selective amplification of BOX sequences within enterobacteria using specialized PCR primers. Ribotyping involves the hybridization of 16S and 23S rDNA sequences as a method of differentiating bacterial subtypes.

Male-specific RNA coliphages have some unique characteristics that permit them to be used for tracking sources of fecal contamination. Phylogenetically, F-specific RNA coliphages fall into four subgroups (Furuse, 1987). Male-specific RNA coliphages are composed of serogroups I through IV. Subgroups I and II are related and form the major group A while subgroups III and IV are very similar and form the major group B. Strains isolated from human feces usually are in group II and III, whereas groups I and IV are usually found in animal feces (Osawa et al., 1981; Furuse, 1987). Cole et al. (2003) have recently reported on the distribution of different subgroups and genotypes of RNA and DNA coliphages. Municipal wastewater samples had high proportions of F^+ DNA coliphages and group II and III F^+ RNA coliphages. Bovine wastewater samples, on the other hand, though containing a large proportion of F^+ DNA coliphages, harbored a majority of group I and IV F+ RNA coliphages. Swine wastewaters harbored equal proportions of F^+ DNA and RNA coliphages. Group I and III F^+ RNA coliphages were the most common types of RNA coliphages in swine wastewaters. The F^+

RNA coliphages (groups I and IV) were present in large numbers in waterfowl feces. Though there was a statistically relevant association between genotypes II and III with human excreta and genotypes I and IV with animal/bird excreta, Schaper et al. (2002) have questioned whether they can be used for absolute distinction between human and animal sources. This was based on the detection of genotype II phages in poultry, cattle, and pig feces and genotype III phages that were reported for the first time in their study. Nevertheless, the understanding of the distribution of genotypes and serotypes in waste streams has paved the way for using male-specific RNA coliphages as indicators for detecting the source of fecal contamination, although antisera for male-specific RNA coliphages are not readily available and some isolates are difficult to serotype. Genotyping of male-specific coliphages with oligonucleotide probes has been found to be a feasible alternative to serotyping (Hsu et al., 1995; Brion et al., 2002). In addition to male-specific coliphages, the use of Bacteroides phages has also been suggested to detect fecal pollution from human sources. Grabow et al. (1995) reported that out of *Bacteroides fragilis* phages were detected in 13% (n = 90) of human fecal samples but were absent in fecal samples from a variety of animals. Thus, the detection of Bacteroides phages is indicative of fecal pollution from human sources.

6.0. SUMMARY

The distribution and occurrence of bacteriophages in source water such as rivers and aquifers have been extensively studied over the past years. There are a number of published articles describing the survival characteristics of bacteriophages in natural and man-made or engineered environments. However, our understanding of the occurrence of bacteriophages in foods is rather limited. Other than a few recent publications, there is a serious lack of understanding of the occurrence, distribution, and survival kinetics of these organisms in foods of animal and plant origin. We are currently unsure of the behavior of these coliphages and Bacteroides phages within foods during food processing and the various pathogen intervention strategies that foods are often subjected to. We are unsure for what types of foods these indicator viruses are robust indicators of fecally associated viral contaminants. We need to determine the food categories and the food processing systems in which coliphages can be used as fecal contamination or process indicators. In addition, to understand the ecology of coliphages on foods we need methods that can effectively recover coliphages from foods. These methods should be easy and efficient to use so that phage levels can be used for the estimation of microbial risk.

The technologies for coliphage detection are relatively mature, but processing protocols to recover coliphages from foods are scant. Although a few published protocols exist for recovering coliphages from certain herbs such as cilantro and parsley, concerted efforts are needed to develop methods for recovering coliphages from various types of vegetables, fruits, salads, and meat and meat products. Rapid methods to characterize the isolated coliphages in terms of their genotype need to be developed so that information about the potential sources of pathogens can also be obtained in parallel. Currently, the methods available for genotyping are restricted to research laboratories. Recent technological advances in micro-array, micro-fluidics, and biosensor technologies need to be exploited to develop user-friendly methods for the specific and sensitive detection and characterization of indicator viruses.

7.0. REFERENCES

Ackermann, H. W., and Nguyen, T. M., 1983, Sewage coliphages studied by electron microscopy. *Appl. Environ. Microbiol.* 45:1049–1059.

Allwood, P. B., Malik, Y. S., Hedberg, C. W., and Goyal, S. M., 2004, Effect of temperature and sanitizers on the survival of feline calicivirus, *Escherichia coli*, and F-specific coliphage MS2 on leafy salad vegetables. *J. Food Prot.* 67:1451–1456.

Ashbolt, N. J., Grabow, W. O. K., and Snozzi, M., 2001, Indicators of microbial water. quality, in: *Water Quality: Guidelines, Standards and Health* (L. Fewtrell, and J. Bartram, eds.), IWA Publishing, London, pp. 289–315.

Araujo, R. A., Puig, J., Lasobras, F., Lucena, F., and Jofre, J., 1997, Phages of enteric bacteria in fresh water with different levels of fecal pollution. *J. Appl. Microbiol.* 82:281–286.

Bell, R. G., 1976, The limitations of the ratio of fecal coliforms to total coliphage as a water pollution index. *Water Res.* 10:745–748.

Berg, G., Dahling, D. R., Brown, G. A., and Berman, D., 1978, Validity of fecal coliforms, total coliforms, and fecal streptococci as indicators of viruses in chlorinated primary sewage effluents. *Appl. Environ. Microbiol.* 36:880–884.

Brion, G. M., Meschke, J. S., and Sobsey, M. D., 2002, F-specific RNA coliphages: occurrence, types, and survival in natural waters. *Water Res.* 36:2419–2425.

Carducci, A., Gemelli, C., Cantiani, L., Casini, B., and Rovini, E., 1999, Assessment of microbial parameters as indicators of viral contamination of aerosol from urban sewage treatment plants. *Lett. Appl. Microbiol.* 28:207–210.

Castillo, G. R., Thiers, R., Dutka, B. J., and El-Shaarawi, A. H., 1988, Coliphage association with coliform indicators: a case study in chile. *Toxicity Assessment: An International Journal* 3:535–550.

Ceballos, B. S., Soares, N. E., Moraes, M. R., Catao, R. M., and Konig, A., 2003, Microbiological aspects of an urban river used for unrestricted irrigation in the semi-arid region of north-east Brazil. *Water Sci. Technol.* 47:51–57.

Chung, H., and Sobsey, M. D., 1993, Comparative survival of indicator viruses and enteric viruses in seawater and sediment. *Water Sci. Technol.* 27:425–429.

Cole, D., Long, S. C., and Sobsey, M. D., 2003, Evaluation of F+ RNA and DNA coliphages as source-specific indicators of fecal contamination in surface waters. *Appl. Environ. Microbiol.* 69:6507–6514.

Cornax, R., Moriñigo, M. A., Paez, I. G., Muñoz, M. A., and Borrego, J. J., 1990, Application of direct plaque assay for detection and enumeration of bacteriophages of *Bacteroides fragilis* from contaminated-water samples. *Appl. Environ. Microbiol.* 56:3170–3173.

Croci, L., De Medici, D., Scalfaro, C., Fiore, A., Divizia, M., Donia, D., Cosentino, A. M., Moretti, P., and Costantini, G., 2000, Determination of enteroviruses, hepatitis A virus, bacteriophages and *Escherichia coli* in Adriatic Sea mussels. *J. Appl. Microbiol.* 88:293–298.

Davies, C. M., Yousefi, Z., and Bavor, H. J., 2003, Occurrence of coliphages in urban stormwater and their fate in stormwater management systems. *Lett. Appl. Microbiol.* 37:299–303.

Dhillon, T. S., Dhillon, E. K. S., Chau, H. C., Li, W. K., and Tsang, A. H. C., 1976, Studies on bacteriophage distribution—virulent and temperate bacteriophage content of mammalian feces. *Appl. Environ. Microbiol.* 32:68–74.

Dore, W. J. and Lees, D. N., 1995, Behavior of *Escherichia coli* and male-specific bacteriophage in environmentally contaminated bivalve molluscs before and after depuration. *Appl. Environ. Microbiol.* 61:2830–2834.

Dowd, S. E., Widmer, K. W., and Pillai, S. D., 1997, Thermotolerant clostridia as an airborne pathogen indicator during land application of biosolids. *J. Environ. Qual.* 26:194–199.

Duran, A. E., Muniesa, M., Moce-Llivina, L., Campos, C., Jofre, J., and Lucena, F., 2003, Usefulness of different groups of bacteriophages as model micro-organisms for evaluating chlorination. *J. Appl. Microbiol.* 95:29–37.

Endley, S. E., Vega, E., Lingeng, L., Hume, M., and Pillai, S. D., 2003a, Male-specific coliphages as an additional fecal contamination indicator for screening fresh carrots. *J. Food Prot.* 66:88–93.

Endley, S., Johnson, E., and Pillai, S. D., 2003b, A simple method to screen cilantro and parsley for fecal indicator viruses. *J. Food Prot.* 66:1506–1509.

Espinosa, I. Y. and Pillai, S. D., 2002, Impaction-based sampler for detecting male-specific bacteriophages in bioaerosols. *J. Rapid Methods Autom. Microbiol.* 10:117–127.

Fiksdal, L., Maki, J. S., LaCroix, S. J., and Staley, J. T., 1985, Survival and detection of Bacteroides spp., prospective indicator bacteria. *Appl. Environ. Microbiol.* 49:148–150.

Flint, L. W., Enquist, R. M., King, V. R., Racaniello and Skalka, A. M., 2000, Principles of virology, in: *Molecular Biology, Pathogenesis and Control* (S. J., ed.), American Society for Microbiology, Washington, DC, p. 804.

Furuse, K., 1987, Distribution of phages in the environment: general considerations, in: *Phage Ecology* (S. M. Goyal, C. P. Gerba, and G. Bitton, eds.), Wiley Interscience, New York, pp. 87–123.

Furuse, K., Ando, A., Osawa, S., and Wantanabe, I., 1981, Distribution of ribonucleic acid coliphages in raw sewage in South and East Asia. *Appl. Environ. Microbiol.* 41:995–1002.

Gantzer, C., Henny, J., and Schwartzbrod, L., 2002, *Bacteroides fragilis* and *Escherichia coli* bacteriophages in human faeces. *Int. J. Hyg. Environ. Health* 205:325–328.

Gerba, C. P., 1987, Phage as indicators of fecal pollution, in: *Phage Ecology* (S. M. Goyal, C. P. Gerba, and G. Bitton, eds.), Wiley Interscience, New York, pp. 197–210.

Gerba, C. P., Goyal, S. M., LaBelle, R. L., Cech, I., and Bodgan, G. F., 1979, Failure of indicator bacteria to reflect the occurrence of enteroviruses in marine waters. *Am. J. Public Health* 69:1116–1119.

Grabow, W. O. K., Neubrech, T. E., Holtzhausen, C. S., and Jofre, J., 1995, *Bacteroides fragilis* and *Escherichia coli* bacteriophages: excretion by humans and animals. *Water Sci. Technol.* 31:223–230.

Goyal, S. M., 1983, Indicators of viruses, in: *Viral Pollution of the Environment* (G. Berg, ed.), CRC Press, Boca Raton, FL, pp. 211–230.

Havelaar, A. H., 1993, Bacteriophages as models of human enteric viruses in the environment. *ASM News*, 59:614–619.

Havelaar, A. H., Furuse, K., and Hogeboom, W. M., 1986, Bacteriophages and indicator bacteria in human and animal feces. *J. Appl. Bacteriol.* 60:255–262.

Havelaar, A. H., Olphen, M., Olphen, Y. C., and van Dorst, Y. C., 1993, F-specific RNA bacteriophages are adequate model organisms for enteric viruses in fresh water. *Appl. Environ. Microbiol.* 59:2956–2962.

Hot, D., Legeay, O., Jacques, J., Gantzer, C., Caudrelier, Y., Guyard, K., Lange, M., and Andreoletti, L., 2003, Detection of somatic phages, infectious enteroviruses and enterovirus genomes as indicators of human enteric viral pollution in surface water. *Water Res.* 37:4703–4710.

Hsu, F. C., Shieh, Y. S., van Duin, J., Beekwilder, M. J., and Sobsey, M. D., 1995, Genotyping male-specific RNA coliphages by hybridization with oligonucleotide probes. *Appl. Environ. Microbiol.* 61:3960–3966.

Hsu, F. C., Shieh, Y. S., and Sobsey, M. D., 2002, Enteric bacteriophages as potential fecal indicators in ground beef and poultry meat. *J. Food Prot.* 65:93–99.

Humphrey, T. J., and Martin, K., 1993, Bacteriophage as models for virus removal from Pacific oysters (*Crassostrea gigas*) during re-laying. *Epidemiol. Infect.* 111:325–335.

IAWPRC Study Group on Health Related Water Microbiology, 1991, Bacteriophages as model viruses in water quality control. *Water Res.* 25:529–545.

Ijzerman, M. M., Falkinham, J. O., and Hagedorn, C., 1994, A liquid, colorimetric presence-absence coliphage detection method. *J. Virol. Methods* 48:349.

ISO (International Standards Organization), 1999, Water Quality. Detection and Enumeration of Bacteriophages—Part 2. Enumeration of bacteriophages infecting *Bacteroides fragilis*. ISO/DIS10705-4, Geneva, Switzerland.

Jiang, S. C., and Chu, W., 2004, PCR detection of pathogenic viruses in southern California urban rivers. *J. Appl. Microbiol.* 97:17–28.

Jiang, S., Noble, R., and Chu, W., 2001, Human adenoviruses and coliphages in urban runoff-impacted coastal waters of Southern California. *Appl. Environ. Microbiol.* 67:179–184.

Jin, G., Englande, A. J., Bradford, H., and Jeng, H. W., 2004, Comparison of *E. coli*, enterococci, and fecal coliform as indicators for brackish water quality assessment. *Water Environ. Res.* 76:245–255.

Jofre, J., Olle, E., Ribas, F., Vidal, A., and Lucena, F., 1995, Potential usefulness of bacteriophages that infect *Bacteroides fragilis* as model organisms for monitoring virus removal in drinking water treatment plants. *Appl. Environ. Microbiol.* 61:3227–3231.

Joklik, W. K., 1988, *Virology*, 3rd ed., Appleton and Lange, Norwalk, OH.

Kibbey, H. J., Hagedorn, C., and McCoy, E. L., 1978, Use of fecal streptococci as indicators of pollution in soil. *Appl. Environ. Microbiol.* 35:711–717.

Lucena, F., Lasobras, J., McIntosh, D., Forcadell, M., and Jofre, J., 1994, Effect of distance from the polluting focus on relative concentrations of *Bacteroides fragilis* phages and coliphages in mussels. *Appl. Environ. Microbiol.* 60:2272–2277.

Meays, C. L., Broersma, K., Nordin, R., and Mazumdar, A., 2004, Source tracking fecal bacteria in water: a critical review of current methods. *J. Environ. Mgt.* 73:71–79.

Mena, K. D., and Pillai, S. D., 2003, Quantitative risk-based microbial standards for irrigation water quality. Abstract Annu. Mtg. Am. Soc. Microbiol. Washington, D.C.

Mocé-Llivina, L., Maite Muniesa, H., Pimenta-Vale, F. L., and Jofre, J., 2003, Survival of bacterial indicator species and bacteriophages after thermal treatment of sludge and sewage. *Appl. Environ. Microbiol.* 69:1452–1456.

Muniain-Mujika, I., Calvo, M., Lucena. F., and Girones, R., 2003, Comparative analysis of viral pathogens and potential indicators in shellfish. *Int. J. Food Microbiol.* 25, 83:75–85.

Muniesa, M., and Jofre, J., 2004a, Abundance in sewage of bacteriophages infecting *Escherichia coli* O157:H7. *Methods Mol. Biol.* 268:79–88.

Muniesa, M., and Jofre, J., 2004b, Factors influencing the replication of somatic coliphages in the water environment. *Antonie Van Leeuwenhoek J. Microbiol.* 86:65–76.

Muniesa, M., Moce-Llivina, L., Katayama, H., and Jofre, J., 2003, Bacterial host strains that support replication of somatic coliphages. *Antony Van Leeuwenhoek J. Microbiol.* 83:305–315.

Nelson, K. L., Cisneros, B. J., Tchobanoglous, G., and Darby, J. L., 2004, Sludge accumulation, characteristics, and pathogen inactivation in four primary waste stabilization ponds in central Mexico. *Water Res.* 38:111–127.

Osawa, S., Furuse, K., and Watanabe, I., 1981, Distribution of ribonucleic acid coliphages in animals. *Appl. Environ. Microbiol.* 41:164–168.

Ottoson, J., and Stenstrom, T. A., 2003, Faecal contamination of greywater and associated microbial risks. *Water Res.* 37:645–655.

Paul, J. H., Rose, J. B., Jiang, S. C., London, P., Xhou, X., and Kellogg, C., 1997, Coliphage and indigenous phage in Mamala Bay, Oahu, Hawaii. *Appl. Environ. Microbiol.* 63:133–138.

Pillai, S. D., and Nwachuku, N., 2000, Field testing of coliphage methods for screening groundwater for fecal contamination. Proc. Am. Water Qual. Technol. Conf., Salt Lake City, UT.

Pillai, S. D., and Nwachuku, N., 2002, Bacteriophage methodologies, in: *Encyclopedia of Microbiology* (G. Bitton, ed.), John Wiley & Sons, New York, pp. 374–384.

Puig, A., Queralt, N., Jofre, J., and Araujo, R., 1999, Diversity of *Bacteroides fragilis* strains in their capacity to recover phages from human and animal wastes and from fecally polluted wastewater. *Appl. Environ. Microbiol.* 65:1772–1776.

Schaper, M., Jofre, J., Uys, M., and Grabow, W. O., 2002, Distribution of genotypes of F-specific RNA bacteriophages in human and non-human sources of faecal pollution in South Africa and Spain. *J. Appl. Microbiol.* 92:657–667.

Shah, P. C., and McCamish, J., 1972, Relative chlorine resistance of poliovirus I and coliphages f2 and T 2 in water. *Appl. Microbiol.* 24:658–659.

Shin, G. A., and Sobsey, M. D., 2003, Reduction of Norwalk virus, poliovirus 1, and bacteriophage MS2 by ozone disinfection of water. *Appl. Environ. Microbiol.* 69:3975–3978.

Skraber, S., Gassilloud, B., and Gantzer, C., 2004, Comparison of coliforms and coliphages as tools for assessment of viral contamination in river water. *Appl. Environ. Microbiol.* 70:3644–3649.

Sobsey, M. D., Schwab, K. J., and Handzel, T. R., 1990, A simple membrane filter method to concentrate and enumerate male-specific RNA coliphages. *J. Am. Water Works Assn.* 82:52–59.

Sobsey, M. D., Yates, M. V., Hsu, F. C., Lovelace, G., Battigelli, D., Margolin, A., Pillai, S. D., and Nwachuku, N., 2004, Development and evaluation of methods to detect coliphages in large volumes of water. *Water Sci. Technol.* 50:211–217.

Suan, S. T., Chuen, H. Y., and Sivaborvorn, K., 1988, Southeast Asian experiences with the coliphage test. *Toxicity Assessment: An International Journal* 3:551–564.
Toranzos, G. A., Gerba, C. P., and Hanssen, H., 1988, Enteric viruses and coliphages in Latin America. *Toxicity Assessment: An International Journal* 3:391–394.
Torrella, F., Lopez, J. P., and Banks, C. J., 2003, Survival of indicators of bacterial and viral contamination in wastewater subjected to low temperatures and freezing: application to cold climate waste stabilisation ponds, *Water Sci. Technol.* 48: 105–112.
USEPA, 1996, ICR (Information Collection Rule). Microbial Laboratory Manual, Office of Research and Development, Washington, DC.
USEPA, 2000a, Method 1601, Male-specific (F+) and somatic coliphage in water by a two-step enrichment procedure. EPA 821-R-00-009, Washington, DC.
USEPA, 2000b, Method 1602, Male-specific (F+) and somatic coliphage in water by single agar layer (SAL) procedure. EPA 821-R-00-009, Washington, DC.

CHAPTER 9

Shellfish-Associated Viral Disease Outbreaks

Gary P. Richards

1.0. INTRODUCTION

Enteric viruses include a broad array of pathogens that enter the host via the fecal-oral route, often through the ingestion of sewage-contaminated food or water. In young children, many illnesses can result from fomites shared among playmates. The enteric viruses include the caliciviruses, which are classified as noroviruses and sapoviruses; picornaviruses, particularly hepatitis A virus, the Aichi virus, and poliovirus; hepatitis E virus; astroviruses; rotaviruses; enteric adenoviruses; coronaviruses and toroviruses; and picobirnaviruses. The most frequently reported food-borne outbreaks are caused by noroviruses, formerly called the agent of winter vomiting disease, Norwalk-like viruses, and small round structured viruses (SRSVs). Hepatitis A virus is also reported as a frequent cause of food-borne illness. Children are infected in early childhood with group A rotaviruses, enteric adenoviruses, astroviruses, and caliciviruses and may develop partial immunity against them (Glass et al., 2001). These so-called childhood viruses may be transmitted easily from child-to-child through casual contact and through fomites.

Enteric viruses have undoubtedly been infecting mankind since the dawn of civilization; however, techniques to isolate and identify these viruses were not developed until the past century. With the advent of sensitive molecular methods, even nonpropagable viruses may now be detected. In spite of these advances, reporting of enteric viral illnesses is poor or nonexistent in many parts of the world today. Noroviruses are believed to constitute the most frequent cause of food-borne illness; however, only major outbreaks are recorded and accurate, quantitative assessment of the number of individuals affected is not available. The best data are available for poliovirus, which has nearly been eradicated through global vaccination and tracking programs. Accountability for hepatitis A and hepatitis E infections is fair in the developed nations, due to the potential seriousness of diseases caused by these agents. However, the incidence of norovirus, sapovirus, rotavirus, astrovirus, and other viral pathogens is not generally recorded, except in outbreaks involving high numbers of individuals or those involving politically or socially important individuals. This is because these viruses seldom cause mortality even though they are the most prevalent causes of food-borne illnesses in the world today. When illnesses are noted, there is seldom epidemiological follow-up to confirm the cause of illness. Most of the illnesses are likely from drinking sewage-contaminated water, followed by the consumption of raw or undercooked foods that were tainted by contaminated water, the hands of food-handlers, or contaminated contact surfaces.

Among the most notable foods that may contain enteric viruses are molluskan shellfish (oysters, clams, mussels, cockles), especially when they are consumed raw or lightly cooked. The shellfish accumulate contaminants, including enteric viruses, from the water and bioconcentrate them within their edible tissues. Consequently, some large outbreaks of hepatitis A and noroviruses have been reported after consumption of contaminated shellfish. Efforts to document such outbreaks have provided some glimmer of the causes and effects of shellfish-borne disease but do not convey the magnitude of the problem (Gerba and Goyal, 1978; Richards, 1985, 1987; Rippey, 1994). The Centers for Disease Control and Prevention have indicated that noroviruses are the most common cause of acute gastroenteritis in the United States, causing an estimated 23 million cases annually with 9.2 million of those cases associated with foods (Mead et al., 1999). The vast majority of illnesses go undiagnosed, and statistics are not maintained on those reported because norovirus illness is not a notifiable disease in most countries including the United States.

Although persons infected with norovirus develop acute vomiting and diarrhea, symptoms are fleeting, lasting only a day or two. Consequently, the patients do not seek medical attention because symptoms resolve rapidly and spontaneously. They may spread the disease to family members through contamination of surfaces or by handling foods with inadequately sanitized hands. Such individuals often miss work for 2 or 3 days, but when they return, they may still carry the virus and be a source of infection to their work mates (White et al., 1986, Iversen et al., 1987; Haruki et al., 1991; Graham et al., 1994; Richards et al., 2004).

The scientific literature is dotted with occasional reports of outbreaks, particularly for hepatitis A and the noroviruses. Epidemiological linkage of an outbreak to a particular source is more difficult for some virus infections due to differences in incubation times. For instance, hepatitis A has an extended incubation period approaching 1 month, and sick individuals may not be able to say with any degree of certainty where or what they ate a month earlier. Larger outbreaks are more likely to reveal the source of infection, be it water, food, or a particular restaurant. Illnesses due to norovirus and sapovirus are easier to track because of their short, 1- to 2-day incubation period. Rotavirus causes diarrhea in infants and young children and, although it may be transmitted by foods, children develop immunity to rotavirus at an early age. Rotavirus diarrhea may lead to dehydration and vascular collapse, particularly when rehydration therapy is not available. Although rotavirus is transmitted by the fecal-oral route, it is likely that most illnesses are from direct contact of children with other children and fomites, rather than through the food-borne route. Astrovirus is another pathogen that has been difficult to track. Molecular diagnostic methods are now available for astroviruses, which may allow some screening of foods for the virus, especially in outbreak investigations.

2.0. CASE STUDIES

Because reporting of viral illnesses and their association with a particular food are inadequate at best, this chapter will not attempt to tabulate and list outbreaks by country or food source; rather, the focus will be to highlight specific, shellfish-related outbreaks in countries around the globe and to indicate sources of contamination, when known. This section highlights outbreaks caused by known shellfish-borne viral pathogens.

2.1. Hepatitis A Virus:

The United States has experienced numerous outbreaks of shellfish-associated hepatitis A. Major outbreaks date back to 1961 with 459 cases in New Jersey and New York from clams, 372 cases in Pennsylvania, Connecticut, and Rhode Island in 1964 from clams, and 293 cases in Georgia, Missouri, New Mexico, Oklahoma, and Texas in 1973 from the consumption of oysters from Louisiana (Richards, 1985). Oysters associated with the 1973 outbreaks were consumed raw but were reportedly obtained from waters that met the standards of the National Shellfish Sanitation Program (Portnoy et al., 1975; Mackowiak et al., 1976). Flooding of polluted water from the Mississippi River into oyster growing waters occurred 2 months earlier and may have been responsible for the outbreaks (Portnoy et al., 1975; Mackowiak et al., 1976). A multistate outbreak of hepatitis A was attributed to the consumption of raw oysters from Florida (Desconclos et al., 1991). The attack rate was calculated at 19 persons per 10,000 dozen oysters consumed in restaurants.

The largest outbreak of hepatitis A occurred in and around Shanghai, China, from January through March, 1988. More than 293,000 individuals became ill after eating clams harvested from recently opened mud flats outside of Shanghai (Xu et al., 1992) with 47 deaths reported (Cooksley, 2000). Most of the cases were reported to have been from direct consumption of the clams, rather than from person-to-person transmission. Since the incubation period for hepatitis A is around 30 days, many people ate the clams before any symptoms appeared. During this same period, factory workers in Shanghai also developed hepatitis A after eating raw and cooked clams (Wang et al., 1990; Halliday et al., 1991; Tang et al., 1991). Because thorough cooking is known to inactivate enteric viruses, it appears that the clams were not fully cooked. Chinese clams imported into the United States were recently found to contain hepatitis A virus using molecular biological methods (Kingsley et al., 2002b). Although import regulations require that clams from China be cooked for importation, these clams were only labeled as cooked, but had the appearance of raw product. They were associated with an outbreak of norovirus in New York State (see below). Clams imported into Japan from China also were associated with outbreaks of hepatitis A (Furuta et al., 2003). Between 1976 and 1985, there were 109 cases of hepatitis A reported in Japan and 11% were believed to be from consuming raw shellfish (Konno et al., 1983; Kiyosawa et al., 1987). Another study reported 225 cases of hepatitis A in Japan and raw oysters were the likely vehicle for infection (Fujiyama et al., 1985).

In 1997, 467 cases of hepatitis A occurred in New South Wales, Australia, from the consumption of oysters harvested from Wallis Lake (Conaty et al., 2000). One person died from hepatitis and a class action suit was filed on behalf of the victim and those who became ill. Before marketing, the government of New South Wales required that all shellfish be subjected to the commercial process of depuration, where shellfish are placed in tanks of clean seawater and are allowed to purge contaminants for 2–3 days. Depuration has been shown to be effective in eliminating many bacterial pathogens and spoilage organisms from molluscan shellfish but not enteric viruses such as hepatitis A and noroviruses (Richards, 1988, 1991; Kingsley and Richards, 2003). Long-term relaying (Richards, 1988) may be a better alternative to the commercial depuration of viruses from shellfish.

Europe too has had its share of hepatitis A associated with contaminated shellfish. Outbreaks of hepatitis A from oysters, cockles, and mussels have been reported in England, Wales, and Ireland (O'Mahony et al., 1983; Polakoff, 1990; Maguire et al., 1992). Shellfish-associated hepatitis A has also been reported in Italy. One outbreak was from imported clams with secondary spread to a public school (Leoni et al., 1998). The total cost of one outbreak of hepatitis A involving 5,889 cases in Italy was estimated at $24 million and costs to a sick individual were estimated at $662 (Lucioni et al., 1998). Spain experienced hepatitis A outbreaks in 1999 with 184 cases from clams meeting European Union standards (Sanchez et al., 2002). Clams imported from Peru also led to 183 cases of hepatitis A in Valencia with hepatitis A virus detected in 75% of the shellfish samples tested (Bosch et al., 2001). A survey of South American imports showed the presence of hepatitis A virus in 4 of 17 lots of molluscs (Romalde et al., 2001).

2.2. Noroviruses

Shellfish-borne outbreaks of norovirus have been widespread. A review of the literature indicates 6,049 documented cases of shellfish-associated gastroenteritis in the United States between 1934 and 1984 (Richards, 1987). Because no bacterial pathogens were associated with these illnesses and symptomatology was consistent with norovirus illness, it seems likely that noroviruses were the causative agents. One outbreak involved 472 cases of gastroenteritis from the consumption of Louisiana oysters. This outbreak resulted in 25% of Louisiana's 250,000 acres of shellfish beds being closed, an estimated loss to the industry of $5.5 million, and disruption of harvesting for 500 licensed oystermen (Richards, 1985). Some outbreaks were small, such as the one in Florida in 1980 involving only six individuals who ate raw oysters (Gunn et al., 1982). In another case, oysters from a defined area in

Louisiana were associated with outbreaks of norovirus illness in at least five states: Louisiana, Maryland, Mississippi, North Carolina, and Florida (Centers for Disease Control, 1993). Although these oysters were distributed throughout the United States, outbreaks were identified only in these five states. Identification of the source of contaminated shellfish was facilitated by tags (labels) on sacks of oysters indicating the location of harvest.

The worst year on record for norovirus outbreaks in the United States was 1983 when New York experienced numerous outbreaks of norovirus illness from raw and steamed clams (Centers for Disease Control, 1982; New York State Department of Health, 1983) and from oysters (Morse et al., 1986). At least 441 people suffered acute gastroenteritis and eight of these individuals subsequently developed hepatitis A as well. Ten outbreaks during the summer were attributed to the illegal harvesting of oysters by an unlicensed digger in polluted waters that were closed to shellfishing along the Massachusetts coast (Morse et al., 1986). Other contaminated shellfish were obtained from Rhode Island waters. Another series of outbreaks in the winter was from clams harvested in New York waters. Negative publicity and the lack of confidence in the safety of local shellfish prompted shellfish dealers to obtain clams that had been depurated in England. Unfortunately, these clams led to more than 2,000 illnesses in 14 separate outbreaks in New York and New Jersey over a 3-month period (Richards, 1985). Clams served at a picnic were responsible for more than 1,100 cases of norovirus illness in one outbreak. The U.S. Food and Drug Administration investigated the outbreaks and concluded that depuration was poorly monitored in plants from which the shellfish were obtained (Food and Drug Administration, 1983).

An outbreak of norovirus illness occurred in 1983 in Rochester, New York. A survey indicated that 84 (43%) of 196 people interviewed contracted norovirus-like illness after eating "cooked" clams served at a clambake. The clams were harvested off the coast of Massachusetts from waters known to be contaminated by untreated municipal sewage (Truman et al., 1987). This outbreak may have been avoided if the clams had been fully cooked or if the shellfish had been obtained from waters meeting the standards of the National Shellfish Sanitation Program. Other U.S. outbreaks of norovirus have been associated with cooked oysters (Kirkland et al., 1996; McDonnell et al., 1997). In an outbreak of norovirus illness that affected 129 individuals in Florida in 1995, surveys indicated that sick individuals had eaten raw, cooked, and thoroughly cooked oysters (McDonnell et al., 1997). Those who reported eating only thoroughly cooked oysters made a subjective judgment and the actual degree of cooking remains unknown. It is unlikely that thoroughly cooked oysters would cause illness unless they were recontaminated after cooking, perhaps with dirty gloves used during shucking or with contaminated shucking knives. There was speculation that the source of norovirus contamination was the overboard dumping of feces during a community-wide outbreak of gastroenteritis (McDonnell et al., 1997). This is not the first instance when overboard disposal of feces or vomit led to contaminated shellfish beds and outbreaks of illness. Kohn et al. (1995) conducted a survey of crew members from oyster harvesting boats and learned that 85% of the boats disposed of sewage overboard. Although this is against regulations, monitoring for compliance is very difficult. Berg et al. (2000) also reported the overboard disposal of sewage by oyster harvesters in Louisiana as the likely source of contaminated oysters in two or more outbreaks. New Zealand experienced a number of oyster-associated outbreaks of norovirus illness, and overboard disposal of sewage from recreational boats was suggested as a likely source of contamination (Simons et al., 2001).

Using molecular biological methods, our laboratory detected both noro- and hepatitis A viruses in clams imported to the United States from China (Kingsley et al., 2002b). These clams were implicated in an outbreak of norovirus illness in New York State. Because the clams were labeled and imported as cooked clams on the half shell, the restaurant served them at their buffet without any heating. Subsequent studies revealed that the clams were raw. No individuals were reported to have developed hepatitis A from these clams. Sequence analysis revealed that both norovirus and hepatitis A virus RNAs contained sequences characteristic of Asian strains of these viruses (Kingsley et al., 2002b).

Other countries have battled with shellfish-associated norovirus outbreaks. A widespread outbreak of norovirus illness infecting more than 2,000 people occurred in Australia in 1978 and was subsequently linked to oyster consumption (Murphy et al., 1979; Grohmann et al., 1980). Another outbreak in Australia affected 25 of 28 people who ate raw oysters at a hotel (Linco and Grohmann, 1980). In response to these outbreaks, in 1981 the government of New South Wales, Australia, implemented regulation requiring that all shellfish be subjected to depuration (Ayres, 1991). A study was undertaken to determine whether depurated oysters from two sites in Australia would cause illness in human volunteers (Grohmann et al., 1981). Oysters from one site produced norovirus illness in 52 people, but none from the second site caused illness. Depuration requirements were recently abandoned in New South Wales, since it has become well recognized that enteric viruses persist within shellfish tissues for periods much longer than the duration of commercial depuration. The extended relay of shellfish offers some hope of reducing or eliminating enteric viruses. Oysters were the presumptive vehicle of norovirus transmission to residents of New South Wales and Queensland in a 1996 outbreak involving 97 cases (Stafford et al., 1997). Although New South Wales required depuration and Queensland did not, these outbreaks demonstrate that depurated oysters can transmit enteric viruses just like nondepurated shellfish.

In Japan, oysters and clams have both been associated with norovirus illness. A study of 80 outbreaks of acute gastroenteritis from 1984 to 1987 revealed that 53 outbreaks were associated with the consumption of oysters (Sekine et al., 1989). Clams imported from China caused 22 cases of norovirus illness and four cases of hepatitis A in Japan (Furuta et al., 2003). Both norovirus and hepatitis A virus were detected in these clams. Another study reported five outbreaks of norovirus illness from eating raw oysters (Otsu, 1999).

Norovirus outbreaks in Europe have also been reported. Cockles were linked to an early outbreak of norovirus illness (Appleton and Pereira, 1977). Raw mussels and clams were the apparent vehicles of transmission for an outbreak of norovirus illness in Italy and a dose-response relationship was observed between the amount of shellfish consumed and illness (Mele et al., 1989). Mussels in a cocktail were responsible for an outbreak at a national convention in the United Kingdom and again a dose-response relationship was noted (Gray and Evans, 1993). English oysters that had been depurated and served at a birthday party caused nine cases of norovirus gastroenteritis (Ang, 1998).

2.3. Hepatitis E Virus

Hepatitis E virus (HEV) is a nonenveloped, positive-strand RNA virus morphologically similar to caliciviruses. Hepatitis E infection occurs via the fecal-oral route and is a major cause of epidemic as well as sporadic viral hepatitis in endemic regions of Asia, the Indian subcontinent, Africa, and the Americas (Velazquez et al., 1990; Arankalle et al., 1994; Clayson et al., 1997; Balayan, 1997). Hepatitis E is less frequently detected in Europe and only a handful of cases have been reported in the United States. In some developing countries, HEV may account for more than 50% of acute viral hepatitis (Balayan, 1997; Clayson et al., 1997). Like hepatitis A virus, HEV normally causes an acute, self-limiting disease with a low mortality rate; however, during pregnancy a mortality rate between 15% and 25% has been reported (Mast and Krawczynski, 1996). Epidemiological studies have shown that HEV transmission occurs mostly by ingestion of contaminated water (Arankalle et al., 1994; Bayalan, 1997), with few significant contributions of person-to-person or food-borne transmission established to date. Shellfish consumption was considered a risk factor for sporadic cases of hepatitis E in Eastern Sicily (Capopardo et al., 1997), and undercooked cockles and muscles were associated with hepatitis E in India (Tomar, 1998). Epidemiological follow-up is difficult with this virus because of a 15- to 60-day incubation period and the sporadic distribution of illnesses. To date, no large outbreaks of shellfish-associated hepatitis E have been reported. Hepatitis E virus should be considered a potential emerging pathogen in the United States and other countries.

3.0. OUTBREAK PREVENTION

3.1. Monitoring and Regulations

The United States and the European Union have implemented criteria for the harvesting and processing of molluscan shellfish. Under the guidelines of the National Shellfish Sanitation Program (NSSP) Model Ordinance (Anon., 1999) and the NSSP Manual of Operations, shellfish harvesting in the United States has been historically based on water quality criteria derived from sanitary surveys of shellfish growing water. The surveys are based on the levels of total or fecal coliforms in water and are determined during periodic water sampling and testing. Water testing has served the country well since its implementation in 1925 (Frost, 1925). Sanitary surveys were originally undertaken to reduce the incidence of typhoid fever among shellfish consumers and a successful outcome was achieved. Today, shellfish growing waters are classified as approved, conditionally approved, restricted, conditionally restricted, or prohibited, depending on the level of coliform contamination.

According to the Model Ordinance (Anon., 1999), shellfish obtained from waters with a most probable number (MPN) of fecal coliforms <14/100 ml are classified as approved for shellfish harvesting and direct sale. Shellfish waters are classified as restricted if the fecal coliform levels are under 88/100 ml, while shellfish are prohibited from harvest when the waters have >88 fecal coliforms/100 ml. Because water classification is an ongoing process and the history of a site can be determined by an examination of past data, some areas with intermittent contamination may be classified as conditionally approved and conditionally restricted. Such waters come under a management plan and shellfish are permitted to be harvested for direct sale or for depuration or relaying when the criteria of the plan are met. Shellfish from restricted areas can be harvested only if they are subjected to depuration or relaying before they enter the marketplace. Shellfish from prohibited areas may never be harvested or marketed.

In contrast, the EU follows Council Directive 91/492/EEC (Anon., 1991), which regulates shellfish based on the levels of fecal coliforms or *E. coli* in the shellfish meats, rather than in shellfish growing waters. Under this system, shellfish meats are classified in any of four categories: A, B, C, or D, as shown in Table 9.1. The numbers of fecal coliforms and *E. coli* are also determined by MPN, but the results are reported per 100 g of shellfish meat. The differences between the U.S. and EU standards are, in large part, due to the fact that there are many shellfish growing waters in the United States that are perceived to be clean enough for direct harvest and sale of shellfish,

Table 9.1 Council Directives for the Production and Marketing of Shellfish According to European Union Standards Based on Fecal Coliform or *E. coli* Levels in the Meats[a]

Classification	*Fecal Coliform Limit*	*E. coli Limit*
A. Sell without processing	<300 MPN/100 g	<230 MPN/100 g
B. Depurate or relay	<6,000 MPN/100 g	<4,600 MPN/100 g
C. Prolonged relay	<60,000 MPN/100 g	N.A.
D. Prohibited	>60,000 MPN/100 g	N.A.

N.A., not applicable; MPN, most probable number.
[a] From Anon., 1991.

whereas, water quality is seldom adequate in Europe for direct shellfish harvest and sale. Hence, most shellfish in the EU must be depurated or relayed before they can be marketed whereas depuration is seldom required in the United States. Regardless of which standard is used, the levels of fecal coliforms are not a good indicator of the virological quality of shellfish, because enteric viruses persist longer than coliforms within shellfish tissues and they depurate poorly. Therefore, reliance on coliforms as a predictive index for virus presence is ineffectual. Only when coliform levels are high do the standards prevent the direct sale of potentially virus-laden shellfish. Viruses tend to be more resilient than coliforms to the effects of sewage treatment processes and environmental stressors; therefore, water containing low or negligible levels of coliforms, because of effective inactivation, may contain high levels of enteric viruses.

Shellfish growing waters are often affected by the disposal of sewage from commercial and recreational vessels (Kohn et al., 1995; McDonnell et al., 1997; Simons et al., 2001), leading to sporadic contamination events that are difficult to assess by either the U.S. or EU methods. Neither method is foolproof. When the incidence of hepatitis A was assessed after an outbreak in Florida, it was established that the attack rate in seafood establishments was 1.9 per 1,000 dozen oysters eaten (Desenclos et al., 1991). Such low-level contamination would likely miss detection using the EU meat standard, because of the low numbers of samples tested, the likely randomness of the contamination, and the lack of correlation between coliforms and enteric viruses within the meats. The utility of the water standard is also limited by the lack of correlation between coliforms and viruses, the generally lower numbers of coliforms (and viruses) in the water compared with the meats, and the lack of homogeneity of the water due to tides, winds, currents, and non-point-source contamination events.

3.2. Enhanced Monitoring and Enforcement

Several areas are in need of better monitoring and enforcement if outbreaks are to be reduced. Tighter enforcement of laws against dumping waste in shellfish harvesting areas would reduce the incidence of enteric virus illness. An area in need of enhanced monitoring is the illegal practice of harvesting shellfish from closed areas, a practice called poaching or bootlegging. Some outbreaks have been attributed to the sale and consumption of poached or bootlegged shellfish (Morse et al., 1986; Desenclos et al., 1991). Typically, the penalties for those who perpetrate such crimes have been relatively small. According to U.S. and EU guidelines, all lots of shellfish must contain tags (U.S.) or health marks (EU), which label the lot with information that allows the shellfish to be tracked to their source. This is important in outbreak investigations as health authorities seek epidemiological evidence to curb the spread of disease. Enhanced monitoring of tags and health marks would serve as a deterrent against poachers.

Tighter enforcement of import laws is needed to restrict the importation of tainted shellfish. Shellfish exported from China, England, Ireland,

Peru, and other countries have been apparent vehicles of enteric virus illness. Exporting countries are required to subscribe to the standards in place for the receiving country. Transactions are often sealed with a memorandum of understanding (MOU) between the exporting and importing nations. Failure to comply with the MOU should impart dire consequences upon the exporting country, including the withdrawal of the MOU in cases that show wanton disregard for the requirements of the agreement. Harvesters, processors, and shippers should meet criteria deemed necessary to ensure the safety of their merchandise. Hazard analysis critical control point (HACCP) plans should be in place to monitor factors that are important in ensuring shellfish safety.

3.3. Improved Sewage Treatment Plants

Another intervention to reduce virus levels in shellfish would be to improve upon sewage treatment plants and septic systems, particularly in coastal regions near rivers, lakes, and shellfish-growing areas. Adequate monitoring and maintenance of treatment facilities are important to reduce viral loads emitted into the environment. The United States routinely chlorinates effluent wastewater and this practice has some penetrating effects on particulate matter that contains potential pathogens. After treatment, the chlorine may be inactivated by sodium thiosulfate treatment. In contrast, the EU often uses ultraviolet irradiation to treat sewage effluent. The lack of penetrating ability, particularly in turbid water or in water containing particulate matter, and the lack of any residual properties imparted by the UV, would be expected to allow some viruses and bacteria to escape inactivation. The technology is available to eliminate or substantially reduce enteric viruses from sewage; however, few if any engineers design sewage treatment facilities with virus reduction in mind. Treatment plant maintenance and operation should be tightly controlled so that the facility works at optimal efficiency.

3.4. Analytical Techniques

Direct monitoring for viruses in water or shellfish should be encouraged using molecular biological methods, namely reverse transcription–polymerase chain reaction (RT-PCR). New RT-PCR protocols continue to be developed along with improved methods to extract the viruses from water and shellfish. Such methods are limited in their practical application because they fail to differentiate infectious from noninfectious viruses (Richards, 1999). Direct assays for infectious viruses would be desirable; however, wild-type hepatitis A and E viruses, noroviruses, sapoviruses, and the astroviruses defy cell culture propagation. Rotaviruses are difficult to assay. Because poliovirus is easily propagated in cell cultures, it was proposed as an indicator for the possible presence of other human enteric viruses when vaccine strains were commonly in use (Richards, 1985). The near eradication of poliovirus and the fear that vaccine strains might revert to wild-type strains have prompted the elimination of vaccine distribution in all but a few select areas. New virus propagation assays are needed to adequately assess shellfish safety from a virological perspective.

3.5. Processing Strategies

Other intervention strategies to reduce or eliminate enteric virus contamination in shellfish should be implemented on multiple fronts. Lessons from previous outbreaks should be heeded. Perhaps the simplest intervention available to consumers is cooking. In most outbreaks, raw or only lightly cooked molluscs appear to be the primary vehicles of infection. Alternative processing strategies, such as irradiation and high hydrostatic pressure processing, have been proposed. The high levels of irradiation required to inactivate enteric viruses from shellfish imparts undesirable flavor characteristics to the meats. On the other hand, high hydrostatic pressure processing for 5 min was shown effective in inactivating 7 $\log_{10}$ of hepatitis A virus and feline calicivirus, a surrogate for the noroviruses (Kingsley et al., 2002a). High pressure inactivates viruses by denaturation of capsid proteins (Kingsley et al., 2002a) and sanitizes the shellfish from bacterial pathogens and spoilage organisms as well. Treated oysters are reported to taste like the raw product.

3.6. Disease Reporting and Epidemiological Follow-Up

Improved reporting and epidemiological follow-up are needed to understand the magnitude of enteric virus illnesses and to reduce the size of outbreaks once they occur. Such reporting has been effective in Italy where 35 participating, local health units link incidence notification with serology and follow-up questionnaires in their surveillance for hepatitis A (Mele et al., 1986, 1997). In a survey of 10 EU countries, eight had national databases for hepatitis A statistics (Lopman et al., 2002). Likewise, the Centers for Disease Control and Prevention have maintained statistics on reported cases of hepatitis A in the United States. Although some countries maintain statistics on the number of cases of hepatitis A reported, few determine the source of the illness due to the high cost for epidemiological follow-up. Norovirus illnesses are not notifiable diseases in most countries, meaning that there are no formal systems to obtain accurate statistics on the number of illnesses.

3.7. Hygienic Practices

Most outbreaks of shellfish-associated viral illness appear to be from shellfish contaminated within their natural environment; however, some cases, particularly those involving cooked shellfish, may actually be from product contamination by shuckers, handlers, or fomites. The contamination of foods by unsanitized hands of food handlers has led to numerous outbreaks of hepatitis A and noroviruses (Richards, 2001). Better enforcement of handwashing practices may prevent some potential outbreaks from becoming a reality. Likewise, all sanitary standards generally applied in the food industry should be enforced in the shellfish industry, especially on harvesting boats, and in processing plants, transport facilities, and restaurants. Better education and monitoring of food handlers are needed to ensure compliance with food sanitation requirements.

4.0. SUMMARY

Numerous outbreaks of shellfish-borne enteric virus illness have been reported worldwide. Most notable among the outbreaks are those involving norovirus illness and hepatitis A. Lessons learned from outbreak investigations indicate that most outbreaks are preventable. Anthropogenic sources of contamination will continue to invade shellfish growing waters, and shellfish, by their very nature, will continue to bioconcentrate these contaminants, including enteric viruses. There is no quick fix for enteric virus contamination of shellfish; however, vigilance on behalf of the industry, regulatory agencies, and the consumer could substantially reduce the incidence of illness. Enhanced monitoring in all areas of shellfish production, harvesting, distribution, and processing would help to reduce viral illnesses. Pollution abatement and improved hygienic practices on behalf of the industry and consumers are needed. New processing and analytical technologies, such as high hydrostatic pressure processing and molecular biological assays, will enhance shellfish safety and continue to provide new avenues to protect the consumer and the industry. Better reporting and epidemiological follow-up of outbreaks are keys to the development of interventions against the foodborne transmission of viral infections.

5.0. REFERENCES

Ang, L. H., 1998, An outbreak of viral gastroenteritis associated with eating raw oysters. *Commun. Dis. Public Health* 1:38–40.

Anon., 1991, Council Directive 91/492/EEC. Laying down the health conditions for the production and placing on the market of live bivalve mollusks. *Official J. Eur. Comm.* L268:1–14

Anon., 1999, National Shellfish Sanitation Program Model Ordinance IV. Shellstock growing areas. Department of Health and Human Services, U.S. Food and Drug Administration, Washington, DC.

Appleton, H., and Pereira, M. S., 1977, A possible virus aetiology in outbreaks of food-poisoning from cockles. *Lancet* 1(8015):780–781.

Arankalle, V. A., Chadha, M. S., Tsarev, S. A., Emerson, S. U., Risbud, A. R., Banerjee, K., and Purcell, R. H., 1994, Seroepidemiology of water-borne hepatitis in India and evidence for a third enterically-transmitted hepatitis agent. *Proc. Natl. Acad. Sci. U. S. A.* 91:3428–3432.

Ayers, P. A., 1991, The status of shellfish depuration in Australia and South-east Asia, in: *Molluscan Shellfish Depuration* (W. S. Otwell, G. E. Rodrick, and R. E. Martin, eds.), CRC Press, Boca Raton, FL, pp. 287–321.

Balayan, M. S., 1997, Epidemiology of hepatitis E virus infection. *J Viral Hepat* 4:155–65.

Berg, D. E., Kohn, M. A., Farley, T. A., and McFarland, L. M., 2000, Multi-state outbreaks of acute gastroenteritis traced to fecal-contaminated oysters harvested in Louisiana. *J. Infect. Dis.* 181(Suppl. 2):S381–386.

Bosch, A., Sanchez, G., Le Guyader, F., Vanaclocha, H., Haugarreau, L., and Pinto, R. M., 2001, Human enteric viruses in Coquina clams associated with a large hepatitis A outbreak. *Water Sci. Technol.* 43:61–65.

Cacopardo, B., Russo, R., Preiser, W., Benanti, F., Brancati, F., and Nunnari, A., 1997, Acute hepatitis E in Catania (eastern Sicily) 1980–1994. The role of hepatitis E virus. *Infection* 25:313–316.

Caredda, F., Antinori, S. R. T., Pastecchia, C., Zavaglia, C., and Moroni, M., 1985, Clinical features of sporadic non-A, non-B hepatitis possibly associated with faecal-oral spread. *Lancet* 2:444–445.

Centers for Disease Control, 1982, Enteric illness associated with raw clam consumption—New York. *Morbid. Mortal. Weekly Rep.* 31:449–450.

Centers for Disease Control, 1993, Multistate outbreak of viral gastroenteritis related to consumption of oysters—Louisiana, Maryland, Mississippi and North Carolina, 1993, *Morbid. Mortal. Weekly Rep.* 42:945–948.

Clayson, E. T., Shrestha, M. P., Vaughn, D. W., Snitbhan, R., Shrestha, K. B., Longer, C. F., and Innis, B. L., 1997, Rates of hepatitis E virus infection and disease among adolescents and adults in Kathmandu, Nepal. *J. Infect. Dis.* 176:763–766.

Conaty, S., Bird, P., Bell, G., Kraa, E., Grohmann, G., and McAnulty, J. M., 2000, Hepatitis A in New South Wales, Australia from consumption of oysters: the first reported outbreak. *Epidemiol. Infect.* 124:121–130.

Cooksley, W. G., 2000, What did we learn from the Shanghai hepatitis A epidemic? *J. Viral Hepat.* 7(Suppl. 1):1–3.

Desenclos, J. C., Klontz, K. C., Wilder, M. H., Nainan, O. V., Margolis, H. S., and Gunn, R. A., 1991, A multistate outbreak of hepatitis A caused by the consumption of raw oysters. *Am. J. Public Health* 81:1268–1272.

Food and Drug Administration, 1983, England Shellfish Program Review. U.S. Food and Drug Administration, U.S. Public Health Service, Washington, DC.

Frost, H. W., 1925, Report of committee on the sanitary control of the shellfish industry in the United States. *Public Health Rep.*, 53(Suppl.):1–17.

Fujiyama, S., Akahoshi, M., Sagara, K., Sato, T., and Tsurusaki, R., 1985, An epidemic of hepatitis A related to ingestion of raw oysters. *Gastroenterol. Jpn.* 20:6–13.

Furuta, T., Akiyama, M., Kato, Y., and Nishio, O., 2003, A food poisoning outbreak caused by purple Washington clams contaminated with norovirus (Norwalk-like virus) and hepatitis A virus. *Kansenshogaku Zasshi* 77:89–94.

Gerba, C. P., and Goyal, S. M., 1978, Detection and occurrence of enteric viruses in shellfish: a review. *J. Food Prot.* 41:743–754.

Glass, R. I., Bresee, J., Jiang, B., Gentsch, J., Ando, T., Fankhauser, R., Noel, J., Parashar, U., Rosen, B., and Monroe, S. S., 2001, Gastroenteritis viruses: an overview. *Novartis Found. Symp.* 238:5–19.

Graham, D. Y., Jiang, X., Tanaka, T., Opekum, A. R., Madore, H. P., and Estes, M. K., 1994, Norwalk virus infection of volunteers: new insights based on improved assays. *J. Infect. Dis.* 170:34–43.

Gray, S. F., and Evans, M. R., 1993, Dose-response in an outbreak of non-bacterial food poisoning traced to a mixed seafood cocktail. *Epidemiol. Infect.* 110:583–590.

Grohmann, G. S., Greenberg, H. B., Welch, B. M., and Murphy, A. M., 1980, Oyster-associated gastroenteritis in Australia: the detection of Norwalk virus and its antibody by immune electron microscopy and radioimmunoassay. *J. Med. Virol.* 6:11–19.

Grohmann, G. S., Murphy, A. M., Christopher, P. J., Auty, E., and Greenberg, H. B., 1981, Norwalk virus gastroenteritis in volunteers consuming depurated oysters. *Aust. J. Exp. Biol. Med. Sci.* 59:219–228.

Gunn, R. A., Janowski, H. T., Lieb, S., Prather, E. C., and Greenberg, H. B., 1982, Norwalk virus gastroenteritis following raw oyster consumption. *Am. J. Epidemiol.* 115:348–351.

Halliday, M. L., Kang, L. Y., Zhou, T. K., Hu, M. D., Pan, Q. C., Fu, T. Y., Huang, Y. S., and Hu, S. L., 1991, An epidemic of hepatitis A attributable to the ingestion of raw clams in Shanghai, China. *J. Infect. Dis.* 164:852–859.

Haruki, K., Seto, Y., Murakami, T., and Kimura, T., 1991, Pattern of shedding small, round-structured virus particles in stools of patients of outbreaks of food-poisoning from raw oysters. *Microbiol. Immunol.* 35:83–86.

Iversen, A. M., Gill, M., Bartlett, C. L., Cubitt, W. D., and McSwiggan, D. A., 1987, Two outbreaks of foodborne gastroenteritis caused by a small round structured virus: evidence for prolonged infectivity in a food handler. *Lancet* 2(8558): 556–558.

Kingsley, D. H., Hoover, D. G., Papafragkou, E., and Richards, G. P., 2002a, Inactivation of hepatitis A virus and a calicivirus by high hydrostatic pressure. *J. Food Prot.* 65:1605–1609.

Kingsley, D. H., Meade, G. K., and Richards, G. P., 2002b, Detection of both hepatitis A virus and Norwalk-like virus in imported clams associated with food-borne illness. *Appl. Environ. Microbiol.* 68:3914–3918.

Kingsley, D. H., and Richards, G. P., 2003, Persistence of hepatitis A virus in oysters. *J. Food Prot.* 66:331–334.

Kirkland, K. B., Meriwether, R. A., Leiss, J. K., and Mac Kenzie, W. R., 1996, Steaming oysters does not prevent Norwalk-like gastroenteritis. *Public Health Rep.* 111:527–530.

Kiyosawa, K., Gibo, Y., Sodeyama, T., Furuta, K., Imai, H., Yoda, H., Koike, Y., Yoshizawa, K., and Furuta, S., 1987, Possible infectious causes in 651 patients with acute viral hepatitis during a 10-year period (1976–1985). *Liver* 7:163–168.

Kohn, M. A., Farley, T. A., Ando, T., Curtis, M., Wilson, S. A., Jin, Q., Monroe, S. S., Baron, R. C., McFarland, L. M., and Glass, R. I., 1995, An outbreak of Norwalk virus gastroenteritis associated with eating raw oysters. Implications for maintaining safe oyster beds. *JAMA* 273:1492.

Konno, T., Chimoto, T., Taneichi, K., Deno, M., Yoshizaya, T., Kimura, O., Sibaki, H., Konno, M., and Kojima, H., 1983, Oyster-associated hepatitis A. *Hokkaido Igaku Zasshi* 58:553–555.

Leoni, E., Bevini, C., Degli Esposti, S., and Graziano, A., 1998, An outbreak of intrafamiliar hepatitis A associated with clam consumption: epidemic transmission to a school community. *Eur. J. Epidemiol.* 14:187–192.

Linco, S. J., and Grohmann, G. S. 1980, The Darwin outbreak of oyster-associated viral gastroenteritis. *Med. J. Aust.* 1:211–213.

Lopman, B., van Duynhoven, Y., Hanon, F. X., Reacher, M., Koopmans, M., Brown, D., and Consortium on Foodborne Viruses in Europe, 2002, Laboratory capability in Europe for foodborne viruses. *Eur. Surveill.* 7:61–65.

Lucioni, C., Cipriani, V., Mazzi, S., and Panunzio, M., 1998, Cost of an outbreak of hepatitis A in Puglia, Italy. *Pharmacoeconomics* 13:257–266.

Mackowiak, P. A., Caraway, C. T., and Portnoy, B. L., 1976, Oyster-associated hepatitis: lessons from the Louisiana experience. *Am. J. Epidemiol.* 103:181–191.

Maguire, H., Heptonstall, J., and Begg, N. T., 1992, The epidemiology and control of hepatitis A. *Commun. Dis. Rep. CDC Rev*. 2:R114–117.

Mast, E. E., and Krawczynski, K., 1996, Hepatitis E: an overview. *Annu. Rev. Med.* 47:257–266.

McDonnell, S., Kirkland, K. B., Hlady, W. G., Aristeguieta, C., Hopkins, R. S., Monroe, S. S., and Glass, R. I., 1997, Failure of cooking to prevent shellfish-associated viral gastroenteritis. *Arch. Intern. Med.* 157:111–116.

Mead, P. S., Slutsker, L., Dietz, V., McCaig, L. F., Bresee, J. S., Chapiro, C., Griffin, P. M., and Tauxe, R. V., 1999, Food-related illness and deaths in the United States. *Emerg. Infect. Dis.* 5:607–625.

Mele, A., Rosmini, F., Zampieri, A., and Gill, O. N., 1986, Integrated epidemiological system for acute viral hepatitis in Italy (SEIEVA): description and preliminary results. *Eur. J. Epidemiol.* 2:300–304.

Mele, A., Rastelli, M. G., Gill, O. N., di Bisceglie, D., Rosmini, F., Pardelli, G., Valtriani, C., and Patriarchi, P., 1989, Recurrent epidemic hepatitis A associated with raw shellfish, probably controlled through public health measures. *Am. J. Epidemiol.* 130:540–546.

Mele, A., Stroffolini, T., Palumbo, F., Gallo, G., Ragni, G., Balocchini, E., Tosti, M. E., Corona, R., Marzolini, A., and Moiraghi, A., 1997, Incidence of and risk factors for hepatitis A in Italy: public health indications from a 10-year surveillance. SEIEVA Collaborating Group. *J. Hepatol.* 26:743–747.

Morse, D. L., Guzewich, J. J., Hanrahan, J. P., Stricof, R., Shayegani, M., Deibel, R., Grabau, J. C., Nowak, N. A., Herrmann, J. E., Cukor, G., and Blacklow, N. R., 1986, Widespread outbreak of clam- and oyster-associated gastroenteritis. Role of Norwalk virus. *N. Engl. J. Med.* 314:678–681.

Murphy, A. M., Grohmann, G. S., Christopher, P. J., Lopez, W. A., Davey, G. R., and Millsom, R. H., 1979, An Australia-wide outbreak of gastroenteritis from oysters caused by Norwalk virus. *Med. J. Aust.* 2:329–333.

New York State Department of Health, 1983, Clam-associated enteric illness in New York May–September 1982: a preliminary report. New York State Department of Health, Albany, NY, pp. 1–10.

O'Mahony, M. C., Gooch, C. D., Smyth, D. A., Thrussell, A. J., Bartlett, C. L., and Noah, N. D., 1983, Epidemic hepatitis A from cockles. *Lancet* 1(8323):518–520.

Otsu, R., 1999, Outbreaks of gastroenteritis caused by SRSVs from 1987 to 1992 in Kyushu, Japan: four outbreaks associated with oyster consumption. *Eur. J. Epidemiol.* 15:175–180.

Polakoff, S., 1990, Report of clinical hepatitis A from public health and hospital microbiology laboratories to the PHLS Comunicable Disease Surveillance Centre during the period 1980–1988. *J. Infect.* 21:111–117.

Portnoy, B. L., Mackowiak, P. A., Caraway, C. T., Walker, J. A., McKinley, T. W., and Klin, C. A. Jr., 1975, Oyster-associated hepatitis. Failure of shellfish certification program to prevent outbreaks. *JAMA* 233:1065–1068.

Richards, G. P., 1985, Outbreaks of enteric virus illness in the United States: requisite for development of viral guidelines. *J. Food Prot.* 48:815–832.

Richards, G. P., 1987, Shellfish-associated enteric virus illness in the United States, 1934–1984. *Estuaries* 10:84–85.

Richards G. P., 1988, Microbial purification of shellfish: a review of depuration and relaying. *J. Food Prot.* 51:218–251.

Richards, G. P., 1991, Shellfish depuration, in: *Microbiology of Marine Food Products* (D. R. Ward and C. R. Hackney, eds.), Van Nostrand Reinhold, New York, pp. 395–428.

Richards, G. P., 1999, Limitations of molecular biological techniques for assessing the virological safety of foods. *J. Food Prot.* 62:691–697.

Richards, G. P., 2001, Enteric virus contamination of foods through industrial practices: a primer on intervention strategies. *J. Indust. Microbiol. Biotechnol.* 17:117–125.

Richards, G. P., 2003, The evolution of molluscan shellfish safety, in: *Molluscan Shellfish Safety* (A. Villalba, B. Reguera, J. L. Romalde, and R. Beira, eds.), Conselleria de Pesca e Asuntos Maritimos, Xunta de Galicia and Intergovernmental Oceanographic Commission of UNESCO, Galicia, Spain, pp. 221–245.

Richards, G. P., Watson, M. A., and Kingsley, D. H. 2004, A SYBR green, real-time RT-PCR method to detect and quantitate Norwalk virus in stools. *J. Virol. Methods* 116: 63–70.

Rippey, S. R., 1994, Infectious diseases associated with molluscan shellfish consumption. *Clin. Microbiol. Rev.* 7:419–425.

Romalde, J. L., Torrado, I., Ribao, C., and Barja, J. L. 2001, Global market: shellfish imports as a source of reemerging food-borne hepatitis A virus infections in Spain. *Int. Microbiol.* 4:223–226.

Sanchez, G., Pinto, R. M., Vanaclocha, H., and Bosch, A., 2002, Molecular characterization of hepatitis A virus isolates from a transcontinental shellfish-borne outbreak. *J. Clin. Microbiol.* 40:4148–4155.

Sekine, S., Okada, S., Hayashi, Y., Ando, T., Terayama, T., Yabuuchi, K., Miki, T., and Ohashi, M., 1989, Prevalence of small round structured virus infections in acute gastroenteritis outbreaks in Tokyo. *Microbiol. Immunol.* 33:207–217.

Simons, G., Greening, G., Gao, W., and Campbell, D., 2001, Raw oyster consumption and outbreaks of viral gastroenteritis in New Zealand: evidence for risk to the public's health. *Aust. N. Z. J. Public Health* 25:234–240.

Stafford, R., Strain, D., Heymer, M., Smith, C., Trent, M., and Beard, J., 1997, An outbreak of virus gastroenteritis following consumption of oysters. *Commun. Dis. Intell.* 21:317–320.

Tang, Y. W., Wang, J. X., Xu, Z. Y., Guo, Y. F., Qian, W. H., and Xu, J. X., 1991, A serologically cofirmed, case-control study, of a large outbreak of hepatitis A in China, associated with consumption of clams. *Epidemiol. Infect.* 107:651–657.

Tomar, B. S., 1998, Hepatitis E in India. *Zhonghua Min Guo Xiao Er Ke Yi Xue Hui Za Zhi* 39:150–156.

Truman, B. I., Madore, H. P., Menegus, M. A., Nitzkin, J. L., and Dolin, R., 1987, Snow Mountain agent gastroenteritis from clams. *Am. J. Epidemiol.* 126:516–525.

Velazquez, O., Stetler, H. C., Avila, C., Ornelas, G., Alvarez, C., Hadler, S. C., Bradley, D. W., and Sepulveda, J., 1990, Epidemic transmission of enterically transmitted non-A, non-B hepatitis in Mexico, 1986–1987. *JAMA* 263:3281–3285.

Wang, J. V., Hu, S. L., Liu, H. Y., Hong, Y. L., Cao, S. Z., and Wu, L. F., 1990, Risk factor analysis of an epidemic of hepatitis A in a factory in Shanghai. *Int. J. Epidemiol.* 19:435–438.

White, K. E., Osterholm, M. T., Mariotti, J. A., Korlath, J. A., Lawrence, D. H., Ristinen, T. L., and Greenberg, H. B., 1986, A foodborne outbreak of Norwalk virus gastroenteritis: evidence for post recovery transmission. *Am. J. Epidemiol.* 124:120–126.

Xu, Z. Y., Li, Z. H., Wang, J. X., Xiao, Z. P., and Dong, D. X., 1992, Ecology and prevention of a shellfish-associated hepatitis A epidemic in Shanghai, China, *Vaccine* 10 (Suppl. 1):S67–68.

CHAPTER 10

Epidemiology of Viral Food-borne Outbreaks

Craig W. Hedberg

1.0. INTRODUCTION

In 1978, Greenberg et al. (1978) reported serologic evidence that Norwalk or Norwalk-like virus was the likely cause of 8 of a series of 25 outbreaks of acute gastroenteritis of unknown etiology for which acute and convalescent sera had been collected. The outbreaks occurred over a period of 12 years in a variety of settings including cruise ships, schools, nursing homes, and the community. Four years later, Kaplan and colleagues demonstrated that Norwalk virus outbreaks had characteristic clinical and epidemiologic features, that a high proportion of outbreaks with these features were caused by Norwalk-like viruses (now known as noroviruses), and that these accounted for most outbreaks of acute nonbacterial gastroenteritis reported in the United States (Kaplan et al., 1982a, 1982b).

During the ensuing two decades, a variety of diagnostic methods were developed and used, culminating in the widespread use of reverse transcription–polymerase chain reaction (RT-PCR) to detect viral RNA (Atmar and Estes, 2001). Sequencing of PCR products has been extremely valuable for classifying these viruses and establishing genetic relationships between virus strains (Ando et al., 2000). However, although these impressive developments in diagnostic methods have broadened our understanding of the epidemiology of noroviruses, their epidemiology has not fundamentally changed. The types of outbreaks currently making headlines reflect the same patterns that were recognized early on. Whether more rigorous investigation and laboratory confirmation of outbreaks will lead to better outbreak control and public health prevention measures remains to be determined.

Although several groups of viruses may be transmitted through the fecal-oral route, norovirus and hepatitis A virus (HAV) are recognized as the most important human food-borne viruses due to the number of outbreaks and people affected by noroviruses and the potential severity of illnesses caused by HAV (Koopmans and Duizer, 2004). An excellent review of food-borne HAV has recently been published (Fiore, 2004), and this chapter will focus on noroviruses.

2.0. OUTBREAK DETECTION, INVESTIGATION, AND SURVEILLANCE

2.1. Outbreak Detection Methods

Outbreaks of gastrointestinal illness are detected by one of two primary means. The first involves recognizing a pattern of illnesses among persons with a common exposure, such as attending a banquet (Hedberg, 2001). Detecting these outbreaks requires that a number of people who become ill have some reason to discuss their illness with others in their group. This allows the group, *a priori*, to attribute the illnesses to the common event and frequently leads them to report the outbreak to public health officials. The primary implications are that large outbreaks are more likely to be detected, and there is also a bias toward detecting outbreaks involving socially cohesive groups. There is a secondary bias toward detecting outbreaks associated with commercial establishments, as many groups are reluctant to report outbreaks in which it is likely that one or more members of the group is the source.

The dependence of outbreak detection on the size of the outbreak and social cohesion of the group is modified by the agent and its characteristic incubation period. For example, contamination of food by chemical agents or large amounts of staphylococcal enterotoxin may cause a high proportion of exposed persons to become ill (high attack rate), and the illnesses may begin while the event is ongoing. Under these circumstances, outbreak detection is unavoidable.

With noroviruses, illnesses typically begin 24–48 hr after the exposure. This reduces the likelihood of detecting outbreaks in restaurants, where most other patrons are anonymous. Individuals, unaware of similar illnesses among other patrons, may write off their experience as "the flu" or attribute it to a more recently eaten meal. Thus, for "one-time" events, being part of a socially cohesive group greatly increases the likelihood of detecting an outbreak of norovirus. A further implication of the above is that being part of a group with multiple or continuous exposures over a time period that exceeds the incubation period for noroviruses also increases the likelihood of detecting an outbreak. Two settings where this has been clearly demonstrated are cruise ships and nursing homes (Centers for Disease Control and Prevention, 2001).

The second means of detecting outbreaks involves identifying an unusual cluster of cases reported through pathogen-specific surveillance. Because there is no routine clinical laboratory testing for noroviruses, there is no pathogen-specific surveillance for them, such as is conducted for *Salmonella enterica* serotypes or *Escherichia coli* O157:H7. The only food-borne virus for which routine laboratory surveillance is conducted is HAV.

Detection of outbreaks of HAV is greatly complicated by both the length and variability of the incubation period. Not only does this require people to individually remember food exposures 2–6 weeks before onset of their illness but also requires linking illnesses that could be separated by as much as a month to a common exposure. Even though most persons diagnosed with HAV are interviewed by public health investigators, most interviewers do not attempt to collect detailed food histories. If several cases identify a common restaurant, it may lead to further investigation. However, in many cases the source remains unknown (Fiore, 2004).

2.2. Public Health Investigation of Outbreaks

Outbreaks associated with events and establishments require prompt and thorough investigation to identify the agent, route of transmission, and source of the outbreak (Hedberg, 2001). Identifying the agent is a complex process that involves collecting information about the occurrence of various symptoms, plotting the distribution of case onsets by time, and collecting stool samples to test for the presence of the agent. Because vomiting, fever, and diarrhea commonly occur with many food-borne diseases, diagnosing an individual illness requires specific laboratory testing. The Centers for Disease Control and Prevention (CDC) have incorporated this logic into their criteria for confirming the etiology of an outbreak (Olsen et al., 2000). Thus, for an outbreak to have a confirmed etiology, two or more cases have to be individually confirmed by laboratory testing.

Taken on face value this seems a reasonable measure. However, the practical implication is that it reduces the effort to identify the agent into one of obtaining stool samples and getting them tested. Unfortunately, in many outbreak investigations, laboratory testing may not be adequate to identify the agent (Hedberg, 2001). In practical terms, this has discouraged investigations of outbreaks with a suspected viral etiology (Bresee et al., 2002). Thus, many outbreaks may be detected and reported to local public health officials but never investigated because of the inability to confirm the agent by laboratory testing. This has created a negative feedback loop in which outbreaks do not get investigated because no agent can be identified. Thus, the outbreaks do not get tabulated in official summaries and do not appear to contribute to the overall burden of illness. One measure of this impact was a survey of public health personnel conducted in Tennessee (Jones and Gerber, 2001). As part of a training program on food-borne disease investigations, public health workers were asked to identify the top three causes of food-borne illness. Only 5% listed Norwalk-like viruses, even though they are estimated to cause two thirds of food-borne illnesses caused by known food-borne agents (Mead et al., 1999).

Careful evaluation of clinical and epidemiological characteristics of outbreaks can allow rapid identification of agents in the absence of laboratory testing and can help guide the public health laboratory to conduct appropriate tests to confirm the presence of noroviruses. Hall et al. (2001) demonstrated that 340 of 712 (48%) outbreaks, for which no pathogen was isolated, fit a Norwalk-like virus epidemiologic profile. During this time period, only seven laboratory-confirmed outbreaks of Norwalk-like virus had been reported.

Table 10.1 Epidemiologic Criteria for Rapid Classification of Norovirus Outbreaks

Source	*Criteria*	*Application*
Kaplan et al., 1982a	1. Stool cultures negative for bacterial pathogens 2. Median incubation period 24–48 hr 3. Median duration of illness 12–60 hr 4. Vomiting in >50% of patients	Criteria as written are appropriate for retrospective evaluation of outbreaks. For rapid prospective evaluation, the incubation period and percent of patients with vomiting are the key determinants.
Hedberg and Osterholm, 1993	1–3. Same as Kaplan's 4. Percent of patients with vomiting > percent of patients with fever	For rapid prospective evaluation, the incubation period and the ratio of patients with vomiting and fever are the key determinants. Because vomiting is less common among outbreaks involving adults, the ratio of vomiting to fever increases the sensitivity of the criteria.

It is frequently claimed that the epidemiology of noroviruses is poorly understood because of a long-standing lack of diagnostic assays (Widdowson et al., 2004), which is certainly true from the standpoint of the molecular epidemiology of virus transmission through populations. However, the availability of a relatively specific epidemiologic profile (Table 10.1) more than 20 years ago should have facilitated national surveillance for norovirus outbreaks (Kaplan et al., 1982a; Hedberg and Osterholm, 1993). A model for what such a system might have looked like could be the state of Minnesota, where the use of epidemiologic criteria to define outbreaks of Norwalk-like viruses was initiated in 1982. From 1981 to 1998, Norwalk-like viruses were the most common cause of food-borne disease outbreaks, accounting for 41% of all food-borne outbreaks reported in Minnesota (Deneen et al., 2000).

The other benefit of using epidemiologic profiles is that it helps train public health investigators to rapidly and carefully collect and analyze detailed information as a routine measure. As a consequence, epidemiologists are better able to assist laboratory staff and environmental health specialists in conducting their evaluations. Rapid epidemiologic investigation of outbreaks requires the ability to interview large numbers of people quickly. Conducting these interviews may be a rate-limiting step for many local public health agencies with limited resources. To address this problem, the Minnesota Department of Health hires public health students to serve as its primary workforce (known as "Team Diarrhea") to conduct interviews. Rapid epidemiologic investigation may also be facilitated by the use of the Internet to send and receive questionnaires (Kuusi et al., 2004).

2.3. Outbreak Surveillance Systems

In the United States, outbreak surveillance usually begins with the detection and investigation of the outbreak at the local level (Olsen et al., 2000). Jurisdiction for investigating an outbreak may depend on the outbreak setting. For example, an outbreak in a restaurant may be investigated by the local environmental health agency that licenses and inspects the establishment. In contrast, an outbreak in a nursing home may be referred to nursing home regulators at the state health department for investigation. In most states, the state health department is responsible for coordinating outbreak investigations across jurisdictions and for compiling and disseminating outbreak reports. On a national level, CDC maintains surveillance only for outbreaks of food-borne and waterborne illnesses (Widdowson et al., 2004). This division of labor and information relating to surveillance of outbreaks in the United States makes it very difficult to develop a comprehensive picture of the public health impact of noroviruses.

In Europe, there is considerable variation in national surveillance systems (Lopman et al., 2003a), and efforts have been made to assess and harmonize surveillance methods (Koopmans et al., 2003). In most countries, outbreaks of gastroenteritis are investigated without regard to outbreak size or possible mode of transmission. Thus, these countries provide the most useful surveillance data on the overall impact of noroviruses. However, there is considerable variation between countries in the use of clinical criteria and laboratory confirmation for inclusion of outbreaks into surveillance databases (Lopman et al., 2003a). Denmark and France primarily investigate outbreaks that appear to be food-borne from the onset. These surveillance systems are more comparable to those in the United States.

3.0. MOLECULAR EPIDEMIOLOGY

The cloning and characterization of the Norwalk virus genome led to the development of RT-PCR, gene sequencing, molecular characterization of noroviruses, and the molecular epidemiology of norovirus outbreaks (Jiang et al., 1990; Ando et al., 1995; Vinje and Koopmans, 1996; Noel et al., 1999). In particular, it has become clear that the noroviruses are genetically diverse viruses, that multiple strains circulate simultaneously, and that individual strains may predominate over a given time period (Hale et al., 2000; Green et al., 2002; Fankhauser et al., 2002; Gallimore et al., 2004b; Lau et al., 2004; Vipond et al., 2004). Furthermore, the emergence of new strains may occur on a global basis accompanied by the increased occurrence of outbreaks in a variety of settings (Noel et al., 1999; Lopman et al., 2004; Widdowson et al., 2004).

It has long been recognized that norovirus outbreaks occur in the context of more widespread illness in the community (Hedberg and Osterholm, 1993). It has also been demonstrated that outbreaks and sporadic cases may be caused by the same virus strains in the community (Buesa et al., 2002;

Gallimore et al., 2004a; Lau et al., 2004). However, the rapid dissemination of new strains leaves open the question of primary mechanisms for their rapid spread (Noel et al., 1999; Lopman et al., 2004; Widdowson et al., 2004). Food-borne outbreaks involving transmission over wide geographic areas have been documented (Ponka et al., 1999; Berg et al., 2000; Anderson et al., 2001; Koopmans et al., 2003). However, extensive investigation of outbreaks on multiple cruise ships caused by identical strains has failed to identify a common source (Widdowson et al., 2004). Furthermore, people infected in one outbreak setting have been identified as the source for outbreaks in other areas (Fretz et al., 2003; Widdowson et al., 2004). Thus, it remains likely that noroviruses, like HAV and rotaviruses, are primarily spread from person-to-person, with outbreaks of food-borne disease serving to periodically amplify transmission (Parashar et al., 1998; Marshall et al., 2003; Fiore, 2004).

The molecular epidemiology of norovirus outbreaks also suggests that norovirus genogroups may differ in pathogenicity or have genogroup-specific characteristics that affect the dynamics of transmission in particular settings. In both the United States and United Kingdom, GII/1,4 strains were more common in nursing home outbreaks than in other settings (Fankhauser et al., 2002; Gallimore et al., 2004a). Conversely, GI strains are more common in other settings. For example, Gallimore et al. (2003) found that GI strains accounted for 38% of cruise ship outbreaks but for <10% of outbreaks in nursing homes or other institutional settings in the United Kingdom. These epidemiologic patterns may be the result of a combination of host, virologic, and environmental factors (Lopman et al., 2003b). Molecular characterization of noroviruses implicated in outbreaks across all settings will be necessary to address these questions.

4.0. MODES OF TRANSMISSION

Outbreaks of norovirus have been reported in virtually every type of institutional and food service setting and in conjunction with various types of water systems (Centers for Disease Control and Prevention, 2001). Most outbreaks appear to be a manifestation of fecal-oral route of exposure, with contaminated food or water serving as a vehicle. However, aerosol transmission of viruses likely contributes to outbreaks in many institutional settings, and environmental contamination has been implicated as well (Cheesebrough et al., 2000; Evans et al., 2002; Kuusi et al., 2002; Marks et al., 2003). The dynamics of person-to-person spread are largely unknown but likely involve combinations of the above.

4.1. Food-borne Disease Transmission

Food-borne transmission of noroviruses occurs because of human fecal contamination of a raw or ready-to-eat food item. This can occur at any point along the food chain; in production, during processing, or at the point of food service. Much of the early literature on food-borne disease transmission of noroviruses focused on contaminated shellfish (Hedberg and Osterholm, 1993). These outbreaks occurred because oysters, in particular, were harvested from waters contaminated by human sewage and consumed raw. Although they continue to have the potential to cause widespread transmission of noroviruses (Berg et al., 2000; Simmons et al., 2001; Koopmans et al., 2003), their relative importance was almost certainly inflated by a bias on the part of public health officials to investigate such outbreaks.

In Minnesota, Norwalk-like viruses accounted for 41% of confirmed food-borne outbreaks reported from 1981 to 1998 (Deneen et al., 2000). However, none of these outbreaks was attributed to shellfish. Fankhauser et al. (2002) identified contaminated food as the cause of 57% of U.S. outbreaks during 1997–2000 in which the source of transmission was investigated; none was related to oysters.

Fresh fruits and vegetables are also susceptible to contamination in the field, at harvest, and during processing. Such contamination has been responsible for large outbreaks of HAV due to produce items ranging from blueberries (Calder et al., 2003) to green onions (Centers for Disease Control and Prevention, 2003b). However, an international outbreak of norovirus associated with frozen raspberries stands out as an exceptional occurrence (Ponka et al., 1999). The lack of laboratory testing to identify specific norovirus strains has limited the ability of investigators to link separate outbreaks to a potential common source (Anderson et al., 2001; Koopmans et al., 2003).

Most food-borne transmission appears to occur as a result of contamination of foods at the point of service. In Minnesota, 62% of confirmed food-borne outbreaks of viral gastroenteritis were likely the result of contamination of foods by contact with bare hands (Deneen et al., 1999). Although such contamination in most restaurant settings is due to food workers, contamination of foods by patrons has also resulted in outbreaks (Marshall et al., 2001).

Food-borne transmission is the most common mode of transmission for outbreaks that occur in restaurants and catered events (Fankhauser et al., 2002; Lopman et al., 2003b). Food-borne transmission also occurs in institutional settings such as schools, nursing homes, camps, and cruise ships. However, multiple modes of transmission occur in most of these settings, and it is frequently difficult to distinguish the role of food-borne disease transmission. It is hoped that the broader use of sensitive assays to both confirm and subtype norovirus strains will result in better understanding of the dynamics of transmission in these settings (Parashar and Monroe, 2001).

4.2. Waterborne Disease Transmission

Most waterborne disease outbreaks have resulted from fecal contamination of private wells and untreated community or noncommunity water systems (Hedberg and Osterholm, 1993; Centers for Disease Control and Prevention,

2001). The route of transmission for most waterborne outbreaks has been identified as a result of epidemiologic investigations and environmental assessments of the facilities. However, isolation of norovirus from water has been uncommon (Carique-Mas et al., 2003; Parshionikar et al., 2003). Waterborne transmission has also occurred on cruise ships, either because of the storage and use of untreated water or because of cross-connections in the ship's plumbing (Gallimore et al., 2003; Widdowson et al., 2004).

Outbreaks of viral gastroenteritis have been associated with swimming in lakes and swimming pools, usually as a result of the presence of one or more infected persons contaminating a crowded swimming area (Hedberg and Osterholm, 1993). Exposure to a recreational water fountain was implicated as the source of an outbreak in The Netherlands (Hoebe et al., 2004). The same norovirus identified from cases was also identified in a water sample from the fountain. Similarly, taking showers with contaminated water has been implicated as an additional route of waterborne transmission in Italy and Norway (Boccia et al., 2002; Nygard et al., 2004).

4.3. Airborne and Environmental Transmission

Most enteric viruses are transmitted by a fecal-oral route, which is reflected in the epidemiologic pattern of outbreaks reported (Fankhauser et al., 2002; Lopman et al., 2003b; Fiore, 2004). Because noroviruses are also expelled in vomit, the aerosolization of vomitus may create opportunities for widespread transmission and environmental contamination with noroviruses that would not occur with HAV (Centers for Disease Control and Prevention, 2001; Fiore, 2004). However, assessing the public health importance of airborne and environmental transmission has been difficult, as such transmission almost always occurs in settings such as cruise ships, nursing homes, and schools where food-borne, waterborne, or person-to-person transmission may also occur.

Evidence for environmental transmission of noroviruses is supported by findings such as the occurrence of illness among hotel employees who did not eat at the hotel during the course of an outbreak at the hotel (Love et al., 2002). Evidence for airborne transmission of noroviruses is supported by findings such as an increased risk of illness among school children after a vomiting event in their classroom (Marks et al., 2003). An increased risk of illness associated with showering in waterborne outbreaks also suggests airborne transmission (Boccia et al., 2002; Nygard et al., 2004).

More definitive evidence for airborne and environmental transmission was an outbreak that occurred in a concert hall after a concert-goer vomited in the auditorium and adjacent toilet (Evans et al., 2002). Illness occurred in 8 of 15 groups of schoolchildren who attended the next day, and risk of illness was associated with whether the group was seated near where the vomiting event occurred. In a protracted outbreak at a hotel, Norwalk-like virus was identified by RT-PCR in multiple environmental swabs (Cheesbrough et al., 2000). Samples collected from areas that were directly contaminated by vomit were more likely to be positive. However, evidence of broader environmental spread was also detected.

Even when noroviruses can be detected in environmental samples, it is necessary to conduct a thorough epidemiological investigation to interpret the significance of the findings. In a prolonged outbreak of Norwalk-like virus at a rehabilitation center in Finland, more than 300 guests and staff members became ill (Kuusi et al., 2002). No food or activity at the center could be associated with illness, and food and water samples tested negative for Norwalk-like viruses. However, Norwalk-like viruses identical to those from patients were detected in three environmental samples. In the context of these findings, it appears that environmental contamination was important to the prolonged occurrence of this outbreak.

4.4. Person-to-Person Transmission

Secondary transmission of noroviruses to household members has been regularly observed after food-borne and waterborne outbreaks (Centers for Disease Control and Prevention, 2001). Such transmission also contributes to the complexity of outbreaks in institutional settings where introduction of the virus through a food vehicle can result in extended person-to-person transmission among persons with continuous or repeated exposures. An unusual example of this is the spread of Norwalk-like virus in 30 daycare centers that shared a common caterer (Gotz et al., 2002). Consumption of a pumpkin salad was implicated as the source for the first cases that occurred with a mean incubation period of 34 hr. The primary attack rate was 27%. The secondary attack rate among daycare and household contacts was 17%. The incubation period for secondary transmission was estimated to be 52 hr (Gotz et al., 2001). Risk factors for spread into households included the occurrence of vomiting and having a child as the primary case.

Person-to-person transmission is frequently identified as the primary mode of transmission in nursing homes, schools, and other institutional settings (Fankhauser et al., 2002; Lopman et al., 2003b). However, this is generally the reflection of an epidemiologic picture in which there is no obvious point source of exposure and cases occur over a prolonged time period. In only a few of these settings do public health investigators actively identify the patterns of personal contact that would be necessary to establish this mode of transmission. In all likelihood, what gets labeled as person-to-person transmission actually represents a complex series of exposures that may result from airborne transmission of vomitus as well as contamination of food, water, and environmental surfaces in the common residential setting (Miller et al., 2002). The occurrence of vomiting as a risk factor for secondary transmission suggests that much of this may be due to undocumented airborne and environmental transmission (Gotz et al., 2001). From a public health standpoint, distinction between direct personal contact and environmental contamination could have implications for the emphasis that is put on specific control measures.

5.0. PREVENTION AND CONTROL

Reducing food-borne transmission of hepatitis A depends on food-handler hygiene and providing pre-exposure prophylaxis to persons at risk of infection. Transmission of HAV will continue to decline with routine vaccination of persons at risk for HAV infection (Fiore, 2004). Prevention and control activities for norovirus transmission need to be targeted to the primary transmission routes, which in turn are dependent on setting (Table 10.2). Untreated community water systems are susceptible to contamination that can lead to large common source outbreaks. Although noroviruses are relatively resistant to chlorine, routine chlorination and filtering of drinking water systems appears to be highly effective at preventing waterborne outbreaks (Centers for Disease Control and Prevention, 2001).

Harvesting shellfish from sewage contaminated waters can lead to large and widespread food-borne outbreaks. Although oysters are not harvested from beds known to be contaminated by municipal sewage outflows, sewage contamination from individual boats can be harder to prevent and track (Simmons et al., 2001). Fresh produce fields and the use of water to cool produce (hydrocooling) at harvest may be similarly susceptible to contamination from human sewage. The development and use of good agricultural practices (GAP) should help prevent transmission by this route.

In restaurant settings, infected food-handlers present the greatest risk of transmission. To reduce the risk of a food-borne outbreak, restaurant managers need to train their workers in proper food-handling techniques and encourage frequent hand-washing. In addition, managers should monitor illnesses in staff and exclude ill food-handlers from working in the restaurant.

Table 10.2 Primary Transmission Routes for Noroviruses by Setting and by Characteristics of the Settings

Setting	*Primary Transmission Route*	*Characteristics of the Setting That Favor Transmission Route*
Facility or community with untreated water system	Waterborne	Fecal contamination of well or water system.
Restaurants	Food-borne	Transient customer base with limited opportunities for environmental contamination and repeated exposures. Resident food workers provide extended source of contamination during outbreaks.
Institutional	Person-to-person, airborne, and environmental	Resident population with many opportunities for environmental contamination and repeated exposures.

Such active managerial control is possible only if managers know what is going on in the restaurant and are able to initiate appropriate control measures, although it probably is not possible to prevent all outbreaks of viral gastroenteritis. Just as with HAV, outbreaks will occasionally occur even when it appears that proper procedures are being followed (Centers for Disease Control and Prevention, 2003a).

In the event of illness occurring among food workers, or if patrons become ill at the restaurant, managers should make sure that all surfaces contaminated by feces, vomit, or hands are cleaned and disinfected thoroughly. If it appears that an outbreak of norovirus is occurring in the community, it may be necessary to modify menus to limit potential for customers to contaminate food (e.g., salad bars), or for food handlers to contaminate ready-to-eat foods. In the event of an outbreak at the restaurant, ill foodworkers should be excluded until they are free of symptoms for 72 hr. If it appears that there is an ongoing risk of transmission to patrons, restaurants should close for 72 hr to allow workers to recover and thoroughly clean and disinfect the establishment (Hedberg and Osterholm, 1993).

Institutional settings, particularly hospitals, nursing homes, and cruise ships, represent the greatest challenge to control norovirus transmission. In such settings, it seems reasonable to encourage frequent hand-washing, exclude ill food workers, clean and disinfect surfaces contaminated by feces or vomit, monitor illnesses in residents and staff, and implement control measures at first sign of the outbreak, including isolation of ill residents, exclusion of ill staff, and aggressive cleaning and disinfection (Centers for Disease Control and Prevention, 2001; McCall and Smithson, 2002; Lynn et al., 2004).

In practice, however, it may be difficult to apply infection control guidelines sufficient to prevent transmission (Miller et al., 2002; Kuusi et al., 2002; Khanna et al., 2003). The challenge in preventing outbreaks on cruise ships is even greater with each cruise bringing a new cohort of passengers representing potential sources of exposure as well as a population at risk from food-borne, waterborne, airborne, and environmental infection (Widdowson et al., 2004).

6.0. PUBLIC HEALTH IMPORTANCE

The landmark paper by Mead et al. (1999) on the burden of food-borne illness did much to establish the public health importance of noroviruses in the United States. For the first time, it was recognized that noroviruses are the leading cause of food-borne illness, accounting for 67% of food-borne illnesses caused by known etiologies; more than 9,000,000 infections with 20,000 hospitalizations and 124 deaths annually. In contrast, while the severity of illness caused by HAV is greater, there are only 4,000 cases with 90 hospitalizations and 4 deaths per year caused by HAV in the United States (Mead et al., 1999). Publication of these estimates has served to stimulate public health interest in surveillance for outbreaks caused by norovirus. In its wake, sensitive diagnostic assays are being widely adopted for use by public health laboratories.

Because of the absence of systematic surveillance for noroviruses in the United States, Mead's estimates were based largely on studies conducted in The Netherlands. Long-standing surveillance for outbreaks in the United Kingdom also presents a broader picture of the impact of norovirus. From 1992 to 2000, 1,877 norovirus outbreaks were reported in England and Wales (Lopman et al., 2003a). Of these, 40% occurred in hospitals and 39% occurred in residential facilities. Because these settings include high proportions of persons at greater risk for serious illness or death, the occurrence of these outbreaks in these settings presents a great public health challenge. Even though food-borne disease surveillance systems in the United States do not typically include reports of outbreaks in these settings, norovirus has been established as the leading cause of outbreaks of gastroenteritis in nursing homes in the United States (Green et al., 2002).

Since the discovery and characterization of Norwalk virus, noroviruses have been demonstrated to be the most frequent known cause of food-borne illness. Furthermore, most illnesses caused by noroviruses are transmitted through other routes, which complicate prevention and control efforts. Although much has been learned about the molecular epidemiology of noroviruses during the past 20 years, this has not yet been translated into more effective prevention and control strategies. More vigorous surveillance and control measures are needed across the public health system.

7.0. SUMMARY AND CONCLUSIONS

Noroviruses are the most common known cause of food-borne illness and outbreaks of food-borne disease in the United States. The clinical and epidemiologic characteristics of these outbreaks were characterized in the early 1980s, but the lack of sensitive diagnostic tests led to their systematic underreporting. The recent development and widespread availability of PCR-based methods throughout the public health laboratory system in the United States has led to a growing awareness of the burden of illness caused by these viruses. Sequencing of PCR gene products is shedding new light on the epidemiology of noroviruses, transmission routes, and global distributions. However, despite these recent advances, our understanding of norovirus epidemiology, prevention, and control is not fundamentally different than it was 20 years ago.

Prevention of norovirus outbreaks relies on the application of infection-control principles: education, surveillance, isolation, and disinfection. Application of these principles needs to be institutionalized throughout the hospitality industry. Encouraging proper hand-washing and excluding ill staff are cornerstones for this effort. The greatest challenge for food service operators is and will continue to be monitoring and managing illnesses among food workers. Finally, despite the increasing availability of diagnostic tests, confirmation of norovirus infections still requires time and effort in obtaining and transporting the sample, running the test, and interpreting the results. Thus, prompt and effective response to norovirus outbreaks cannot depend on laboratory confirmation but should be initiated at the first sign that the outbreak appears consistent with the epidemiology of norovirus.

8.0. REFERENCES

Anderson, A. D., Garrett, V. D., Sobel, J., Monroe, S. S., Fankhauser, R. L., Schwab, K. J., Bresee, J. S., Mead, P. S., Higgins, C., Campana, J., and Glass, R. I., 2001, Multistate outbreak of Norwalk-like virus gastroenteritis associated with a common caterer. *Am. J. Epidemiol.* 154:1013–1019.

Ando, T., Monroe, S. S., Gentsch, J. R., Jin, Q., Lewis, D. C., and Glass, R. I., 1995, Detection and differentiation of antigenically distinct small round-structured viruses (Norwalk-like viruses) by reverse-transcription-PCR and Southern hybridization. *J. Clin. Microbiol.* 33:64–71.

Ando, T., Noel, J. S., and Fankhauser, R. L., 2000, Genetic classification of "Norwalk-like viruses." *J. Infect. Dis.* 181:S336–S348.

Atmar, R. L., and Estes, M. K., 2001, Diagnosis of noncultivable gastroenteritis viruses, the human caliciviruses. *Clin. Microbiol. Rev.* 14:15–37.

Berg, D. E., Kohn, M. A., Farley, T. A., and McFarland, L. M., 2000, Multi-state outbreaks of acute gastroenteritis traced to fecal-contaminated oysters harvested in Louisiana. *J. Infect. Dis.* 181(Suppl 2):S381–S386.

Billgren, M., Christenson, B., Hedlund, K. O., and Vinje, J., 2002, Epidemiology of Norwalk-like human caliciviruses in hospital outbreaks of acute gastroenteritis in the Stockholm area in 1996. *J. Infect.* 44:26–32.

Boccia, D., Tozzi, A. E., Cotter, B., Rizzo, C., Russo, T., Buttinelli, G., Caprioli, A., Marziano, M. L., and Ruggeri, F. M., 2002, Waterborne outbreak of Norwalk-like virus gastroenteritis at a tourist resort, Italy. *Emerg. Infect. Dis.* 8:563–568.

Bresee, J. S., Widdowson, M. A., Monroe, S. S., and Glass, R. I., 2002, Foodborne viral gastroenteritis: challenges and opportunities. *Clin. Infect. Dis.* 35:748–753.

Buesa, J., Collado, B., Lopez-Andujar, P., Abu-Mullouh, R., Rodriguez Diaz, J., Garcia Diaz, A., Prat, J., Guix, S., Llovet, T., Prats, G., and Bosch, A., 2002, Molecular epidemiology of caliciviruses causing outbreaks and sporadic cases of acute gastroenteritis in Spain. *J. Clin. Microbiol.* 40:2854–2859.

Calder, L., Simmons, G., Thornley, C., Taylor, P., Pritchard, K., Greening, G., and Bishop, J., 2003, An outbreak of hepatitis A associated with consumption of raw blueberries. *Epidemiol. Infect.* 131:745–751.

Carrique-Mas, J., Anderson, Y., Petersen, B., Hedlund, K. O., Sjogren, N., and Giesecke, J., 2003, A norwalk-like virus waterborne community outbreak in a Swedish village during peak holiday season. *Epidemiol. Infect.* 131:737–744.

Centers for Disease Control and Prevention, 2001, "Norwalk-like viruses": public health consequence and outbreak management. *Morb. Mort. Wkly. Rpt.* 50(RR-9): 1–34.

Centers for Disease Control and Prevention, 2003a, Foodborne transmission of hepatitis A—Massachusetts, 2001. *Morbid. Mortal. Weekly Rep.* 52:565–567.

Centers for Disease Control and Prevention, 2003b, Hepatitis A Outbreak Associated with Green Onions at a Restaurant—Monaca, Pennsylvania, 2003. *Morbid. Mortal. Weekly Rep.* 52:1155–1157.

Cheesbrough, J. S., Green, J., Gallimore, C. I., Wright, P. A., and Brown, D. W., 2000, Widespread environmental contamination with Norwalk-like viruses (NLV) detected in a prolonged hotel outbreak of gastroenteritis. *Epidemiol. Infect.* 125:93–98.

Deneen, V. C., Hunt, J. M., Paule, C. R., James, R. I., Johnson, R. G., Raymond, M. J., and Hedberg, C. W., 2000, The impact of foodborne calicivirus disease: the Minnesota experience. *J. Infect. Dis.* 181(Suppl 2):S281–S283.

Evans, M. R., Meldrum, R., Lane, W., Gardner, D., Ribeiro, C. D., Gallimore, C. I., and Westmoreland, D., 2002, An outbreak of viral gastroenteritis following environmental contamination at a concert hall. *Epidemiol. Infect.* 129:355–360.

Fankhauser, R. L., Monroe, S. S., Noel, J. S., Humphrey, C. D., Bresee, J. S., Parashar, U. D., Ando, T., and Glass, R. I., 2002, Epidemiologic and molecular trends of "Norwalk-like viruses" associated with outbreaks of gastroenteritis in the United States. *J. Infect. Dis.* 186:1–7.

Fiore, A. E., 2004, Hepatitis A transmitted by food. *Clin. Infect. Dis.* 38:705–715.

Fretz, R., Schmid, H., Kayser, U., Svoboda, P., Tanner, M., and Baumgartner, A., 2003, Rapid propagation of norovirus gastrointestinal illness through multiple nursing homes following a pilgrimage. *Eur. J. Clin. Microbiol. Infect. Dis.* 22: 625–627.

Gallimore, C. I., Richards, A. F., and Gray, J. J., 2003, Molecular diversity of noroviruses associated with outbreaks on cruise ships: comparison with strains circulating within the UK. *Commun. Dis. Publ. Health* 6:285–293.

Gallimore, C. I., Green, J., Lewis, D., Richards, A. F., Lopman, B. A., Hale, A. D., Eglin, R., Gray, J. J., and Brown, D. W., 2004a, Diversity of noroviruses co-circulating in the north of England from 1998 to 2001. *J. Clin. Microbiol.* 42:1396–1401.

Gallimore, C. I., Green, J., Richards, A. F., Cotterill, H., Curry, A., Brown, D. W., and Gray, J. J., 2004b, Methods for the detection and characterization of noroviruses associated with outbreaks of gastroenteritis: outbreaks occurring in the northwest of England during two norovirus seasons. *J. Med. Virol.* 73:280–288.

Gotz, H., Ekdahl, K., Lindback, J., de Jong, B., Hedlund, K. O., and Giesecke, J., 2001, Clinical spectrum and transmission characteristics of infection with Norwalk-like virus: findings from a large community outbreak in Sweden. *Clin. Infect. Dis.* 33: 622–628.

Gotz, H., de J. B., Lindback, J., Parment, P. A., Hedlund, K. O., Torven, M., and Ekdahl, K., 2002, Epidemiological investigation of a food-borne gastroenteritis outbreak caused by Norwalk-like virus in 30 day-care centers. *Scand. J. Infect. Dis.* 34: 115–121.

Green, K. Y., Belliot, G., Taylor, J. L., Valdesuso, J., Lew, J. F., Kapikian, A. Z., and Lin, F. Y., 2002, A predominant role for Norwalk-like viruses as agents of epidemic gastroenteritis in Maryland nursing homes for the elderly. *J. Infect. Dis.* 185: 133–146.

Greenberg, H. B., Valdesuso, J., Yolken, R. H., Gangarosa, E., Gary, W., Wyatt, R. G., Konno, T., Suzuki, H., Chanock, R. M., and Kapikian, A. Z., 1978, Role of Norwalk virus in outbreaks of nonbacterial gastroenteritis. *J. Infect. Dis.* 139:564–568.

Hale, A., Mattick, K., Lewis, D., Estes, M., Jiang, X., Green, J., Eglin, R., and Brown, D., 2000, Distinct epidemiological patterns of Norwalk-like virus infection. *J. Med. Virol.* 62:99–103.

Hall, J. A., Goulding, J. S., Bean, N. H., Tauxe, R. V., and Hedberg, C. W., 2001, Epidemiologic profiling: evaluating foodborne outbreaks for which no pathogen was isolated by laboratory testing: United States, 1982–1989. *Epidemiol. Infect.* 127:381–387.

Hedberg, C. W., 2001, Epidemiology of foodborne illnesses, in: *Food Microbiology: Fundamentals and Frontiers*, 2nd ed. (M. P. Doyle, L. R. Beuchat, and T. J. Montville, eds.), ASM Press, Washington, DC, pp. 435–447.

Hedberg, C. W., and Osterholm, M. T., 1993, Outbreaks of food-borne and waterborne viral gastroenteritis. *Clin. Microbiol. Rev.* 6:199–210.

Hoebe, C. J., Vennema, H., de Roda Husman, A. M., and van Duynhoven, Y. T., 2004, Norovirus outbreak among primary schoolchildren who had played in a recreational water fountain. *J. Infect. Dis.* 189:699–705.

Jiang, X., Graham, D. Y., Wang, K., and Estes, M. K., 1990, Norwalk virus genome cloning and characterization. *Science* 250:1580–1583.

Jones, T. F., and Gerber, D. E., 2001, Perceived etiology of foodborne illness among public health personnel. *Emerg. Infect. Dis.* 7:904–905.

Kaplan, J. E., Feldman, R., Campbell, D. S., Lookabaugh, C., and Gary, G. W., 1982a, The frequency of a Norwalk-like pattern of illness in outbreaks of acute gastroenteritis. *Am. J. Pub. Health* 72:1329–1332.

Kaplan, J. E., Gary, G. W., Baron, R. C., Singh, N., Schonberger, L. B., Feldman, R., and Greenberg, H. B., 1982b, Epidemiology of Norwalk gastroenteritis and the role of Norwalk virus in outbreaks of acute nonbacterial gastroenteritis. *Ann. Intern. Med.* 96:756–761.

Khanna, N., Goldenberger, D., Graber, P., Battegay, M., and Widmer, A. F., 2003, Gastroenteritis outbreak with norovirus in a Swiss university hospital with a newly identified virus strain. *J. Hosp. Infect.* 55:131–136.

Koopmans, M., and Duizer, E., 2004, Foodborne viruses: an emerging problem. *Int. J. Food Microbiol.* 90:23–41.

Koopmans, M., Vennema, H., Heersma, H., van Strien, E., van Duynhoven, Y., Brown, D., Reacher, M., and Lopman, B., 2003, Early identification of common-source foodborne outbreaks in Europe. *Emerg. Infect. Dis.* 9:1136–1142.

Kuusi, M., Nuorti, J. P., Maunula, L., Minh, N. N., Ratia, M., Karlsson, J., and von Bonsdorff, C. H., 2002, A prolonged outbreak of Norwalk-like calicivirus (NLV) gastroenteritis in a rehabilitation centre due to environmental contamination. *Epidemiol. Infect.* 129:133–138.

Kuusi, M., Nuorti, J. P., Maunula, L., Miettinen, I., Pesonen, H., and von Bonsdorff, C. H., 2004. Internet use and epidemiologic investigation of gastroenteritis outbreak. *Emerg. Infect. Dis.* 10:447–450.

Lau, C. S., Wong, D. A., Tong, L. K., Lo, J. Y., Ma, A. M., Cheng, P. K., and Lim, W. W., 2004, High rate and changing molecular epidemiology pattern of norovirus infections in sporadic cases and outbreaks of gastroenteritis in Hong Kong. *J. Med. Virol.* 73:113–117.

Lopman, B. A., Reacher, M. H., Van Duijnhoven, Y., Hanon, F. X., Brown, D., and Koopmans, M., 2003a, Viral gastroenteritis outbreaks in Europe, 1995–2000. *Emerg. Infect. Dis.* 9:90–96.

Lopman, B. A., Adak, G. K., Reacher, M. H., and Brown, D. W., 2003b, Two epidemiologic patterns of norovirus outbreaks: surveillance in England and Wales, 1992–2000. *Emerg. Infect. Dis.* 9:71–77.

Lopman, B., Vennema, H., Kohli, E., Pothier, P., Sanchez, A., Negredo, A., Buesa, J., Schreier, E., Reacher, M., Brown, D., Gray, J., Iturriza, M., Gallimore, C., Bottiger,

B., Hedlund, K. O., Torven, M., von Bonsdorff, C. H., Maunula, L., Poljsak-Prijatelj, M., Zimsek, J., Reuter, G., Szucs, G., Melegh, B., Svennson, L., van Duijnhoven, Y., and Koopmans, M., 2004, Increase in viral gastroenteritis outbreaks in Europe and epidemic spread of new norovirus variant. *Lancet* 363:682–688.

Love, S. S., Jiang, X., Barrett, E., Farkas, T., and Kelly, S., 2002, A large hotel outbreak of Norwalk-like virus gastroenteritis among three groups of guests and hotel employees in Virginia. *Epidemiol. Infect.* 129:127–132.

Lynn, S., Toop, J., Hanger, C., and Millar, N. 2004, Norovirus outbreaks in a hospital setting: the role of infection control. *N. Z. Med. J.* 117:U771.

Marks, P. J., Vipond, I. B., Regan, F. M., Wedgwood, K., Fey, R. E., and Caul, E. O., 2003, A school outbreak of Norwalk-like virus: evidence for airborne transmission. *Epidemiol. Infect.* 131:727–736.

Marshall, J. A., Yuen, L. K., Catton, M. G., Gunesekere, I. C., Wright, P. J., Bettelheim, K. A., Griffith, J. M., Lightfoot, D., Hogg, G. G., Gregory, J., Wilby, R., and Gaston, J., 2001, Multiple outbreaks of Norwalk-like virus gastro-enteritis associated with a Mediterranean-style restaurant. *J. Med. Microbiol.* 50:143–151.

Marshall, J., Botes, J., Gorrie, G., Boardman, C., Gregory, J., Griffith, J., Hogg, G., Dimitriadis, A., Catton, M., Bishop, R., 2003, Rotavirus detection and characterization in outbreaks of gastroenteritis in aged-care facilities. *J. Clin. Virol.* 28:331–340.

McCall, J., and Smithson, R., 2002, Rapid response and strict control measures can contain a hospital outbreak of Norwalk-like virus. *Commun. Dis. Pub. Health* 5:243–246.

Mead, P. S., Slutsker, L., and Dietz, V., 1999, Food-related illness and death in the United States. *Emerg. Infect. Dis.* 5:607–625.

Miller, M., Carter, L., Scott, K., Millard, G., Lynch, B., and Guest, C., 2002, Norwalk-like virus outbreak in Canberra: implications for infection control in aged care facilities. *Commun. Dis. Intell.* 26:555–561.

Noel, J. S., Fankhauser, R. L., Ando, T., Monroe, S. S., and Glass, R. I., 1999, Identification of a distinct common strain of "Norwalk-like viruses" having a global distribution. *J. Infect. Dis.* 179:1334–1344.

Nygard, K., Vold, L., Halvorsen, E., Bringeland, E., Rottingen, J. A., and Aavitsland, P., 2004, Waterborne outbreak of gastroenteritis in a religious summer camp in Norway, 2002. *Epidemiol. Infect.* 132:223–229.

Olsen, S. J., MacKinnon, L. C., Goulding, J. S., Bean, N. H., and Slutsker, L., 2000, Surveillance for foodborne-disease outbreaks—United States, 1993–1997. *CDC Surveillance Summaries, Morbid. Mortal. Weekly Rep.* 49:1–64.

Parashar, U. D., Bresee, J. S., Gentsch, J. R., and Glass, R. I., 1998, Rotavirus. *Emerg. Infect. Dis.* 4:561–570.

Parashar, U. D., and Monroe, S. S., 2001, "Norwalk-like viruses" as a cause of foodborne disease outbreaks. *Rev. Med. Virol.* 11:243–252.

Parshionikar, S. U., Willian-True, S., Fout, G. S., Robbins, D. E., Seys, S. A., Cassady, J. D., and Harris, R., 2003, Waterborne outbreak of gastroenteritis associated with a norovirus. *Appl. Environ. Microbiol.* 69:5263–5268.

Ponka, A., Maunula, L., von Bonsdorf, C. H., and Lyytikainene, O., 1999, Outbreak of calicivirus gastroenteritis associated with eating frozen raspberries. *Epidemiol. Infect.* 123:469–474.

Simmons, G., Greening, G., Gao, W., and Campbell, D., 2001, Raw oyster consumption and outbreaks of viral gastroenteritis in New Zealand: evidence for risk to the public's health. *Aust. N. Z. J. Pub. Health* 25:234–240.

Vinje, J., and Koopmans, M. P. G., 1996, Molecular detection and epidemiology of small round structured viruses in outbreaks of gastroenteritis in the Netherlands. *J. Infect. Dis.* 174:610–615.

Vipond, I. B., Caul, E. O., Hirst, D., Carmen, B., Curry, A., Lopman, B. A., Pead, P., Pickett, M. A., Lambden, P. R., and Clarke, I. N., 2004, National epidemic of Lordsdale Norovirus in the UK. *J. Clin. Virol.* 30:243–247.

Widdowson, M. A., Cramer, E. H., Hadley, L., Bresee, J. S., Beard, R. S., Bulens, S. N., Charles, M., Chege, W., Isakbaeva, E., Wright, J. G., Mintz, E., Forney, D., Massey, J., Glass, R. I., and Monroe, S. S., 2004, Outbreaks of acute gastroenteritis on cruise ships and on land: identification of a predominant circulating strain of norovirus-United States, 2002. *J. Infect. Dis.* 190:27–36.

CHAPTER 11

Role of Irrigation Water in Crop Contamination by Viruses

Charles P. Gerba and Christopher Y. Choi

1.0. INTRODUCTION

Foods traditionally eaten raw or receiving minimal processing provide an ideal route for the transmission of viruses. Fruits and vegetables can potentially become contaminated before harvesting by irrigation water, water used for spray application of pesticides, or water used in processing (e.g., washing, hydrocooling with ice, etc.). An increase in the number of produce-associated outbreaks corresponds with the increased consumption of fresh fruits and vegetables and with the expanded global sources of these products over the past two decades (Sivapalasingam et al., 2004). Produce-associated outbreaks have increased from 0.7% of all outbreaks in the 1970s to more than 6% in the 1990s in the United States. In 2002, the number of cases of produce-associated illnesses was almost equal to all of those reported for beef, poultry, and seafood combined (Center for Science in the Public Interest, 2002). Several known and suspected food-borne outbreaks have been ascribed to crops contaminated in the field, suggesting contamination by irrigation or during harvesting (Dentinger et al., 2001; CDC, 2003). Perhaps more significant is the low-level transmission of viruses by food contaminated with irrigation water. Quantitative microbial risk analysis has suggested that low levels of virus in irrigation water can result in a significant level of risk to consumers (Petterson et al., 2001). Stine et al. (2005c) estimated that less than one hepatitis A virus per 10 L of irrigation water could result in a risk exceeding 1:10,000 per year considering the efficiency of transfer of the virus to crop and its survival till harvest time. The 1:10,000 risk of infection per year is currently the acceptable level used by the United States Environmental Protection Agency for Drinking Water (Regli et al., 1991).

The largest use of freshwater in the world is in agriculture with more than 70% being used for irrigation. About 240 million ha, 17% of the world's cropland, are irrigated, producing one third of the world's food supply (Shanan, 1998). Nearly 70% of this area is in developing countries. In the United States, California and Arizona are the major producers of lettuce, carrots, broccoli, and cantaloupe (Arizona Farm Bureau, 2003). All of these crops are grown almost entirely by irrigated agriculture. It is thus surprising that we know little about the microbial quality of irrigation water. Most studies have dealt with the occurrence and fate of enteric pathogens in reclaimed water used for irrigation and not the quality of surface waters currently in use.

Almost no data exist on the occurrence of enteric viruses in irrigation waters, which do not intentionally receive sewage discharges.

2.0. WATER QUALITY STANDARDS FOR IRRIGATION WATER

For more than 100 years, irrigation or fertilization of food crops with feces or fecally contaminated water has been known to play a role in the transmission of enteric microorganisms. For this reason, the use of night soil or irrigation with untreated domestic wastewater is not allowed in the United States and is not recommended by the World Health Organization (Mara and Cairncross, 1989). Most of the research on enteric pathogen contamination of vegetables and fruits during production has been done to evaluate the safety of reclaimed wastewater irrigation. Many states in the United States have standards for the treatment of reclaimed water to be used for food crop irrigation (Asano, 1998), and the World Health Organization has also made recommendations in this regard (Mara and Caincross, 1989). The state of California requires advanced physical-chemical treatment and extended disinfection to produce "virus-free" effluent. A coliform standard of <2/100 ml must also be met (Asano, 1998). The state of Arizona has a virus standard of 1 plaque forming unit (pfu)/40 L and *Giardia* cysts of 1 per 40 L in addition to a fecal coliform standard of 25/100 ml. Although standards for the use of reclaimed wastewater exist for food crops eaten raw in the United States, irrigation using reclaimed water for crop irrigation is seldom practiced. In developing countries, raw or partially treated wastewater is often used to irrigate crops, especially in the arid regions.

There are a few published studies on the quality of nonreclaimed wastewater used as an irrigation source (Steele and Odumeru, 2004). Irrigation agriculture requires approximately 2 acre-feet of water per acre of growing crop. The frequency and volume of application must be carefully programmed to compensate for deficiencies in rainfall distribution and soil moisture content during the growing season. Rivers and streams are tapped by large dams and then diverted into extensive canal systems. In addition, groundwater may also be pumped from wells into canals (which puts this water at risk from surface contamination). Because water availability is often critical, little attention is paid to the microbial quality of the irrigation water. In water-short areas, available sources are subjected to contamination by sewage discharge from small communities (unplumbed housing along canals in developing countries is common), cattle feedlot drainage, grazing animals along canal embankments, storm-water events, and return irrigation water (noninfiltrated water from the field being irrigated is returned to the irrigation channels). Because irrigation channels are frequently small, these changes in pollution discharges can result in rapid deterioration of water quality.

One of the few early studies conducted on irrigation waters documented a wide range in the microbial quality of this water (Geldreich and Bordner,

1971). The wide variation was attributed to the discharge of domestic sewage into streams from which the irrigation water was obtained. This study was conducted in the western United States (Wyoming, Utah, and Colorado). Median fecal coliform values ranged from 70 to 450,000/100 ml. Based on results obtained with the occurrence of *Salmonella* in the same waters, these authors recommended a fecal coliform standard for irrigation waters of 1,000/100 ml.

Guidelines for the microbial quality of surface water tend to be more lenient than those for wastewater because of the belief that enteric viruses and other human pathogens are less likely to be present or are less numerous (Steele and Odumeru, 2004). The criteria range from <100 to <1,000 fecal coliforms per 100 ml. Other criteria for *Escherichia coli* and fecal streptococcus are also used by some regulatory agencies (Steele and Odumeru, 2004). The United States Environmental Protection Agency guidelines for surface water recommend fewer than 1,000 fecal coliforms per 100 ml of surface water, including river water, for irrigation of crops (EPA, 1973). The differences among the guidelines reflect widespread uncertainty about the actual risk of disease transmission by irrigation water. Obviously, data on the occurrence of pathogens in irrigation waters would aid in the development of a risk-based approach to the development of standards.

3.0. OCCURRENCE OF VIRUSES IN IRRIGATION WATER

The microbial quality of irrigation water depends on the source of the water. Sources of human enteric viruses may involve sewage discharges into source water, septic tanks, recreational bathers, and so forth. Although groundwater is often considered a microbially safe source for irrigation water, recent studies in the Unites States have indicated that 8% to 31% of the groundwaters may contain viruses (Abbaszadegan et al., 2003; Borchardt et al., 2003). These viruses may originate from septic discharges, leaking sewer lines, or infiltration from lakes, rivers, and oxidation ponds.

Currently, only one study is available on the occurrence of enteric viruses in irrigation waters (Kayed, 2004). In this study, the occurrence of protozoan parasites, indicator bacteria, and noroviruses in irrigation waters in central and western Arizona was investigated. The irrigation waters in central Arizona are derived from a series of dammed reservoirs. The water is then channeled through a series of canals traveling distances as great as 40 miles before reaching the fields to be irrigated. In western Arizona, the water comes from a reservoir on the Colorado River. Noroviruses were detected in 20.7% of the canal samples from western Arizona and 18.2% of the canal samples from central Arizona. Geometric averages of *E. coli* were 6.4/100 ml in western Arizona canals and 18/100 ml in central Arizona. *Salmonella* and *Campylobacter* spp. were also detected in the irrigation water, especially after rainfall events. Because polymerase chain reaction was used to detect noroviruses, the infectivity of the viruses could not be determined. However, the results demonstrate that contamination of irrigation water by enteric viruses does occur even when there is no intentional discharge of sewage into the system.

4.0. CONTAMINATION OF PRODUCE DURING IRRIGATION

The likelihood of the eatable parts of a crop becoming contaminated during irrigation depends upon a number of factors including growing location, type of irrigation application, and nature of the produce surface. If the eatable part of the crop grows in or near the soil surface, it is more likely to become contaminated than a fruit growing in the aerial parts of a plant. Some produce surfaces are furrowed or have other structures that may retain water (e.g., a tomato vs. a cantaloupe). There are three distinct methods of irrigation: sprinkler systems, gravity-flow (furrow) systems, and microirrigation systems. Microirrigation includes surface drip and subsurface irrigation methods. In 2000, approximately 63 million acres of farmland was irrigated in the United States, 31.5 million acres (50%) with sprinkler irrigation systems, 28.4 million acres (45%) with gravity-flow systems, and the remainder (5%) with microirrigation systems (Anon., 2001).

Studies have been done on contamination of crops by enteric bacteria present in irrigation water, but only a few have evaluated the degree of contamination by viruses. Stine et al. (2005a) quantified the transfer of virus (coliphage PRD-1) and enteric bacteria in water used to prepare pesticide spray to the surface of cantaloupe, iceberg lettuce, and bell peppers. The average transfer of bacteria from the water to the surface of fruit was estimated to range from 0.00021% to 9.4%, while the average viral transfer ranged from 0.055% to 4.2% depending on the type of produce.

Oron et al. (1995) applied irrigation water containing up to 1,000 pfu/ml of poliovirus on tomato plants by subsurface drip irrigation in an outdoors setting using both surface water and wastewater. Some virus was detected in the leaves of the plants but not in the fruits. The authors stated that the high virus content of the water might explain the occurrence of virus in the leaves. No virus was detected in plants irrigated with wastewater containing the same level of virus as the surface water. The authors suggested that this was due to the interaction of the virus with particulate or soluble matter present in the wastewater preventing their entrance into the roots.

Alum (2001) studied the effectiveness of drip irrigation in the control of viral contamination of salad crops (lettuce, tomato, cucumber) in a greenhouse in potted plants. The plants were irrigated with secondary effluent using surface drip and subsurface irrigation. Irrigation water was periodically seeded with coliphage MS-2, phage PRD-1, poliovirus type 1, adenovirus 40, or hepatitis A virus. Surface irrigation always resulted in surface contamination of both above-ground parts and the underground parts of the plants. In lettuce, it was observed that only the outer leaves of the plant became contaminated. No contamination of the plants occurred when subsurface drip irrigation was used. No systemic uptake of viruses was observed in any of the crops.

Choi et al. (2004) assessed viral contamination of lettuce by surface and subsurface drip irrigation using coliphages MS-2 and PRD-1. A greater number of coliphages was recovered from lettuce grown in subsurface plots as compared with those in the furrow-irrigated plots. Shallow drip tape installation and preferential water paths through cracks on the soil surface appeared to be the main cause of high viral contamination. In subsurface drip irrigation plots, penetration of the water to the soil surface was observed, which led to the direct contact of the lettuce stems with the irrigation water. Thus, drip tape depth can influence the probably of produce contamination. Greater contamination by PRD-1 was observed, which might be due to its longer survival time.

Stine et al. (2005b) compared furrow irrigation and subsurface irrigation on the contamination of cantaloupe, iceberg lettuce, and bell peppers when the water was seeded with coliphage PRD-1 under field conditions in Arizona. No coliphage was detected on bell peppers. The maximum virus transfer was 0.046% to lettuce and 0.02% to cantaloupe by furrow irrigation. No viruses were detected on lettuce when subsurface irrigation was used, and the transfer to the cantaloupe was reduced to 0.00039%.

5.0. SURVIVAL OF VIRUSES ON PRODUCE IN THE FIELD

Studies on the survival of viruses on produce postharvest indicate little virus inactivation because of the low temperatures of storage (Seymour and Appleton, 2001). Few studies are available on the survival of viruses on growing crops preharvest. Tierney et al. (1977) found that poliovirus survived on lettuce for 23 days after flooding of outdoor plots with wastewater. The virus persisted in the soil for 2 months during the winter and 2 to 3 days during the summer months. Sadovski et al. (1978) spiked wastewater and tap water used to irrigate cucumbers with high titers of poliovirus. The cucumbers were grown with either (i) surface drip irrigation or with (ii) the soil and drip lines covered with polyethylene sheets to reduce contact of the irrigation water with the plants. Virus was detected on cucumbers by both methods of application. Virus was detected only occasionally on cucumbers that were irrigated with the drip lines covered by plastic sheets. Viruses survived on the cucumbers for at least 8 days after irrigation (hepatitis A virus and coliphage PRD-1 survival on growing produce was found to be similar under high and low humidity conditions) (Stine et al., 2005b). In general, the inactivation rates of the viruses were less than those of *E. coli* 0157:H7, *Shigella sonnei*, and *Salmonella enterica* on cantaloupe, lettuce, and bell peppers. The hepatitis A virus levels were reduced to about 90% after 14 days, indicating that enough viruses could survive from an irrigation event to harvest time to pose a potential risk.

6.0. SUMMARY

Risks from the use of virus-contaminated irrigation water are poorly understood. Information on the occurrence of viruses in irrigation water and potential sources of viruses in irrigation water is sorely needed. Sources of viral contamination from irrigation need to be defined and their transport and survival need to be determined. Irrigation methods and the type of produce also affect the degree of contamination. Although the percent virus transfer from irrigation water to produce is low, it should be realized that ingestion of low numbers of viruses can result in a significant risk of infection (Peterson and Ashbolt, 2001). It appears that temperature and the nature of the produce surface are the most important factors in virus persistence on produce surfaces. Limited studies suggest that enteric viruses survive longer than enteric bacteria and may survive from the time of irrigation to harvesting.

7.0. REFERENCES

Anon., 2001, 2000 Irrigation Survey. *Irrigation J.* 51:1–17.

Abbaszadegan, M., LeChevallier, M., and Gerba, C. P., 2003, Occurrence of viruses in US groundwater. *J. Am. Water Works Assoc.* 95:107–120.

Alum, A., 2001, Control of viral contamination of reclaimed water irrigated vegetables by drip irrigation. Ph.D. diss., University of Arizona, Tucson, AZ.

Arizona Farm Bureau, 2003, *Arizona Agricultural Statistics*. Arizona Farm Bureau, Phoenix, AZ.

Asano, T., 1998, *Wastewater and Reclamation Reuse*. Technomic Publishing, Lancaster, PA.

Borchardt, M. A., Bertz, P. D., Spencer, S. K., and Battiagelli, D. A., 2003, Incidence of enteric viruses in groundwater from household wells in Wisconsin. *Appl. Environ. Microbiol.* 69:1172–1180.

CDC, 2003, Hepatitis A outbreak associated with green onions at a restaurant-Monaca, Pennsylvania, 2003. *Morbid. Mortal. Weekly Rep.* 52:1155–1157.

Center for Science in the Public Interest, 2002, *Outbreak Alert*. CSPI, Washington, DC.

Choi, C., Song, I., Stine, S., Pimentel, J., and Gerba, C. P., 2004, Role of irrigation and wastewater: comparison of subsurface irrigation and furrow irrigation. *Water Sci. Technol.* 50:61–68.

Dentinger C., Bower, W. A., Nainan, O. V., Cotter, S. M., Myers, G., Dubusky, L. M., Fowler, S., Salehi, E. D., and Bell, B. P., 2001, An outbreak of hepatitis A associated with green onions. *J. Infect. Dis.* 183:1273–1276.

EPA, 1973, *Water Quality Criteria*. Ecological Research Series, EPA R3-73-033, Washington, DC.

Geldreich, E. E., and Bordner, R. H., 1971, Fecal contamination of fruits and vegetables during cultivation and processing for market. *J. Milk Food Technol.* 34:184–198.

Kayed, D., 2004, Microbial quality of irrigation water used in the production of fresh produce in Arizona. Ph.D. diss., University of Arizona, Tucson, AZ.

Mara, D., and Cairncross, S., 1989, *Guidelines for the Safe use of Wastewater and Excreta Agriculture and Aquaculture.* World Health Organization, Geneva, Switzerland.

Oron, G., Goemans, M., Manor, Y., and Feyen, J., 1995, Poliovirus distribution in the soil-plant system under reuse of secondary wastewater. *Water Res.* 29:1069–1078.

Petterson, S. R., Teunis, P. F., and Ashbolt, N. J., 2001, Modelling virus inactivation on salad crops using microbial count data. *Risk Anal.* 21:1097–1108.

Regli, S., Rose, J. B., Haas, C. N., and Gerba, C. P., 1991, Modeling the risk from *Giardia* and viruses in drinking water. *J. Am. Water Works Assoc.* 88:76–84.

Sadovski, A. Y., Fattal, B., Goldberg, D., Katzenelson, E., and Shuval, H. I., 1978, High levels of microbial contamination of vegetables irrigated with wastewater by the drip method. *Appl. Environ. Microbiol.* 36:824–830.

Seymour, I. J., and Appleton, H., 2001, Foodborne viruses and fresh produce. *J. Appl. Microbiol.* 91:759–773.

Shanan, L., 1998, Irrigation development: proactive planning and interactive management, in: *The Arid Frontier* (H. Bruins and L. Harvey, eds.), Kluwer Academic Press, London.

Sivapalasingam, S., Friedman, C. R., Cohen, L., and Tauxe, R. V., 2004, Fresh produce: a growing cause of outbreaks of foodborne illness in the United States, 1973 through 1997. *J. Food Protect.* 67:2342–2353.

Steele, M., and Odumeru, J., 2004, Irrigation water as source of foodborne pathogens on fruit and vegetables. *J. Food Protect.* 67:2839–2849.

Stine, S. W., Song, I., Choi, C. Y., and Gerba, C. P., 2005a, Microbial risks from application of pesticide sprays to fresh produce. *Int. J. Food Microbiol.* 68:1352–1358.

Stine, S. W., Song, I., Pimentel, J., Choi, C. Y., and Gerba, C. P., 2005b, The effect of relative humidity on pre-harvest survival of bacterial and viral pathogens on the surface of cantaloupe, lettuce, and bell pepper. *J. Food Protect.* 68:913–918.

Stine, S. W., Song, I., Choi, C. Y., and Gerba, C. P., 2005c, Application of microbial risk assessment to the development of standards for enteric pathogens in water used to irrigate fresh produce. *J. Food. Protect.* (in press).

Tierney, J. T., Sullivan, R., and Larkin, E. P., 1977, Persistence of poliovirus 1 in soil and on vegetables grown in soil previously flooded with inoculated sewage sludge or effluent. *Appl. Environ. Microbiol.* 33:109–113.

USDA, 1998, Available at http://www.nass.usda.gov/census/census97/fris/fris.htm.

CHAPTER 12

Chemical Disinfection Strategies Against Food-borne Viruses

Syed A. Sattar and Sabah Bidawid

1.0. INTRODUCTION

The impact of food-borne viral pathogens on human health can be substantial (Lüthi, 1997; Meade et al., 1999; Appleton, 2000; Sair et al., 2002) even though significantly fewer viruses than bacteria can spread via foods (Cliver, 1997). General difficulties in recovering and identifying viruses from foods and clinical specimens collected during food-borne outbreaks grossly underestimate the true role of viruses as food-borne pathogens (Collins, 1997; Bresee et al., 2002; Koopmans et al., 2002), reinforcing the need for proper inactivation of food-borne viruses. The main focus of this chapter is on the testing and application of chemicals (microbicides) that can be used to inactivate viruses on inanimate and animate food contact surfaces as well as for the decontamination of foods consumed raw or with minimal processing. Table 12.1 defines the common terms to be used in this chapter.

2.0. BASIC CONSIDERATIONS

Whereas the use of microbicides in reducing the risk from food-borne infections is widespread, there are many aspects of this practice that require reevaluation, especially those for interrupting the spread of food-borne viruses. A clear understanding of the following factors is necessary for the development of any successful strategy for the use of microbicides in preventing and controlling the spread of food-borne viral infections:

1. Hepatitis A virus (HAV), an important food-borne pathogen, can survive better than many enteric bacteria on inanimate and animate surfaces (Sattar et al., 2000); recent studies have shown this to be the case with caliciviruses as well (Bidawid et al., 2003).
2. A microbicide shown to be effective against vegetative bacteria may not be suitable in inactivating viruses, particularly the nonenveloped ones (Ansari et al., 1989).
3. Unlike many types of bacteria, viruses cannot replicate in contaminated foods; thus, holding of foods at an inappropriate temperature as such is not a risk factor in case of viral contamination. But viruses may remain viable in contaminated foods for several days, especially under refrigeration.

Table 12.1 Glossary of Common Terms Used in this Chapter

Term	*Explanation*
Antimicrobial agent	A physical or chemical agent that kills microorganisms or suppresses their growth.
Antiseptic	An agent that destroys pathogenic or potentially pathogenic microorganisms on living skin or mucous membranes.
Carrier	An inanimate surface or object inoculated with a test organism.
Cleaning (precleaning)	Removing, by physical and/or chemical means, visible soil, dirt, or organic debris from a surface or object.
Microbial contamination	The presence of viable microorganisms in or on a given material or object.
Decontamination	Freeing a person, object, or surface of harmful microorganisms, chemicals, or radioactive materials.
Disinfectant	A physical or chemical agent that destroys pathogenic or potentially pathogenic microorganisms in or on inanimate surfaces or objects.
EBSS	Earle's balanced salt solution.
Eluate	An eluent that contains microorganism(s) recovered from a carrier.
Eluent	Any solution that is harmless to the test organism(s) and that is added to a carrier to recover the organism(s) in or on it.
Chemical microbicide	A chemical that kills pathogenic or potentially pathogenic microorganisms in or on inanimate surfaces/objects and on living skin/mucous membranes.
High-level disinfectant	A chemical or a mixture of chemicals that is bactericidal, fungicidal, mycobactericidal, and virucidal; may also be sporicidal with an extended contact time.
Intermediate-level disinfectant	A chemical or a mixture of chemicals that is bactericidal, fungicidal, mycobactericidal, and virucidal, but not sporicidal.
Label	Written, printed, or graphic matter on, or attached to, the microbicide containers or wrappers.
Low-level disinfectant	A chemical or a mixture of chemicals that kills only vegetative bacteria and enveloped viruses.
Microbicide (microbiocide)	A physical or chemical agent that kills microorganisms.
Neutralization	Quenching the antimicrobial activity of a test formulation by dilution of the organism/test formulation mixture and/or addition of one or more chemical neutralizers to the mixture.
OTC	Over-the-counter topicals.
Pathogen	Any disease-producing microorganism.
Pesticide	Any substance or mixture of substances intended for preventing, destroying, repelling, or mitigating any pest.

Table 12.1 *Continued*

Term	*Explanation*
Potency	The degree of strength or power of a microbicide to render disease-causing microorganisms noninfectious.
QCT-1	Quantitative carrier test—tier 1.
QCT-2	Quantitative carrier test—tier 2.
Sanitization	A process that reduces the microbial load on a surface or object.
Soil load	A solution of one or more organic and/or inorganic substances added to the suspension of the test organism to simulate the presence of body secretions, excretions, or other extraneous substances.
Sterile	Free from living microorganisms.
Sterilization	A process that kills all forms of microbial life.
Stringency of test method	The level of rigor, strictness, or severity built into the method to reflect factors the test formulation may encounter under in-use conditions.
Test formulation	A formulation that incorporates antimicrobial ingredients.
$TCID_{50}$ (50% tissue culture infective dose)	The dilution at which 50% of all infected cell cultures show evidence of virus infection.
Test organism	An organism that has readily identifiable characteristics. It also may be referred to as a surrogate or a marker organism.
Use-dilution	The level to which a concentrated microbicide is diluted for use.
Virucide (viricide)	An antimicrobial agent that kills (inactivates) viruses.
Water hardness	The measure of the amount of metallic (e.g., calcium) salts in water.

4. Foods such as shellfish harvested from fecally polluted waters do not lend themselves readily to decontamination by chemicals.

5. Hands can readily acquire or donate infectious virus particles under conditions encountered during the handling and preparation of foods (Sattar and Springthorpe, 1996).

6. Suitable microbicides, when properly used, can interrupt the transfer of viruses from contaminated surfaces to foods (Bidawid et al., 2000).

7. Safety considerations exclude the use of certain types of microbicides (e.g., phenolics) on food contact surfaces (Gulati et al., 2001).

8. Microbicides often used on food contact surfaces are neither required to nor are tested against common types of food-borne viruses.

9. In the United States, there are no officially accepted methods for evaluating the virucidal activity of handwash and handrub agents; nor is there any regulatory framework to allow such products to make claims against viruses (Sattar et al., 2002).

10. Recognized flaws in current methods to assess microbicidal activity can compromise the label claims of disinfectants in general.

3.0. TEST METHODOLOGY TO DETERMINE VIRUCIDAL ACTIVITY

The virucidal potential of microbicides is normally assessed by "suspension" or "carrier" tests (Springthorpe et al., 1986; Springthorpe and Sattar, 1990; Quinn and Carter, 1999). In suspension tests, a known volume of the challenge virus, with or without a soil load, is mixed with a 5- to 10-fold larger volume of the test microbicide. For control, the virus is suspended in an equivalent volume of a liquid known to be harmless to the virus. The mixtures are held for a defined contact time at a specified temperature, neutralized to stop virucidal activity, titrated for infectious virus, and the degree of loss in virus viability calculated (ASTM, 2002a). While suspension tests are easier to perform, they are also easier to pass (Sattar et al., 1986; Abad et al., 1997) and are thus suitable for screening the activity of microbicides under development. Regulatory agencies in North America do not accept claims of virucidal activity based on suspension tests for product registration purposes.

Under most field conditions, the target virus is present on an animate or inanimate surface. In view of this, carrier tests, where the challenge virus is first dried on a representative surface and then exposed to the test formulation, are considered more suitable in assessing the potential of microbicides under in-use conditions (Springthorpe and Sattar, 1990).

4.0. FACTORS IN TESTING FOR VIRUCIDAL ACTIVITY

4.1. Test Viruses

With the exception of certain blood-borne viruses, the U.S. Environmental Protection Agency (EPA) so far does not accept surrogates in tests for virucidal activity of microbicides but requires that a given product be tested against each virus to be listed on the product label. On the other hand, Health Canada currently allows for a general virucidal claim when a given product shows the required level of activity against the Sabin strain of poliovirus type 1 (CGSB, 1997). The use of this nonenveloped virus, which is also safe to handle and is relatively resistant to microbicides, makes product development easier and label claims simpler and reliable. However, one or more substitutes for poliovirus may be needed soon in view of the anticipated eradication of poliomyelitis and the expected ban on the laboratory use of all types of polioviruses (Aylward et al., 2003).

What should one look for in selecting viruses to assess the activity of microbicides against food-borne viruses? Fortunately, the list of major food-borne viruses is short, and identification of potential surrogates for them is easy. The two most suitable viruses in this regard would be cell culture–adapted strains of HAV (e.g., HM-175) and the F9 strain of feline calicivirus (FCV). Indeed, investigations in the past decade have already demonstrated the feasibility of using such strains in testing disinfectants and antiseptics to be used in settings where foods are processed and handled (Doultree et al., 1999; Gulati et al., 2001; Sattar et al., 2000). Rotaviruses can also cause food-borne outbreaks (Sattar et al., 2001), and it is feasible to use the cell culture–adapted Wa strain of human rotavirus (HRV) to evaluate microbicides against them (Sattar et al., 1994).

HAV shows the highest level of microbicide resistance of the food-borne viruses tested so far (Mbithi et al., 1992), and it would thus make a good surrogate if the selection were to be based on this factor alone. Working with this virus has become safer because an effective vaccine against it is now available. The possible drawbacks in the use of HAV are that the turnaround time for test results is at least 1 week and that many formulations in current use may fail against this virus. This points to the need for further discussions on the justification of using one or more surrogates in testing microbicides against food-borne viruses, and a consensus between major stakeholders is needed on which virus(es) may be the most suitable for this purpose.

4.2. Nature and Design of Carriers

The three categories of surfaces to be discussed here are inanimate non-porous environmental items that may contact foods during storage, preparation, and serving; fruits and vegetables that are consumed raw or with minimal processing; and hands of food handlers.

4.2.1. Environmental Surfaces

Food contact surfaces vary widely in their nature, usage, and level of cleanliness. The microtopography of a given surface may also change with the type and extent of use, which may provide either more or less protection to viruses deposited on it (Springthorpe and Sattar, 1990). Because it is impractical to test microbicides on all types of food contact surfaces prior to product registration, it would be logical to develop and use a "surrogate" surface. The selection of such a surrogate, inanimate, food contact surface should take the following into consideration: (a) how frequently it contacts foods and hands of food handlers; (b) how readily it releases infectious viruses it carries; (c) it must not inactivate the test virus or irreversibly bind or sequester it such that virus elution from it becomes difficult; (d) its surface should be uneven enough to represent those in the field; (e) if meant for reuse, it should readily withstand repeated decontamination and sterilization; and (f) it should be resistant to microbicides commonly used in decontamination of food contact surfaces.

Further, any carriers made out of such a surrogate material should allow the convenient deposition of the desired volume of the test virus as well as the test microbicide, and the entire carrier should be submersible in a reasonably small volume of the eluent without any wash-off. The need for keeping the eluent volume per carrier as small as possible is particularly relevant when working with viruses, because unlike tests against bacteria, membrane filtration cannot be readily used to trap viruses from large volumes of eluates. The need for cell cultures for detection and quantitation of infectious virus in test samples also restricts the eluate volumes that can be easily and economically processed.

Disks (~1 cm diameter) of brushed stainless steel offer almost all the desired attributes of a surrogate surface in testing microbicides against food-borne viruses (Springthorpe and Sattar, 1990; Sattar and Springthorpe 2001a, 2001b). The microtopography of the disk surface is sufficiently uneven, and the carriers can be handled in a closed system so that wash-off of the test virus does not occur. If needed, disks similar in size to those described above can be readily prepared from other types of food contact surfaces (Lloyd-Evans et al., 1986). Porous materials can also be made into disks as carriers (Traoré et al., 2002), but they are generally more difficult to work with in testing microbicides because their absorbent nature reduces the efficiency of recovery of test organisms. Besides, such materials are rarely meant to be decontaminated using microbicides.

4.2.2. Food Items

In view of the potential of fresh produce to spread viruses (Seymour and Appleton, 2001), such items may be treated with microbicides before consumption (Beuchat, 2001). The use of microbicides for this purpose requires that they be evaluated for their virucidal efficacy on representative types of vegetables and fruits that are eaten raw or after minimal processing. A carrier test using small disks or pieces of items such as lettuce or strawberries represents a feasible approach (Bidawid et al., 2000).

4.2.3. Hands

Although virucides intended for use on human skin are often tested using hard inanimate surfaces, comparative testing has found skin to present a stronger challenge to microbicides (Woolwine and Gerberding, 1995). This reinforces the need for using carriers of a suitable animate surface for evaluating the virucidal activity of formulations for the decontamination of hands. Virucides can be tested using the entire surface of both hands of an adult subject (ASTM, 2002a), but the disadvantages of such an approach include high variability in results, inability to run controls and test samples simultaneously, lack of statistical power, and the need for larger volumes of high-titered virus pools (Sattar and Ansari, 2002). The fingerpad method (Ansari et al., 1989), which is a standard of ASTM International (2002a), avoids these drawbacks by using the thumb- and fingerpads of adult subjects as *in vivo* carriers. In this method, the test virus is placed on targeted areas on the hands, allowed to dry, and then exposed to a handwash or handrub formulation for a suitable contact time. It also allows for the determination of reduction in virus infectivity after exposure to the test formulation alone or after post-treatment water rinsing and with or without drying using cloth, paper, or warm air (Ansari et al., 1991). The fingerpad method has already been applied to assess the microbicidal activity of food-borne viruses such as HAV (Mbithi et al., 1992), FCV (Bidawid et al., 2003), and HRV (Ansari et al., 1988).

4.3. Nature and Level of Soil Loading

The organic matrix or "*soil load*" surrounding viruses, whether they are in body fluids or sewage/sludge, enhances their survival in the environment. Normal precleaning of surfaces and items to be disinfected may reduce the amount of such soil, but enough of it remains and can interfere with the activity of the applied microbicide by either binding to it or by preventing its access to the target virus. Any good method for virucidal activity must, therefore, simulate the presence of such soil by incorporating in the test virus suspension a certain amount of organic and inorganic material, and this is now a requirement in several standard protocols (ASTM, 2002a, 2002b; CGSB, 1997).

Although many different types and levels of substances are used as the soil load in testing microbicides, extra precautions are needed in their selection and use when working with viruses. For example, animal sera may contain specific antibodies or nonspecific inhibitors against viruses such as rotaviruses. Fecal suspensions, which have been used in testing microbicides against HAV (Mbithi et al., 1990), are inherently variable and thus unsuitable as a soil load for standardized test protocols. To overcome these difficulties, a "universal" soil load has been developed for testing microbicides against viruses as well as other pathogens (Springthorpe and Sattar, 2003); it consists of a mixture of bovine mucin, tryptone, and bovine albumin in phosphate buffer. The concentrations and ratios of the three ingredients are designed to provide a challenge roughly equivalent to that in 5–10% bovine serum. This soil has been found to be compatible with all viruses as well as other types of pathogens tested thus far (Sattar and Springthorpe, 2003).

4.3.1. Diluent for Test Microbicide

Many microbicides are tested by manufacturers using distilled water as the product diluent, and because this is not clearly specified in label directions, most users use tap water instead. Formulations with marginal virucidal activity may work with distilled water but fail when tap water is used as the diluent (Sattar et al., 1983). This highlights the importance of choosing the right diluent during product development and to clearly specify it on the label.

Although tap water is commonly used in the field and may represent a stronger challenge to microbicides under test, it is unsuitable as a diluent in standardized tests for virucidal activity. This is because the quality of tap water as well as the nature and levels of disinfectants in tap water vary both temporally and geographically. In view of this, water with a standard hardness level of 200–400 parts per million $CaCO_3$ is considered a more desirable diluent in such tests (Sattar et al., 2001b).

4.3.2. Dried Virus Inoculum as the Challenge

As stated above, a carrier with the test inoculum dried on it presents a stronger challenge to the microbicide being evaluated. Although this may be possible with some viruses, certain commonly used surrogates (e.g., polioviruses) lose high levels of infectivity on drying (Mbithi et al., 1991) especially at low levels of relative humidity (RH). A fine balance may therefore be required to achieve the right degree of drying of the virus inoculum on carriers or by selecting a surrogate that is more stable during the drying of the test inoculum. HAV, FCV, and HRV are all more resistant to drying than enteroviruses in general (Sattar and Ansari, 2002). Suitable controls must be included to determine the loss in the infectivity of the test virus during the drying process, and the level that survives becomes the baseline for measuring the extent of virus inactivation by the test formulation (Springthorpe and Sattar, 2003).

4.4. Time and Temperature for Virus-Microbicide Contact

Except for products that are meant for prolonged soaking of items to be decontaminated, the contact between the target virus(es) on an environmental surface and microbicide under in-use conditions is generally very brief. This should be properly reflected in the design of a carrier test for virucidal activity, and such contact times should not be longer than about 3 min to allow for relatively slow-acting but commonly used actives such as ethanol. This is in contrast with currently accepted microbicide test protocols that incorporate a minimum contact of 10 min, which is much too long to simulate the use of environmental surface disinfectants in the field (AOAC, 1998).

Formulations to be used on environmental surfaces are tested at an air temperature of 20°C; this is lower than the ambient temperature indoors in many work settings and requires the use of suitable climate control chambers to maintain the desired temperature. Air temperatures higher than 20°C may enhance the activity of microbicides while also accelerating the rate of their evaporation from the carrier surface. Products to be used outdoors during winter months or indoors in refrigerators must be shown to be effective against viruses at lower temperatures.

4.5. Elimination of Cytotoxicity

Cytotoxicity of the test formulation to host cells is an important consideration in virucidal tests (Quinn and Carter, 1999) because it can interfere with the reading and interpretation of test results. In addition, any material(s) and procedure used to remove and/or neutralize cytotoxicity must itself be safe for the test virus.

A 10- to 100-fold dilution of the virus-microbicide mixture at the end of the contact time is one simple and potentially viable approach to reducing cytotoxicity (Lloyd-Evans et al., 1986). This approach, however, requires relatively high titered pools of the test virus and may not work on its own for chemicals that are highly cytotoxic. Microbicides such as formaldehyde can effectively kill host cells without detaching them or producing any apparent damage to them. Such cytotoxicity can be misleading because host cell monolayers may appear to be undamaged but are unable to support virus replication. Moreover, one should note that even when toxicity appears to be visibly removed, subtle effects on the cells and potentially on their ability to support virus replication may remain. This needs to be examined through a low-level virus challenge (Sattar et al., 2003).

Gel filtration (ASTM, 1998) or high-speed centrifugation (Doultree et al., 1999) of virus-microbicide mixtures may be effective in the removal of cytotoxicity, but such steps invariably extend the contact of the virus with the test microbicide by several minutes or more and bring into question the accuracy and relevance of claims of virucidal activity for many applications. Other considerations in the selection and use of procedures for the elimination of cytotoxicity have been described before (Sattar and Springthorpe, 2001a).

4.6. Neutralization of Virucidal Activity

For accurate and reproducible results, the microbicidal activity of the test formulation must be arrested immediately at the end of the contact time (Sutton, 1996). This can be achieved by either the addition of a neutralizer or dilution of the virus-microbicide mixture or a combination of both. Whichever approach is adopted, its effectiveness must be properly validated before the test results can be considered as meaningful.

The difficulties in choosing a suitable chemical neutralizer are somewhat similar to those enumerated above for cytotoxicity removal. Although a 100-fold dilution of the virus-microbicide mixture soon after the end of the contact time has proved effective in dealing with most types of microbicides (Lloyd-Evans et al., 1986), this procedure requires that the volume of the diluent be kept relatively small to allow for the titration of most of the eluate.

4.7. Quantitation of Virus Infectivity

The availability of a simple and reproducible method for assaying infectious virus in the test and control samples is absolutely essential for determining virucidal activity. Indirect measures of virus infectivity based on assays for antigens, enzymes, or nucleic acids are not recommended because of the lack of demonstrated correspondence between their concentrations and those of infectious virus in the samples being assayed.

It is noteworthy that the presence of microbicide residues, even in diluted eluates, may increase or decrease the susceptibility of the host cells to the test virus. In case of decreased susceptibility, the host system could overestimate the activity of the tested microbicides by not being able to detect the presence of low levels of infectious virus in the inoculum. An increase in the level of infectivity could possibly be due to any one or a combination of (a) unmasking of more viral receptors on the host cell surface, (b) inactivation of specific or nonspecific virus inhibitors, (c) altering the electrostatic charges on the virus and/or the cell surface, and (d) deaggregation of viral clumps. Controls must, therefore, be included in virucidal tests to rule out the presence of such interference and for the results to be considered valid. The best way to approach this is to first expose the cell monolayer to a non-cytotoxic level of the test microbicide and subsequently challenge the cells to the test virus diluted to yield countable infectious foci such as plaques. If the number of infectious foci in such pre-exposed monolayers is not statistically different from those in the monolayers treated with a control fluid, the product can be assumed to be free from such interference.

4.8. Number of Test and Control Carriers

Enough test and control carriers must be included to make the results statistically meaningful. This requires some knowledge of the degree of reproducibility of the assay methods; because viruses require a host system, the results tend to be inherently more variable than those observed for bacteria and fungi. In general, methods that determine virus plaque- or focus-forming units are more accurate than the most probable number (MPN) techniques. Each measure of reduction in virus infectivity by a microbicide is obtained by comparison with controls not exposed to microbicide. Therefore, it is important that sufficient numbers of such controls are included to obtain an accurate mean value against which each test carrier can be assessed.

4.9. Product Performance Criteria

For government registration, microbicidal products must meet a performance criterion that is based on practical considerations rather than on sound public health science. A 3–4 $\log_{10}$ reduction in virus infectivity titer after exposure to the test formulation is regarded as satisfactory virucidal activity. The CGSB (1997) standard, for example, requires the tested product to show a >3 $\log_{10}$ reduction (beyond the level of cytotoxicity) in the level of infectious virus to meet its requirements. This criterion is lower than the minimum 5–6 $\log_{10}$ reductions required for other classes of pathogens because of the general difficulties in generating high-titered virus pools.

5.0. CURRENTLY AVAILABLE TESTS

Table 12.2 lists the methods currently accepted or under consideration as standard test protocols for testing the virucidal activity of microbicides.

5.1. Quantitative Suspension Tests

The ASTM suspension test for virucidal activity (E-1052) has recently been revised, and the current version incorporates several changes. This test is for special applications of virucides such as inactivation of viruses in contaminated wastes and as a first step in determining the virucidal potential of liquid chemical microbicides, liquid hand soaps, over-the-counter (OTC) topical products, or other skin care products.

Another quantitative suspension test for virucidal activity of chemical disinfectants and antiseptics is being drafted by CEN (Comité Européen de Normalisation) Technical Committee (TC) 216. An adenovirus and a vaccine strain of poliovirus are listed as test viruses. The contact time at 20 ± 1°C ranges from 30 to 60 min depending on the intended use of the product. The formula being evaluated is tested with and without an added protein load in

Table 12.2 Standard Test Methods for Evaluating the Virucidal Activity of Microbicides Designed to Be Used on Environmental Surfaces or Human Hands

Organization	*Title of Standard*	*Document No.*
ASTM International	Standard Test Method for Efficacy of Antimicrobial Agents Against Viruses in Suspension	E-1052
	Standard Test Method for Efficacy of Virucidal Agents Intended for Inanimate Environmental Surfaces	E-1053
	Standard Test Method for Neutralization of Virucidal Agents in Virucidal Efficacy Evaluations	E1482
	Standard Test Method for Determining the Virus-Eliminating Effectiveness of Liquid Hygienic Handwash and Handrub Agents Using the Fingerpads of Adult Volunteers	E-1838
	Standard Test Method for Evaluation of Handwashing Formulations for Virus-Eliminating Activity Using the Entire Hand	E-2011
	Standard Quantitative Disk Carrier Test Method for Determining the Bactericidal, Virucidal, Fungicidal, Mycobactericidal, and Sporicidal Activities of Liquid Chemical Germicides	E-2197
Canadian General Standards Board	Assessment of Efficacy of Antimicrobial Agents for Use on Environmental Surfaces and Medical Devices (Canadian national standard)	CAN/ CGSB-2.161-M97

the form of either 0.3% bovine serum albumin or 5% defibrinated sheep blood. The product performance criterion is a minimum $4\log_{10}$ reduction in the infectivity titer of the test virus.

5.2. Quantitative Carrier Tests

There are four methods in this category in North America. The first is an ASTM (2002a) standard, which has also been revised recently (E-1053). It is meant for evaluating the activity of liquid or pressurized antimicrobials against viruses on inanimate, nonporous, environmental surfaces. This standard lists 10 different viruses with varying degrees of resistance to liquid chemical microbicides. It recommends, however, that the test formulation be evaluated at least against a poliovirus, a herpesvirus, and an adenovirus to qualify for a general virucidal claim. The test virus suspension is first dried on a glass Petri plate and then overlaid with a known volume of the test formulation for a predetermined contact time at ambient temperature. At the end of the contact time, a diluent is added to the virus-product mixture, and the test surface is scraped to resuspend the virus film. The eluates and controls are assayed for infectious virus to determine the loss in virus titer due to the test formulation's virucidal activity. Calf serum is recommended as organic soil (except for rotaviruses), and water with a specific level of hardness is to be used if the product requires dilution in water prior to use.

The second carrier test is a part of the Canadian General Standard Board's document Assessment of Efficacy of Antimicrobial Agents for Use on Environmental Surfaces and Medical Devices (CAN/CGSB-2.161-M97). This test permits the use of glass Petri plates, glass slides, or disks of glass, metal, or plastic. The recommended test virus is the Sabin strain of poliovirus type 1 (ATCC VR-192) to permit a general virucidal claim when testing is carried out with 5% fetal bovine serum as the organic load and, when necessary, water with a hardness of 200 ppm $CaCO_3$ as diluent for the test product.

The third carrier test is the *in vivo* fingerpad method to assess the virus-eliminating activity of topicals. Full details of this method are given in the literature already cited above or in the ASTM standard itself (ASTM E-1838). The fourth method is a standard for virucidal evaluation of formulations when these are performed on the whole hand (ASTM E-2011).

The second tier of a quantitative carrier test (QCT-2) method developed by us under the sponsorship of the U.S. Environmental Protection Agency (EPA) can be applied to all major classes of microorganisms including viruses (Springthorpe and Sattar, 2003, 2005). This would allow direct comparison of results among different classes of microorganisms. The test uses flat stainless steel disk carriers (approximately 1 cm diameter) and a microbial inoculum of 10 μl. After drying of the inoculum, the contaminated carrier is exposed to approximately 50 μl of the test microbicide at 20°C for the manufacturer's recommended contact time. The reaction is terminated by neutralization of the microbicide. In most cases, neutralization is achieved by simple dilution with a physiological saline, but in some instances a chemical neutralizer is required prior to dilution. The test virus can then be titrated by standard methods. This method, which is now a standard of ASTM International (E-2197), has been used to test for virucidal activity against several types of viruses (Lloyd-Evans et al., 1986; Sattar et al., 1989, 2003; Mbithi et al., 1990).

6.0. PRACTICAL ASPECTS OF TESTING MICROBICIDES

We describe in this section the key practical considerations in evaluating microbicides against food-borne viruses. The methods described here are standards of ASTM International and have been used extensively in working with a variety of viruses (Springthorpe and Sattar, 2003; Sattar et al., 2003; Ansari and Sattar, 2002). Figure 12.1 summarizes the basic steps in the quantitative carrier test method (ASTM 2002a), and Figure 12.2 presents the basic steps in the fingerpad method (ASTM 2002c). Although no nationally or internationally accepted methods are available to assess the activity of microbicides used in the decontamination of fruit and vegetables, published studies

Inoculate each disk with 10 µl of test virus with the soil load; allow inoculum to dry. Place one carrier (disk), inoculated side up, on the inside bottom surface of a sterile holder/vial

↓

Place 50 µl of the test formulation on the surfaces of 3–10 test carriers. Place an equivalent volume of normal saline or Earle's balanced salt solution (EBSS) on each of at least 3 control carriers

↓

Hold the carriers for the desired contact time

↓

Add 950 µl of EBSS, with or without a neutralizer, to each vial containing disk carriers

↓

Vortex contents of vials for 45–60 seconds

↓

Transfer eluate to a 2-ml tube and make 10-fold dilutions in EBSS as necessary

↓

Inoculate dilutions to be tested onto monolayers of host cells and incubate

↓

Examine cell cultures after appropriate incubation and determine $\log_{10}$ reduction of the inoculated virus

↓

Determine if the test formulation meets the specified performance criterion

Figure 12.1 Basic steps in disk-based quantitative carrier test for virucidal activity.

(Bidawid et al., 2000, 2003) should serve as a guide in the design and performance of such testing.

The three viruses described below have been selected based on their (a) relevance as food-borne pathogens, (b) relative resistance to microbicides, (c) ability to withstand drying on environmental surfaces and human skin, (d) availability of cell culture–based infectivity assays, and (e) safety for work in experimental settings and for placement on the intact skin of adult subjects.

The need for cell cultures adds an extra layer of difficulty when working with viruses. Also, procedures that work perfectly in one laboratory do not always work in another. This may be due to even slight variations in the quality of water for making media and reagents or to procedures for the clean-up and sterilization of labware, and so forth. Each laboratory must develop and document its own standard operating procedures for each host cell type and test virus to be used. However, regardless of the methods used for cell culture, preparation of virus pools, and quantitation of virus infectivity, the procedures for testing the activity of microbicides must adhere to

Panelist washes hands with non-germicidal soap and water and dries them

↓

Five milliliters of 70–75% (v/v) ethanol is rubbed on hands till they are dry

↓

Ten microliters of virus with soil load is placed at the center of each thumb and fingerpads. Inoculum from thumbpads is eluted immediately to act as "input" control for virus.

↓

Inoculum on the fingerpads is allowed to become visibly dry (20–25 minutes)

↓

Two randomly selected fingerpads are eluted at the end of drying ("baseline" control)

↓

Dried inoculum on two or more fingerpads is exposed to 1 ml of test or control fluid for desired contact time (for waterless agents or to test virus elimination after exposure to product alone, fingerpads can be eluted without further treatment)

↓

To simulate post-treatment rinsing of hands, fingerpads are exposed to 1–15 ml of water for 5–10 seconds. Virus can be eluted at this stage or after drying of hands

↓

To determine virus removal after the drying of washed hands, they can be dried in air or with paper or cloth towel for specified time and virus recovered from them

↓

One milliliter of eluent is used to recover virus from each thumb or fingerpad. Eluates and controls are titrated for infectious virus and $\log_{10}$ reductions calculated.

Figure 12.2 Basic steps in the fingerpad method (ASTM, 2002) for testing handwash or handrub agents against viruses.

the basic requirements as described above to ensure a sufficient level of stringency and reproducibility and to allow the comparison of the results from tests using different viruses and test formulations.

Although ultracentrifugation may sometimes be needed to increase the virus titer, the use of highly purified virus pools is not recommended for testing microbicides because such purification is likely to enhance susceptibility of the virions to microbicides. Described below are some viruses that can be used in such tests.

6.1. Strain HM-175 (ATCC VR-1402) of HAV

HAV, an important food-borne pathogen, affects the liver and is excreted in the feces of infected individuals. It is relatively resistant to drying and mechanical damage and is also generally more resistant to microbicides than other nonenveloped viruses of human origin. Immunization of lab workers with the recently available vaccines makes the handling of this virus much safer. The recommended cell line for making HAV pools and for performing infectivity titrations is FRhK-4 (ATCC CRL-1688). No less than 7 days are needed to complete an infectivity assay due to the relatively slow rate of growth of the virus.

6.2. Strain F9 (ATCC VR-782) of FCV

FCV, pathogenic to cats but believed harmless to humans, belongs to a group of small round viruses. FCV, which is nonenveloped, is closely related to Norwalk or norovirus (NV), a major cause of acute gastroenteritis in humans and also a significant food-borne pathogen. Because NV cannot be grown *in vitro*, FCV is generally accepted as its surrogate (Doultree et al., 1999) and has been used in testing microbicides in settings where foods may be handled (Gulati et al., 2001; Bidawid et al., 2003). The cell line recommended for work with FCV is CrFK (ATCC CCL-94), and a plaque assay system based on these cells has been developed (Bidawid et al., 2002). This virus grows to high titers (~10^8 infective units/ml) within 28–36 hr and produces visible CPE or plaques in less than 36 hr. This is helpful in making test results available relatively rapidly.

6.3. Human Rotavirus

6.3.1. Wa Strain (ATCC VR-2018)

Rotaviruses, which are a common cause of acute gastroenteritis in humans, are excreted in diarrheic feces in numbers generally higher than those for other enteric viruses (Ward et al., 1991). Food-borne spread of rotaviruses has been documented (Sattar et al., 2001a). Recommended cell lines for growing rotaviruses are MA-104 (CRL-2378) and CV-1 (ATCC CCL-70; Sattar et al., 2000). Rotaviruses are safe for normal healthy adults as most adults have acquired immunity against them. The ability of rotaviruses to withstand drying also adds to their attraction as surrogates in testing microbicides. Two important factors to note when working with rotaviruses are that (a) many of them are inhibited by fetal bovine sera often used in cell culture, and (b) the presence of proteolytic enzymes such as trypsin is needed to promote rotavirus infection of host cells.

6.4. Additional Controls in Virucidal Tests

The use of cell cultures requires the incorporation of the following additional controls in tests for virucidal activity (Sattar et al., 2002, 2003) because either the test substance or the neutralizer or both could alter the susceptibility of host cells to the virus in the test. These controls must be run initially at least once and may not need to be included in subsequent tests as long as the same cell line, virus, test formulation, neutralizer, and method are used in testing.

6.4.1 Cytotoxicity Control

This control (a) determines the dilution of the test substance that causes no apparent degeneration (cytotoxicity) of the cell line to be used for measuring virus infectivity and (b) assesses whether the neutralizer reduces or enhances such cytoxicity. For this control, make a 1:20 dilution, then a 1:200 dilution of the test substance in Earle's balanced salt solution (EBSS) with and without the neutralizer. Remove the culture medium from the monolayers of the host cell line(s) and put into each test monolayer separately the same volume of inoculum as used in virus titration; control monolayers receive an equivalent amount of EBSS only (without any neutralizer). Use at least three monolayers for controls as well as for each dilution of the test substance being assessed. Hold the cultures at room temperature for the same length of time used for virus adsorption, then examine under an inverted microscope for any visible cytotoxicity.

If cytotoxicity is observed, a different neutralizer or alternative approaches to the removal/reduction of cytotoxicity may be needed. It is sometimes advisable to use gel filtration to remove the disinfectant, although this procedure may lengthen the exposure time of the test organism to the disinfectant. If no cytotoxicity is observed at either dilution, the test substance and the neutralizers should be subjected to the following interference test.

6.4.2. Control for Interference with Virus Infectivity

Levels of the test substance that show no obvious cytotoxicity could still reduce or enhance the ability of the challenge virus to infect or replicate in host cells, thus interfering with the estimation of its virucidal activity (Sattar et al., 2003). An interference control must, therefore, be included to rule out such a possibility. For this purpose, remove the culture medium from monolayers of the host cell line(s) and add a 1:20 dilution, or a dilution greater than the one that demonstrated cytotoxicity, of the test substance in EBSS to each of the test monolayers with and without neutralizer, using the same volume as that of the inoculum used in virus titration. Controls receive EBSS alone (without the neutralizer). Hold the monolayers at room temperature for the same length of time as used for virus adsorption and inoculate each with a low number (approximately 10–20) of infective units of the challenge virus. Incubate the monolayers for virus adsorption, place maintenance medium in the cultures, incubate them for the time required for virus replication, and then examine for cytopathology or foci of virus infection.

Any significant difference in virus infectivity titer is indicative of the ability of the test material or the neutralizer to affect the virus susceptibility of the host cells. In such case, a different neutralizer or alternative approaches to the removal of the residues of the test product may be needed. Both the cytotoxicity and interference controls must be included even when virus infectivity is titrated using the $TCID_{50}$ method.

6.4.3. Control Carriers

The minimum number of control carriers to be used in each test is three regardless of the number of test carriers. For control carriers, add 50 µl of EBSS instead of the test formulation. The contact time and temperature for the control carriers must be the same as those for the test carriers.

7.0. MICROBICIDES IN ENVIRONMENTAL CONTROL OF FOOD-BORNE VIRUSES

Table 12.3 summarizes recent data on the activity of various concentrations of sodium hypochlorite (bleach) against FCV as tested using QCT-2 and stainless steel disks as carriers. The product effectiveness criterion was arbitrary set as 3 $\log_{10}$ or greater reduction in virus infectivity. A minimum contact time of 10 min under ambient conditions was needed to meet this requirement with 500 ppm of available chlorine; when the available chlorine level was doubled, the contact time could be reduced to as low as 1 min.

The results of similar tests with other environmental surface disinfectants are presented in Table 12.4. Except for 75% ethanol, the remaining four formulations are commercially available environmental surface disinfectants; chlorine dioxide was prepared just before testing using the two solutions provided by the manufacturer. The trade names of the products tested are not given because formulations with identical types and levels of ingredients may be sold under a different trade name elsewhere. Only chlorine dioxide could meet this criterion in a contact time of 1 min, while at least 3 min were required for accelerated hydrogen peroxide and the two commercial alcohol-based sprays to do so. The quat-based formulation and 75% ethanol met the criterion after a minimum contact time of 10 and 5 min, respectively.

The findings of tests with three commercial alcohol-based hand rubs are summarized in Table 12.5. When the ASTM (2002a) fingerpad protocol was used with a contact time of 20 s, the $\log_{10}$ reduction in the infectivity titer of FCV ranged from 1.20 to 1.49. In this regard, FCV proved to be more resistant than adeno-, rota-, and rhinoviruses (Sattar et al., 2000) and less resistant than HAV (Mbithi et al., 1993).

Table 12.3 Activity of Domestic Bleach Against Feline Calicivirus[a,b]

Available Chlorine in Parts per Million	*Contact Time in Minutes*	*Log_{10} Reduction in Virus Infectivity*
100	5	0.56
500	5	1.39
500	10	>4.7
1,000	1	>4.7
1,000	2	>4.7
1,000	3	>4.7

[a] Sattar et al. unpublished data.

[b] The bleach tested was a commercially available product with 5.25% sodium hypochlorite. The metal disk-based test method is a standard ASTM method (E-2197). All testing was done at 25°C in the presence of a soil load and water with a hardness of 400 ppm as calcium carbonate (used to dilute the product).

Table 12.4 Activity of Selected Microbicides Against Feline Calicivirus

Active(s) in Microbicide	*Conc. Tested (ppm)*	*Contact Time (min)*	*Log_{10} Reduction*
0.5% accelerated H_2O_2	Undiluted (5,000)	1	1.55
	Undiluted (5,000)	3	>4.5
A mixture of four quaternary ammonium compounds (octyl decyl dimethyl NH_4Cl (4.6%); dioctyl dimethyl NH_4Cl (1.84%); didectyl dimethyl NH_4Cl (2.76%); dimethyl benzyl NH_4Cl (6.14%)	Diluted 1:62 (2,470)	5	2.3
	Diluted 1:62 (2,470)	10	4.0
79% (v/v) ethanol + 0.1% alkyl dimethyl benzyl ammonium saccharinate	Undiluted	1	2.2
	Undiluted	3	3.8
Each 100 g contains ethanol 25.92 g, 2-propanol 11.50 g, and polyhexanide 0.054 g	Undiluted	1	0.8
	Undiluted	3	3.0
	Undiluted	5	3.5
Ethanol	75% (v/v)	5	3.8
	75% (v/v)	10	4.7
Liquid chlorine dioxide	1,000	1	4.5

[a] Sattar et al. unpublished data.
[b] The metal disk-based test method used is a standard of ASTM (E-2197); all testing was done at 23 ± 2°C in the presence of a soil load and water with a hardness of 400 ppm as calcium carbonate (used to dilute the product).

Table 12.5 Activity of Commercially Available Ethanol-Based Handrub Agents Against Feline Calicivirus Using the Fingerpad Method[a,b]

Ethanol in Handrub (%)	*Log_{10} Reduction in pfu After Contact with Test Formulation*
60	1.20
70	1.42
80	1.49

pfu, plaque-forming unit.
[a] Sattar et al. unpublished data.
[b] Each fingerpad was contaminated with 10 µl of the test virus suspended in a soil load and the inoculum allowed to dry under ambient conditions. Virus from two fingerpads was eluted to determine the pfu remaining after the drying period; this figure (8.9×10^4) was used as the "baseline" to determine log_{10} reduction in virus titer after exposure to the formulation under test. Each formulation was tested on at least three adult subjects, and no less than two fingerpads were exposed to the test formulation in each test. The contact time in all experiments was 20 s.

8.0. CONCLUSIONS

Viruses continue to be important pathogens in general and as food-borne pathogens in particular, but our understanding of the actual sources of viral contamination in many food-borne outbreaks remains incomplete, making it difficult to design and apply proper strategies to prevent and control the spread of such pathogens. However, hands are universally recognized as vehicles for the spread of a number of viruses. Successful strategies to prevent virus spread through these vehicles involve a sound hand-decontamination protocol, diligently applied with a good topical agent. A lack of compliance with hand antisepsis guidelines and, perhaps, the use of ineffective agents continue to undermine the full potential of infection-control measures in this regard. The ease with which washed hands can pick up infectious viruses upon contact with contaminated environmental surfaces and objects suggests that the emphasis on hand antisepsis should be combined with an awareness of the need for proper and regular cleaning and decontamination of those surfaces and objects that come in frequent contact with decontaminated hands.

Standardization of virucide tests, nationally and internationally, will promote confidence among microbicide users and the general public. This chapter provides the basis for general understanding of the potential pitfalls in testing virucides and suggests the basic protocols and controls that should be present in generic methods. This should allow the reader to better understand this field and to be able to critique the published literature independently.

Standard tests for virucides are now available. These tests provide improved carrier design, better methods for cytotoxicity removal, a universal soil load, and other improvements. However, regulatory agencies, especially in the United States, must soon decide on accepting surrogates in tests for virucidal activity and label claims and also set product performance standards. Some jurisdictions already have one or both of these in place (CGSB, 1997). Any such discussion must consider activity against one or more carefully selected nonenveloped viruses representative of food-borne viral pathogens. Many products currently on the market list only enveloped viruses among the organisms on the label. Persons unfamiliar with virus classification can be easily misled by this, especially if the enveloped viruses listed are among those most feared.

Our current knowledge does not allow, with any degree of certainty, the determination of the desired level of reduction in virus load in a given setting to significantly reduce disease transmission. There are also obvious practical limitations to how high a level of challenge virus(es) one can present to the product under evaluation. By the same token, what would one regard as too low a level of challenge? Experience accumulated over the past two decades clearly indicates that if test viruses are chosen carefully, it is feasible to determine a 3–4 $\log_{10}$ reduction in virus infectivity titer after its exposure to a test microbicide in a proper carrier test. The viruses selected for QCT-2 are based on their (a) relative safety for the laboratory staff, (b) ability to grow to titers sufficiently high for testing, (c) property to produce cytopathic effects or plaques, or both, in cell cultures, (d) potential to spread through contaminated environmental surfaces and (e) relatively high resistance to a variety of chemicals.

Given these considerations and the fact that enveloped viruses in general do not survive well on environmental surfaces and are relatively more susceptible to chemical microbicides, all viruses included here are nonenveloped viruses. Other strains or types of nonenveloped viruses may be substituted in the test provided they meet the preceding criteria.

9.0. REFERENCES

Abad, F. X., Pinto, R. M., and Bosch, A., 1997, Disinfection of human enteric viruses on fomites. *FEMS Microbiol. Lett.* 156:107–111.

Ansari, S. A., and Sattar, S. A., 2002, The need and methods for assessing the activity of topical agents against viruses, in: *Handbook of Topical Antimicrobials: Industrial Applications in Consumer Products and Pharmaceuticals* (D. Paulson, ed.), Marcel Dekker, New York, pp. 411–445.

Ansari, S. A., Sattar, S. A., Springthorpe, V. S., Wells, G. A., and Tostowaryk, W., 1988, Rotavirus survival on human hands and transfer of infectious virus to animate and non-porous inanimate surfaces. *J. Clin. Microbiol.* 26:1513–1518.

Ansari, S. A., Sattar, S. A., Springthorpe, V. S., Wells, G. A., and Tostowaryk, W., 1989, In vivo protocol for testing efficacy of hand-washing agents against viruses and bacteria: experiments with rotavirus and Escherichia coli. *Appl. Environ. Microbiol.* 55:3113–3118.

Ansari, S. A., Springthorpe, V. S., and Sattar, S. A., 1991, Survival and vehicular spread of human rotaviruses: possible relationship with seasonality of outbreaks. *Rev. Infect. Dis.* 13:448–461.

Ansari, S. A., Springthorpe, V. S., and Sattar, S. A., 1991, Comparison of cloth-, paper- and warm air-drying in eliminating viruses and bacteria from washed hands. *Am. J. Infect. Control* 19:243–249.

AOAC International, 1998, *Disinfectants—Official Methods of Analysis.* AOAC International, Gaithersburg, MD.

Appleton, H., 2000, Control of food-borne viruses. *Br. Med. Bull.* 56:172–183.

ASTM International, 1998, Standard test method for neutralization of virucidal agents in virucidal efficacy evaluations, E1482. ASTM, West Conshohocken, PA.

ASTM International, 1999, Standard test method for evaluation of handwashing formulations for virus-eliminating activity using the entire hand, E2011. ASTM, West Conshohocken, PA.

ASTM International, 2002a, Standard quantitative disk carrier test method for determining the bactericidal, virucidal, fungicidal, Mycobactericidal and sporicidal activities of liquid chemical microbicides, E2197. ASTM, West Conshohocken, PA.

ASTM International, 2002b, Standard test method for determining the virus-eliminating effectiveness of liquid hygienic handwash and handrub agents using the fingerpads of adult volunteers, E1838. ASTM, West Conshohocken, PA.

ASTM International, 2002c, Standard test method for efficacy of antimicrobial agents against viruses in suspension, E1052. ASTM, West Conshohocken, PA.

ASTM International, 2002d, Standard Test Method for Efficacy of Virucidal Agents Intended for Inanimate Environmental Surfaces, E1053, ASTM, West Conshohocken, PA.

Aylward, R. B., Acharya, A., England, S., Agocs, M., and Linkins, J., 2003, Global health goals: lessons from the worldwide effort to eradicate poliomyelitis. *Lancet* 362:909–914.

Beuchat, L.R., 2001, Surface decontamination of fruit and vegetables eaten raw: a review. Document No. WHO/FSF/FOS/98.2. World Health Organization, Geneva, Switzerland.

Bidawid, S., Farber, J. M., and Sattar, S. A., 2000, Contamination of foods by food handlers: experiments on hepatitis A virus transfer to food and its interruption. *Appl. Environ. Microbiol.* 66:2759–2763.

Bidawid, S., Malik, N., Adegbunrin, O., Sattar, S. A., and Farber, J. M., 2002, A feline kidney cell line-based plaque assay for feline calicivirus, a surrogate for Norwalk virus. *J. Virol. Methods* 107:163–167.

Bidawid, S., Malik, N., Adegbunrin, O., Sattar, S. A., and Farber, J. M., 2003, Norovirus cross-contamination during food handling and interruption of virus transfer by hand antisepsis: experiments with feline calicivirus as a surrogate. *J. Food Protect.* (in press).

Bresee, J. S., Widdowson, M. A., Monroe, S. S., and Glass, R. I., 2002, Foodborne viral gastroenteritis: challenges and opportunities. *Clin. Infect. Dis.* 35:748–753.

Canadian General Standards Board (CGSB), 1997, Assessment of the efficacy of antimicrobial agents for use on environmental surfaces and medical devices. National Standards of Canada Document No. CAN/CGSB-2.161.97. Canadian General Standards Board, Ottawa, Ontario, Canada.

Comité Européen de Normalization, 2001, CEN/TC 216 "Chemical disinfectants and antiseptics" listing of efficacy methods for disinfectants. CEN, Brussels, Belgium.

Cliver, D. O., 1997, Virus transmission via food. *World Health Stat. Q.* 50:90–101.

Collins, J. E., 1997, Impact of changing consumer lifestyles on the emergence/reemergence of foodborne pathogens. *Emerg. Infect. Dis.* 3:471–479.

Doultree, J. C., Druce, J. D., Birch, C. J., Bowden, D. S., and Marshall, J. A., 1999, Inactivation of feline calicivirus—a Norwalk virus surrogate. *J. Hosp. Infect.* 41: 51–57.

Gulati, B. R., Allwood, P. B., Hedberg, C. W., and Goyal, S. M., 2001, Efficacy of commonly used disinfectants for the inactivation of calicivirus on strawberry, lettuce, and a food-contact surface. *J. Food Prot.* 64:1430–1434.

Koopmans, M., von Bonsdorff, C. H., Vinje, J., de Medici, D., and Monroe, S., 2002, Foodborne viruses. *FEMS Microbiol. Rev.* 26:187–205.

Lloyd-Evans, N., Springthorpe, V. S., and Sattar, S. A., 1986, Chemical disinfection of human rotavirus-contaminated inanimate surfaces. *J. Hyg.* 97:163–173.

Lüthi, T. M., 1997, Food and waterborne viral gastroenteritis—a review of agents and their epidemiology, Mitt. Gebete. *Lebensm. Hyg.* 88:119–150.

Mbithi, J. N., Springthorpe, V. S., and Sattar, S. A., 1990, Chemical disinfection of hepatitis A virus on environmental surfaces. *Appl. Environ. Microbiol.* 56:3601–3604.

Mbithi, J. N., Springthorpe, V. S., and Sattar, S. A., 1991, The effect of relative humidity and air temperature on the survival of hepatitis A virus on environmental surfaces. *Appl. Environ. Microbiol.* 57:1394–1399.

Mbithi, J. N., Springthorpe, V. S., Boulet, J. R., and Sattar, S. A., 1992, Survival of hepatitis A virus on human hands and its transfer on contact with animate and inanimate surfaces. *J. Clin. Microbiol.* 30:757–763.

Mbithi, J. N., Springthorpe, V. S., and Sattar, S. A., 1993, Comparative in vivo efficiency of hand-washing agents against hepatitis A virus (HM-175) & poliovirus type 1 (Sabin). *Appl. Environ. Microbiol.* 59:3463–3469.

Mead, P. S., Slutsker, L., Dietz, V., McCaig, L. F., Bresee, J. S., Shapiro, C., Griffin, M., and Tauxe, R. V., 1999, Food-related illness and death in the United States. *Emerging. Infect. Dis.* 5:607–625.

Quinn, P. J., and Carter, M. E., 1999, Evaluation of virucidal activity, in: *Principles and Practices of Disinfection, Preservation and Sterilization* (A. D. Russell, W. B. Hugo, and G. A. J. Ayliffe, eds.), Blackwell, London, pp. 197–206.

Sair, A. I., D'Souza, D. H., and Jaykus, L. A., 2002, Human enteric viruses as causes of foodborne disease. *Comp. Rev. Food Sci. Technol.* 1:73–89.

Sattar, S. A., and Ansari, S. A., 2002, The fingerpad protocol to assess hygienic hand antiseptics against viruses. *J. Virol. Methods* 103:171–181.

Sattar, S. A., and Bidawid, S., 2001, Environmental considerations in preventing the foodborne spread of hepatitis A, in: *Foodborne Disease Handbook* (Y. H. Hui, S. A. Sattar, K. D. Murrell, W. K. Nip, and P. S. Stanfield, eds.), Marcel Dekker, New York, pp. 205–216.

Sattar, S. A., and Springthorpe, V. S., 1996, Transmission of viral infections through animate and inanimate surfaces and infection control through chemical disinfection, in: *Modeling Disease Transmission and its Prevention by Disinfection* (C. J. Hurst, ed.), Cambridge University Press, New York, pp. 224–257.

Sattar, S. A., and Springthorpe, V. S., 1999, Viricidal activity of biocides—activity against human viruses, in: *Principles and Practice of Disinfection, Preservation and Sterilization*, 3rd ed. (A. D. Russell, W. B. Hugo, and G. A. J. Ayliffe, eds.), Blackwell, London, pp. 168–186.

Sattar, S. A., and Springthorpe, V. S., 2001a, Methods for testing the virucidal activity of chemicals, in: *Disinfection, Sterilization and Preservation*, 5th ed. (S. S. Block, ed.), Lippincott Williams & Wilkins, Philadelphia, pp. 1391–1412.

Sattar, S. A., and Springthorpe, V. S., 2001b, New methods for efficacy testing of disinfectants and antiseptics, in: *Disinfection, Sterilization, and Antisepsis: Principles and Practices in Healthcare Facilities* (W. A. Rutala, ed.), Association of Professionals in Infection Control (APIC), Washington, DC, pp. 173–186.

Sattar, S. A., Raphael, R. A., Lochnan, H., and Springthorpe, V. S., 1983, Rotavirus inactivation by chemical disinfectants and antiseptics used in hospitals. *Can. J. Microbiol.* 29:1464–1469.

Sattar, S. A., Lloyd-Evans, N., Springthorpe, V. S., and Nair, R. C., 1986, Institutional outbreaks of rotavirus diarrhea, potential role of fomites and environmental surfaces as vehicles for virus spread. *J. Hyg.* 96:277–289.

Sattar, S. A., Springthorpe, V. S., Karim, Y., and Loro, P., 1989, Chemical disinfection of non-porous inanimate surfaces experimentally contaminated with four human pathogenic viruses. *Epidemiol. Infect.* 102:493–505.

Sattar, S. A., Jacobsen, H., Rahman, H., Rubino, J., and Cusack, T., 1994, Interruption of rotavirus spread through chemical disinfection. *Infect. Contr. Hosp. Epidemiol.* 15:751–756.

Sattar, S. A., Tetro, J., and Springthorpe, V. S., 1999, Impact of changing societal trends on the spread of infectious diseases in American and Canadian homes. *Am. J. Infect. Control* 27:S4–S21.

Sattar, S. A., Abebe, M., Bueti, A., Jampani, H., and Newman, J., 2000a, Determination of the activity of an alcohol-based hand gel against human adeno-, rhino-, and rotaviruses using the fingerpad method. *Infect. Control Hosp. Epidemiol.* 21: 516–519.

Sattar, S. A., Tetro, J., Bidawid, S., and Farber, J., 2000b, Foodborne spread of hepatitis A: recent studies on virus survival, transfer and inactivation, *Can. J. Infect. Dis.* 11:159–163.

Sattar, S. A., Springthorpe, V. S., and Tetro, J., 2001, Rotavirus, in: *Foodborne Disease Handbook* (Y. H. Hui, S. A. Sattar, K. D. Murrell, W-K. Nip, and P. S. Stanfield, eds.), Marcel Dekker, New York, pp. 99–125.

Sattar, S. A., Springthorpe, V. S., Tetro, J., Vashon, B., and Keswick, B., 2002, Hygienic hand antiseptics: should they not have activity and label claims against viruses? *Am. J. Infect. Control* 30:355–372.

Sattar, S., Springthorpe, V. S., Adegbunrin, O., Zafer, A. A., and Busa, M., 2003, A disc-based quantitative carrier test method to assess the virucidal activity of chemical microbicides. *J. Virol. Methods.* 112:3–12.

Seymour, I. J., and Appleton, H., 2001, Foodborne viruses and fresh produce. *J. Appl. Microbiol.* 91:759–779.

Springthorpe, V. S., and Sattar, S. A., 1990, Chemical disinfection of virus-contaminated surfaces. *CRC Crit. Rev. Environ. Control* 20:169- 229.

Springthorpe, V. S., and Sattar, S. A., 2003, Quantitative carrier tests to assess the germicidal activities of chemicals: rationales and procedures. Centre for Research on Environmental Microbiology (CREM), University of Ottawa, Ottawa, ON, Canada.

Springthorpe, V. S., and Sattar, S. A., 2005, Carrier tests to assess microbicidal activities of chemical disinfectants for use on medical devices and environmental surfaces. *J. AOAC Int.* 88:182–201.

Springthorpe, V. S., Grenier, J. L., Lloyd-Evans, N., and Sattar, S. A., 1986, Chemical disinfection of human rotaviruses, efficacy of commercially available products in suspension test. *J. Hyg.* 97:139–161.

Sutton, S. V. W., 1996, Neutralizer evaluations as control experiments for antimicrobial efficacy tests, in: *Handbook of Disinfectants and Antiseptics* (J. M. Ascenzi, ed.), Marcel Dekker, New York, pp. 43–62.

Traoré, O., Springthorpe, V. S., and Sattar, S. A., 2002, A quantitative study of the survival of two species of Candida on porous and non-porous environmental surfaces and hands. *J. Appl. Microbiol.* 92:549–555.

Ward, R. L., Bernstein, D. I., Knowlton, D. R., Sherwood, J. R., Young, E. C., Cusack, T. M., Rubino, J. R., and Schiff, G. M., 1991, Prevention of surface-to-human transmission of rotavirus by treatment with disinfectant spray. *J. Clin. Microbiol.* 29:1991–1996.

Woolwine, J. D., and Gerberding, J. L., 1995, Effect of testing method on apparent activities of antiviral disinfectants and antiseptics. *Antimicrob. Agents Chemother.* 39:921–923.

CHAPTER 13

Food-borne Viruses: Prevention and Control

Efstathia Papafragkou, Doris H. D'Souza, and Lee-Ann Jaykus

1.0. INTRODUCTION

Epidemic and sporadic gastroenteritis is an important public health problem in both developed and developing countries. In the United States, as many as 67% of all food-borne illnesses, 33% of the associated hospitalizations, and 7% of deaths attributable to food-borne disease may be caused by viruses, resulting in approximately 30.9 million cases each year (Mead et al., 1999). Costs of illness are high simply by virtue of the frequent occurrence and high transmissibility of enteric viruses (Koopmans et al., 2002). The total burden of enteric viral disease can only be estimated as most of the illnesses are mild, go unreported, and routine testing of patients for specific virus infections is not usually done (Richards, 2001). Year-round outbreaks have affected adults and children in various settings. These viruses circulate readily in families, communities, and in places where individuals are in close proximity or are using a common source of food or water. Consequently, many documented outbreaks of enteric viral illness have occurred in schools, recreational camps, hotels, hospitals, orphanages, and nursing homes and among individuals consuming a common food item served in a restaurant or banquet setting. Hepatitis A virus (HAV), human caliciviruses, and group A rotaviruses are among the most important food-borne enteric viruses.

Unlike bacteria, viruses cannot replicate in food or water. As a result, when virus contamination of food occurs, the number of infectious virions will not increase during processing and storage. The ability of contaminated food to serve as a vehicle of infection, therefore, depends on virus stability, degree of initial contamination, the method of food processing and storage, viral dose needed to produce infection, and the susceptibility of the host (Koopmans et al., 2002). Virus particles are stable to environmental extremes including pH, low temperatures, and some enzymes, particularly those found in the human gastrointestinal tract (Jaykus, 2000a). Consequently, human enteric viruses can withstand a wide variety of food storage and processing conditions making virtually any kind of food product a potential vehicle for virus transmission (Jaykus, 2000b). Because food-borne viruses are transmitted via the fecal-oral route through contact with human feces and because infected individuals can shed millions of virus particles in their stools, the role of infected food-handlers cannot be underestimated.

From an epidemiological perspective, human caliciviruses are the most significant by virtue of the sheer number of cases caused by this virus group. Caliciviruses are composed of multiple genera; however, the noroviruses and the saporoviruses are the primary cause of human disease and are responsible for up to 2.3 million infections, 50,000 hospitalizations, and 300 deaths per year in the United States alone (Mead et al., 1999). Of all the gastroenteritis outbreaks reported in England and Wales, nearly 50% were due to noroviruses, a figure that is similar to those reported for other European countries including Finland, Sweden, The Netherlands, and Germany (Lopman et al., 2002). Indeed, noroviruses are the most common cause of acute, nonbacterial gastroenteritis worldwide. Although the disease caused by this virus group is short-lived and recovery is usually complete, immunity is poorly understood and there is significant antigenic and genetic diversity within the genus. As a result, some individuals may remain susceptible and/or become infected multiple times during the course of their lives (Jaykus, 2000b).

Noroviruses are transmitted directly by person-to-person contact or indirectly via contaminated food, water, or fomites. Person-to-person transmission can occur by two routes: fecal-oral and aerosol formation after projectile vomiting (Patterson et al., 1997). A total of 348 outbreaks of norovirus gastroenteritis were reported to the U.S. Centers for Disease Control and Prevention (CDC) between January 1996 and November 2000. Of these, food was implicated in 39%, person-to-person contact in 12%, and water in 3% of the outbreaks. Interestingly, 18% of the outbreaks could not be linked to a specific transmission mode (Centers for Disease Control, 2001). Frequently during an outbreak, secondary cases occur as a consequence of contact with the primary cases (Koopmans and Duizer, 2004). A low infectious dose (10–100 virus particles; Koopmans et al., 2002) and a propensity for long-term virus excretion (up to 2 weeks postinfection) are the likely reasons for the high attack rates (~50%) seen in norovirus outbreaks (Koopmans et al., 2002). Although attention has recently focused on high-profile cruise ship outbreaks (Centers for Disease Control, 2002), an estimated 60–89% of all acute viral gastroenteritis outbreaks occur on land (Centers for Disease Control, 2003b). Although norovirus infection predominates during cold-weather months (Mounts et al., 2001), hence the term "winter vomiting disease," it has been diagnosed year-round (Bresee et al., 2002).

Hepatitis A virus (HAV) is another enteric virus that can be transmitted by contaminated food. More than 95% of HAV infections are transmitted by the fecal-oral route (Ciocca, 2000) and person-to-person contact is considered to be the primary mode of transmission (Koopmans et al., 2002). Therefore, outbreaks of HAV are common in high population density settings such as schools, prisons, and military bases (Koopmans et al., 2002). Approximately 50% of the reported HAV cases do not have a recognized source of infection, and only 5% have a clear food- or waterborne route. In the United States alone, HAV causes an estimated 83,000 illnesses per year (Mead et al., 1999). Improved sanitary conditions have caused a decline in the prevalence of this disease in the developed world, but increases in international travel and trade have brought a naïve population in contact with endemic disease, resulting in a reemergence of HAV infection in developed countries (Romalde et al., 2001). For example, a recent, large outbreak of HAV was attributable to imported green onions. This incident alone resulted in more than 600 infections and three deaths (Centers for Disease Control, 2003c).

Rotavirus is the most common cause of severe infant diarrhea (Lopman et al., 2002), estimated to be responsible for 130 million illnesses and more than 600,000 deaths per year throughout the world (Mead et al., 1999). Rotaviruses are most often transmitted by the waterborne route but have occasionally been implicated in outbreaks attributed to contaminated food (Bresee et al., 2002). It is estimated that only 1% of rotavirus cases are food-borne (Sair et al., 2002). Spread of the disease is mainly through the fecal-oral route, although aerosol transmission has also been suggested (Caul, 1994).

In reviewing food-borne transmission of human enteric viruses, three major at-risk food categories stand out: (i) shellfish contaminated by fecally impacted growing waters, (ii) human sewage pollution of drinking and irrigation water, and (iii) ready-to-eat (RTE) and prepared foods contaminated by infected food-handlers as a result of poor personal hygiene (Jaykus, 2000b). Among the more common foods that have been implicated as vehicles for enteric virus transmission are shellfish, fruits, vegetables, salads, sandwiches, and bakery items. Indeed, any food that has been handled manually and is not further heated prior to consumption has the potential to be virally contaminated (Richards, 2001). In further discussion of prevention and control strategies for enteric viruses, we categorize the discussion into these three distinct commodity groups because prevention and control differ for each.

2.0. SHELLFISH

Bivalve mollusks (including mussels, clams, cockles, and oysters) have been implicated as vectors in the transmission of bacterial and viral enteric diseases for many decades. The most commonly implicated bivalves are oysters, followed by clams (Potasman et al., 2002). It is estimated that human enteric viruses are actually the most common human disease agents transmitted by molluskan shellfish (Lees, 2000; Formiga-Cruz et al., 2002). Although more than 100 different types of enteric viruses can be excreted in human feces, only a few (HAV, noroviruses, astroviruses) have been epidemiologically linked to shellfish-associated viral disease (Richards, 2001). Of these, infectious hepatitis caused by HAV is probably the most serious. Immunosuppressed patients are at high risk for serious disease and as a precaution should be advised to avoid this particular cuisine (Potasman et al., 2002). In fact, the largest viral food-borne disease outbreak occurred in Shanghai,

China, in 1988, where 300,000 people were infected with HAV after consumption of clams harvested from fecally impacted growing waters (Halliday et al., 1991). In the 1990s, noroviruses were implicated as the primary etiological agents among reported cases of infectious diseases associated with shellfish consumption (Centers for Disease Control, 2001). Shellfish contamination with noroviruses is of great significance to food safety not only for its direct implications but also for the secondary cases that readily occur after a primary food-borne outbreak (Beuret et al., 2003). Significant and recent outbreaks of viral food-borne diseases associated with shellfish consumption are summarized in Table 13.1.

Unlike illnesses caused by naturally occurring *Vibrio* spp., enteric viral illnesses originate from human fecal wastes only. Even though most sewage treatment processes cannot completely eliminate viruses, adequate sewage treatment remains the first line of defense in protecting shellfish and their harvesting waters (Sorber, 1983). Factors contributing to human sewage pollution of marine waters include the illegal dumping of human waste directly into shellfish harvesting areas, failing septic systems along shorelines, sewage treatment plants overloaded with storm water, and discharges of treated and untreated municipal wastewater and sludge (Jaykus et al., 1994; Shieh et al., 2000). For instance, a recent outbreak in France and Italy was associated with norovirus-contaminated oysters harvested from a pond that was polluted from an overflowing water purification plant (Doyle et al., 2004).

Shellfish are at particular risk of transmitting human enteric viruses because (i) they are frequently eaten whole and raw or only lightly cooked; (ii) there are no good methods to ascertain whether the shellfish or their harvest waters contain infectious viruses; and (iii) shellfish can bioconcentrate viruses within their edible tissues to levels much higher than in the water itself (Richards, 2001). This bioaccumulation is probably assisted by the ionic binding of virus particles to mucopolysaccharide moieties of the shellfish mucus (DiGirolamo et al., 1977). The degree of virus uptake and survival in shellfish depends on several factors, including exposure time, virus concentration in overlay water, presence of particulate matter (and/or excess turbidity), temperature, interspecies differences, individual shellfish differences, type of virus, food availability, and pH and salinity of water (Sobsey and Jaykus, 1991; Jaykus, 1994). Because uptake of virus by shellfish is dependent on active feeding, any factor affecting the physiological activity of the animal can influence virus accumulation. Likewise, the elimination kinetics of enteric viruses by bivalve mollusks can vary with the type of shellfish, type of microorganism, and environmental conditions and season (Burkhardt and Calci, 2000; Mounts et al., 2001).

Preharvest contamination is the most common source of contamination and occurs in shellfish exposed to human fecal pollution of waters from which they are harvested. Once contaminated, the viruses can persist in both the overlay waters and the shellfish. Enriquez et al. (1992) reported rapid uptake (in less than 24 hr) of HAV in the mussel *Mytilus chilensis*, with virus persistence for about 7 days. The ability of HAV to persist in Eastern oysters

Table 13.1 Epidemiological studies summarizing recent enteric virus outbreaks associated with molluscan shellfish

Agent	*Food*	*Samples tested*	*Detection Methods*	*Conclusions*	*Reference*
Norovirus	Oysters	Clinical samples	Electron Microscopy (EM), RT (reverse transcription)-PCR, Sequencing	Overboard disposal of sewage from a harvesting boat into the oyster beds	Aristeguieta et al., 1995
Norovirus	Oysters	Outbreak (clinical and food)	IgG antibodies, EM, RT-PCR, Sequencing	Contamination in the oyster beds from overboard disposal of sewage from handlers	Berg et al., 2000
Norovirus	Oysters	Clinical samples and oyster samples	Bacteriological sampling, Screen for F+ and somatic phages	Virus particles can persist in the oysters for many weeks after depuration	Chalmers and Mcmillan, 1995
Norovirus, HAV	Oysters	Clinical samples	RT-PCR, Sequencing	Imported contaminated clams, not properly steamed	Furuta et al., 2003; Kohn et al., 1995
HAV	Clams	Clinical samples	Enzyme Immunoassay (ELISA), EM	Untreated sewage from fishing vessels and the surrounding residential area	Halliday et al., 1991
HAV	Clams	Outbreak (clinical)	ELISA	Secondary infection through person to person contagion	Leoni et al., 1998

(*Crassostrea virginica*) was recently investigated by Kingsley and Richards (2003), who detected virus in the animal as little as 16 hr after exposure, with the oysters remaining infectious for up to 3 weeks thereafter. Of the factors that influence the survival and persistence of enteric virus in seawater, temperature is particularly important (Muniain-Mujika et al., 2003). For instance, it has been demonstrated that the time necessary to inactivate 90% of HAV in seawater was 671 days at 4°C but only 25 days at 25°C (Gantzer et al., 1998). Similarly, poliovirus type 1 at a concentration of 10^5 pfu/ml lost its infectivity in ocean water within 27 days during the summer but within 65 days during the winter (Lo et al., 1976). Recent research has indicated that about 10% (17/191) of Japanese oysters sampled and intended for raw consumption harbored noroviruses, and in some of these samples, the virus level was quite high (Nishida et al., 2003).

2.1. Preharvest Control Strategies

2.1.1. Harvest Water Quality: Preventing Illegal Sewage Discharge
The most effective and reliable approach to control viral contamination of shellfish is to harvest them from areas with good water quality, most notably from estuarine environments that are free of human sewage contamination. Unfortunately, recent virus outbreaks have been caused by the dumping of untreated human sewage from boats, resulting in contamination of shellfish beds. Enforcement of proper waste disposal in certain discharge locations may be difficult as harvesting areas are frequently remote from the shore and there may be many harvesting and recreational vehicles present at any one time. A preemptive measure would be to provide dockside receptacles for waste disposal. Alternatively, mandating the use of a waste container that cannot be easily dumped or flushed into the harvesting waters could protect water quality (Berg et al., 2000). Furthermore, imposing severe monetary penalties for violation of overboard waste dumping and improper onboard waste receptacles may also be an effective deterrent. Lastly, educating harvesters about the public health risks associated with overboard sewage disposal, as well as the need for compliance with regulations for waste disposal, are integral steps to preharvest control.

2.1.2. Harvest Water Quality: Microbiological Indicators
Historically, the fecal coliform index has been employed as an indicator of the sanitary quality of shellfish and their harvesting waters. This index has been considered appropriate as fecal coliforms are normal inhabitants of the gastrointestinal tract of warm-blooded animals and are excreted in the feces in large numbers (Jaykus, 1994). Standards exist in the United States for shellfish and shellfish-harvesting waters based on the enumeration of total and fecal coliforms. The National Shellfish Sanitation Program (NSSP), a federal/state cooperative program recognized by the U.S. Food and Drug Administration (FDA) and the Interstate Shellfish Sanitation Conference (ISSC), is responsible for the promotion of sanitary quality of shellfish sold for human consumption. NSSP sponsors numerous programs to promote

shellfish safety, including evaluation of state program elements, dealer certification, and state growing area classification, including sanitary surveys of shellfish-growing waters (National Shellfish Sanitation Program, 2000). Accordingly, a fecal coliform standard of less than 14 MPN (most probable number) per 100 ml of water, with not more than 10% of samples exceeding 43–49 MPN/100 ml (depending on the microbiological method), is required for classification of harvesting waters as approved. Outside of these standards, waters can be classified as conditionally approved, restricted, or prohibited for shellfish harvesting (Somerset, 1991). Current regulations also stipulate fecal coliform counts of less than 45 MPN per 100 g of shellfish meat for fresh product to be commercially marketed.

The fecal coliform standards for shellfish, although effective in controlling bacterial disease transmission, do not necessarily prevent virally contaminated shellfish from reaching the marketplace (Kingsley et al., 2002a). In fact, these standards may offer no indication of viral contamination, as viruses can persist for a month or longer within shellfish or estuarine sediments, long after coliform counts have reached acceptable levels (Kingsley and Richards, 2001). As a result, there is no clear and consistent relationship between the occurrence of bacteriological indicators and viruses in water or shellfish (Sobsey and Jaykus, 1991; Jaykus et al., 1994). Romalde et al. (2002) examined European shellfish contaminated with HAV and human enteroviruses and confirmed that there was no correlation between the presence of the traditional bacterial indicators and enteric viruses. Power and Collins (1989) arrived at similar conclusions when investigating depuration of poliovirus, *Escherichin. coli*, and a coliphage by the common mussel, *Mytilus edulis*. Apparently, *E. coli* was not a good indicator of the efficiency of virus reduction during depuration, while the coliphage appeared to be a more reliable one. These data and others confirm that compliance with the fecal coliform end-product standards does not provide a guarantee of the absence of enteric viruses, even in depurated shellfish. In fact, in areas where the current European microbiological standards characterizing waters as suitable for harvesting shellfish were in place, the infectivity of HAV in mussels after depuration was recorded to be reduced by only 98.7% (Abad et al., 1997b). Another study done with mussels harvested in Italy revealed that although the product's microbiological quality was in accordance with the European bacteriological standards, HAV was detected in 13 of 36 specimens (Croci et al., 2000).

As bacterial indicators generally fail to signal the potential for viral contamination, bacteriophages, enteroviruses, and adenoviruses have all been proposed as alternative indicators (Lees, 2000; Muniain-Mujika et al., 2002). The feasibility of using human adenoviruses as indicators of human enteric viruses in environmental and shellfish samples was suggested by Pina et al. (1998) who reported that these viruses were easily detected and seemed to be more abundant and stable in environmental samples. In fact, the presence of human adenoviruses appears to correlate with the presence of other human viruses, and they have been proposed to monitor viral contamination in shellfish harvested from Greece, Spain, Sweden, and the United Kingdom (Formiga-Cruz et al., 2002). On the contrary, no solid correlation could be demonstrated between the occurrence of human enteroviruses and noroviruses in estuarine waters in Switzerland (Beuret et al., 2003).

Bacteriophages, especially F-specific coliphages, somatic coliphages, and phages of *Bacteroides fragilis*, have long been proposed as possible indicators of viral contamination of the environment. Of these, the distribution and survival of F-specific coliphages (also known as FRNA phage or male-specific coliphage) are considered to be more similar to human enteric viruses in the environment. The survivability of FRNA phage in seawater was found to be similar to that of a variety of enteric viruses, including HAV, poliovirus, and rotavirus (Chung and Sobsey, 1993). A study examining the distribution of FRNA phages in shellfish harvesting areas revealed that these phages were more resistant to environmental stresses and that their numbers in shellfish were consistently higher than those of *E. coli*, indicating that their presence in shellfish may be a more representative assessment of virus contamination (Dore et al., 2003). In another study, Sinton et al. (2002) provided evidence that FRNA phages were more resistant to sunlight inactivation than were fecal coliforms, *E. coli*, enterococci, and somatic coliphages, making them potentially more useful for monitoring the virological quality of freshwaters. In a recent study examining the comparative survival of feline calicivirus (FCV, a norovirus surrogate), *E. coli* and FRNA phage in dechlorinated water at 4°C, 25°C, and 37°C, a correlation was found between the survival of the phage and FCV, indicating that the phage may be a potential environmental surrogate for noroviruses (Allwood et al., 2003).

FRNA phages have also been suggested as a complementary parameter for evaluating the efficiency of depuration in heavily contaminated mussels. In a study by Muniain-Mujika et al. (2002), a 5-day depuration period was considered an adequate decontamination treatment because neither human enteric viruses nor FRNA phages could be detected in the shellfish after that time. The applicability of F-specific coliphage as an indicator of depuration efficacy was confirmed by Dore and Lees (1995), who monitored the elimination patterns of *E. coli* and FRNA phage in contaminated mussels and oysters. After a 48-hr depuration period, it was found that *E. coli* was completely eliminated but FRNA phages were still largely retained in the digestive gland. The F-specific RNA bacteriophage have also been suggested as an alternative indicator of potential norovirus contamination in depurated, market-ready oysters (Dore et al., 2000). Formiga-Cruz et al. (2003) studied the correlation between FRNA phages, somatic coliphages, and bacteriophages of *B. fragilis* as indicators of viral contamination in shellfish and found that FRNA phages were better predictors of norovirus contamination than for adenovirus, enterovirus, or HAV contamination.

Evidence that phages infecting *B. fragilis* RYC2056 may be a suitable group of indicators for viral pollution in shellfish has recently been provided; however, further research is needed to develop the appropriate methodology (Muniain-Mujika et al., 2003). Earlier research from the same group suggested that phages infecting *B. fragilis* RYC2056 are preferable to phages infecting *B. fragilis* HSP40 as potential indicators of viral contamination in shellfish, as the former are more abundant is shellfish as well as in sewage (Muniain-Mujika et al., 2000).

Callahan et al. (1995) compared the inactivation rates of HAV, poliovirus 1 (PV-1), F-specific coliphage, and somatic *Salmonella* bacteriophages (SS phages) in seawater and found that SS phages survived significantly longer at 20°C (Callahan et al., 1995). Although the final concentration of all four viruses was reduced to similar levels at the end of the study, the rates of inactivation were different. For instance, it took 10 and 4 weeks to achieve a $4\log_{10}$ reduction of SS phages and HAV, respectively, whereas PV1 and FRNA phages were reduced to the same extent within 1 week. The study concluded that because SS phage persisted longer in seawater environments, it may be a more reliable indicator of enteric virus contamination.

Somatic coliphages have also been suggested as reliable indicators of the efficiency of shellfish depuration as they can be easily and rapidly assayed (Power and Collins, 1989; Muniain-Mujika et al., 2002). However, Legnani et al. (1998) observed no significant differences between the occurrence of somatic coliphages and fecal indicator bacteria in seawater. In another study, phages of *B. fragilis* and *Salmonella* were found not to be adequate to indicate fecal contamination (Chung et al., 1998), whereas FRNA phage and *Clostridium perfringens* spores were more reliable indicators of human enteric viruses in oysters and the best predictors of fecal contamination in water and oysters. In conclusion, the efficacy of the various bacteriophages as indicators of enteric viruses is still under investigation.

2.1.3. Depuration and Relaying

Two preharvest control strategies rely on extending the natural filter-feeding process of the animal in clean seawater in order to purge out microbial contaminants (Lees, 2000). Both methods are based on the ability of the shellfish to eliminate contaminating microorganisms from their digestive tracts through normal feeding, digestion, and excretion activities. Once in clean waters, shellfish can purge at least some of their contaminants, provided that the water and feeding conditions (primarily temperature, salinity, and dissolved oxygen) are favorable (Richards, 2001).

Depuration, or controlled purification, is the process of reducing the levels of bacteria and viruses in contaminated, live shellfish by placing them in a controlled water environment. In order to produce a safe and wholesome depurated product, specific growing area classification, process approval, and process controls are required (Somerset, 1991). Depuration usually takes place in tanks provided with a supply of clean, often disinfected, seawater under specific operating conditions (Sobsey and Jaykus, 1991). The more common methods of water disinfection for use in shellfish depuration are ultraviolet light, chlorine, or ozonation (De Leon and Gerba, 1990). Ozone, although not a new technology, has generated renewed interest for use in molluskan, shellfish depuration systems (Garrett et al., 1997).

Because the environmental conditions during depuration are tightly controlled, the process usually takes only 2–3 days (Jaykus et al., 1994; Richards, 2001). The process does not suffer from possible recontamination due to changing environmental conditions but is not recommended for shellfish harvested from heavily contaminated (prohibited) waters (Roderick and Schneider, 1994).

Relaying, or natural purification, refers to the transfer of shellfish from contaminated growing areas to approved areas (Sobsey and Jaykus, 1991). Although relaying has lower initial costs compared with depuration, its drawbacks include a lower yield of marketable product and a less steady supply due to environmental variations (Blogoslawski, 1991). In addition, relaying requires extended periods (often 10 days to 2 weeks or more) (Richards, 2001) and is sometimes limited by the availability of suitable pristine coastal areas (Lees, 2000).

Depuration is used extensively around the world and has been successful in reducing bacterial illnesses associated with shellfish consumption in the United Kingdom (Sobsey and Jaykus, 1991). Generally, nonindigenous (enteric) bacteria are rapidly (usually within 48 hr) reduced to nondetectable levels by depuration, while viruses are purged more slowly and may persist for several days (Richards, 2001). It is important to note that the efficiency of both depuration and relaying processes is influenced by numerous factors such as type of shellfish, individual variation in feeding rates, initial level of virus contamination, temperature, turbidity, availability of particulate matter, salinity, pH and oxygen availability (Cook and Ellender, 1986; Sobsey and Jaykus, 1991).

Nevertheless, depurated shellfish can also cause enteric viral illness either as a result of inadequate process control or due to insufficiency of the process itself. Consequently, only lightly contaminated shellfish should be subjected to depuration, whereas the more heavily contaminated ones should be relayed for extended periods of time (Richards, 2001). Most food-borne outbreaks associated with depurated shellfish have been caused by HAV, as this virus does not appear to be as readily eliminated during depuration as do other virus types (Richards, 2001). In an Italian study of 290 mussels collected from various sources, HAV RNA was detected in 20% (20/100) of the nondepurated mussels, 11.1% (10/90) of the depurated samples, and 23% (23/100) of the mussels sampled from different seafood markets (Chironna et al., 2002). These authors concluded that this high prevalence of contamination could be due to the practice of keeping shellfish alive for prolonged periods in possibly contaminated waters or else due to inefficient and/or inadequate depuration. Another Italian study examined the effectiveness of depuration on the decontamination of mussels contaminated with HAV (De Medici et al., 2001). By using a closed-circuit depuration system with constant levels of salinity and temperature and with both ozone and UV disinfection of water, the initial levels of HAV were reduced significantly after 48 hr, but extending depuration to 120 hr allowed for detectable virus reconcentration.

2.2. Postharvest Control Strategies

2.2.1. Temperature

Temperature control, particularly the use of refrigeration temperatures, is a long-accepted method to control the growth of bacterial spoilage microorganisms and pathogens in food. Unfortunately, because enteric viruses do not grow in foods and are in fact quite persistent in low-temperature environments, this approach is not very effective for their control. Indeed, the common practice of icing and freezing are likely to facilitate the survival of viruses, as these are widely used as laboratory preservation techniques (Lees, 2000). With respect to temperature control, a study done with Olympic oysters contaminated with poliovirus showed that even after 15 days of storage at 5°C, the virus titer was reduced by only 60%, while after extended storage for 30 days at 5°C, 13% of the input virus remained infectious (DiGirolamo et al., 1970). The same group of researchers studied the survival of poliovirus in Pacific oysters kept at −17.5°C and concluded that after 4 weeks of storage, the virus titer was reduced by little more than $0.5 \log_{10}$, while further storage for 12 weeks resulted in a $1 \log_{10}$ reduction (DiGirolamo et al., 1970). Tierney et al. (1982) reported survival of infectious poliovirus after a 28-day period of storage at 5°C (Tierney et al., 1982). Greening et al. (2001) found poliovirus type 2 to persist in green-lipped mussels, *Perna canaliculus*, even after 2 days of refrigeration (81% of the original titer was recovered), as well as after 28 days of storage at −20°C (44% of the initial titer was detected) (Greening et al., 2001). When T4 coliphage was used as a surrogate for enterovirus contamination in West Coast crabs (*Cancer magister* and *C. antennarius*), less than $1 \log_{10}$ reduction in virus titer was obtained when the crabs were kept for 120 hr at 8°C, and about 25% of the input coliphage could still be recovered when the crabs were stored at −20°C for 30 days (DiGirolamo and Daley, 1973).

Early thermal inactivation study focused on HAV in steamed clams showed that it took 4 to 6 min of steaming for complete virus elimination, at which point the internal temperature of the clam tissue was 100°C (Koff and Sear, 1967). The authors suggested that it would not be safe to consume steamed clams when they first open, as opening of the shells usually happens within the first minute of steaming. In mussels contaminated with HAV and subjected to steaming for 5 min, 0.14% of the initial HAV could still be recovered (Abad et al., 1997). Pacific oysters artificially contaminated with poliovirus were heat processed by stewing, frying, baking, and steaming, with virus inactivation barely exceeding 90% after conventional cooking times for each treatment (DiGirolamo et al., 1970). Studies in artificially contaminated cockles revealed that HAV was only partially reduced when the shellfish were immersed for 1 min in water at 85°C, 90°C, or 95°C or when steamed for the same period. For complete inactivation of HAV, the internal temperature of the shellfish had to reach 85–90°C and be maintained there for 1 min (Millard et al., 1987). Similarly, another study reported that heat treatment at 100°C for 2 min (internal temperature of 90°C) was needed to assure complete inactivation of HAV in artificially contaminated mussels (Croci et al., 1999).

The failure of several cooking methods (grilling, stewing, and frying) to prevent a large oyster-associated gastroenteritis outbreak was reported in Florida in January 1995 (McDonnell et al., 1997). However, experimental data from the inactivation of feline calicivirus (FCV), a norovirus surrogate, showed that the previous heat-processing recommendations (internal temperature of 90°C for 1.5 min) for the elimination of HAV in cockles could also successfully eliminate FCV in shellfish (Slomka and Appleton, 1998). The survival of FCV was studied at 56°C, 70°C, and 100°C and it was found that 7.5 $\log_{10}$ titer of FCV could be completely inactivated (i.e., nondetectable by infectivity assay) by heating for 5 min at 70°C or for 1 min at 100°C (Doultree et al., 1999). Apart from this study, there has been little work done to determine the thermal inactivation kinetics of the noroviruses. Although not studied *per se*, a 1988 outbreak of HAV linked to a fast-food restaurant in Tennessee suggested that microwave heating may partially inactivate the virus, as customers who reheated their sandwiches before consumption did not develop clinical illness (Mishu et al., 1990).

Not unlike bacteria, the type of virus and the matrix in which the viruses are suspended play a significant role in virus sensitivity to heat (Millard et al., 1987; Croci et al., 1999). For instance, a longer heat treatment was necessary to inactivate HAV suspended in shellfish homogenate compared with the same amount of virus suspended in buffer (Croci et al., 1999). Because shellfish tissue is dense and the virus is likely to be concentrated in the digestive diverticula, heat penetration of this product is of particular concern. It must also be recognized that variability of shellfish species, size, time after harvest, contamination level, and cooking conditions account for the difficulty in establishing a minimum cooking time for complete virus inactivation (De Leon and Gerba, 1990). Moreover, standardization of conditions of commercial heat treatment of shellfish may be difficult because excessive heating may result in undesirable organoleptic changes such as toughening of meat texture.

2.2.2. Ionizing Radiation

Although not a promising technology, ionizing radiation has also been tested as a means to inactivate viruses in shellfish. Oysters (*Crassostrea virginica*), hard-shelled clams (*Mercenaria mercenaria*), and soft-shelled clams (*Mya arenaria*) were contaminated with HAV and rotavirus strain SA-11 and treated with irradiation doses ranging from 1 to 7 kGy. Although a 3-kGy dose resulted in a 95% reduction in virus load, the organoleptic properties of the shellfish also deteriorated at this dose. Using a lower dose of 2 kGy resulted in less than 95% virus inactivation and a product with adequate sensory quality (Mallet et al., 1991). The authors suggested that combining depuration with radiation doses of 2 kGy may effectively decontaminate shellfish, although it is recognized that such an approach would likely be very costly.

2.2.3. High Hydrostatic Pressure Processing

Recently, alternative technologies, particularly high hydrostatic pressure (HPP), have been proposed for the inactivation of HAV and noroviruses in shellfish (Kingsley et al., 2002b). The shellfish industry is very interested in HPP as it has previously been shown to eliminate *Vibrio* species in oysters while maintaining the organoleptic properties of the raw shellfish meat (Berlin et al., 1999). In model studies, HAV and FCV (a norovirus surrogate) suspended in tissue culture medium were eliminated after exposure to 450 MPa for 5 min and 275 MPa for 5 min, respectively. However, model studies with poliovirus showed a general failure of high pressure to inactivate the virus, even at pressures as high as 600 MPa for 15 min (Wilkinson et al., 2001). Extending the pressure treatment for 1 hr had no significant effect on reducing virus infectivity. The authors proposed that perhaps the pressure resistance of poliovirus is correlated with capsid composition. It is clear that further research is needed before HPP can be considered as a viable option for the inactivation of viruses in raw molluskan shellfish.

3.0. PRODUCE

A number of viral food-borne disease outbreaks associated with the consumption of contaminated raw produce have occurred over the past several years, presumably due to the combined effect of increased consumption and better epidemiological surveillance (Centers for Disease Control, 2003c). For example, between 1988 and 1997, the U.S. CDC reported 130 food-borne outbreaks linked to the consumption of fresh produce. A report published by the Public Health Laboratory Service (PHLS) in England and Wales indicated that 83 outbreaks between 1992 and 1999 were associated with the consumption of contaminated salad vegetables or fruit (O'Brien et al., 2000). Of the viral agents, HAV and noroviruses are most commonly documented as contaminating fruits and vegetables. Although fresh produce can certainly serve as a vehicle for the transmission of viral food-borne disease, the exact attribution of this commodity group to the overall burden of this set of diseases is unknown. Furthermore, we know relatively little about the persistence of enteric viruses when they contaminate produce, and the data regarding the efficacy of various virus inactivation methods intended for use on fresh produce are limited and variable (Seymour and Appleton, 2001; Lukasik et al., 2003). Taken together, this means that there is much to learn about viruses in this food commodity.

In most instances, contamination of fruits and vegetables with enteric viruses is believed to occur before the product reaches food service establishments (Koopmans et al., 2002). Sources of such contamination include contaminated soil, contaminated irrigation or washing water, or infected food-handlers who harvest and handle the produce (Lopman et al., 2002). Treatment of sewage sludge by drying, pasteurization, anaerobic digestion, and composting can reduce but not eliminate viruses, especially the more thermoresistant ones (Metcalf et al., 1995). Therefore, using recycled sewage effluent and sludge to irrigate or fertilize crops intended for human consumption carries with it the risk of virus contamination (Ward et al., 1982). Likewise, soils can also become contaminated by land disposal of sewage sludge and through the use of fecally impacted irrigation water. Viruses can survive in contaminated soil for long periods of time depending on factors such as growing season, soil composition, temperature, rainfall, resident microflora, and virus type (Yates et al., 1985; Seymour and Appleton, 2001).

It is believed that most virus contamination occurs mainly on the surface of fresh produce, although a few studies have reported on the potential for uptake and translocation of virus within damaged plant tissue (Seymour and Appleton, 2001). Use of wastewater for spray irrigation may be particularly risky as this may facilitate virus attachment to produce surfaces (Richards, 2001). Green onions and other select produce items may be particularly prone to virus contamination because their surfaces are complex, allowing fecal matter and other organic materials to adhere tenaciously (Centers for Disease Control, 2003c). The survival of viruses on vegetables has been shown to be dependent on pH, moisture content, and temperature (Harris et al., 2002). Because noroviruses and HAV have been associated with a number of produce-associated outbreaks, it seems possible, though not yet supported by studies, that these viruses may be resistant to some of the virucidal substances naturally found in produce such as organic acids, and phenolic and sulfur compounds (Seymour and Appleton, 2001).

As items implicated in outbreaks are usually picked and processed long before consumption, it is often difficult to identify the point at which contamination occurred (Hutin et al., 1999). Moreover, locating the growing site of a particular produce item may be complicated. For example, in the case of an HAV outbreak linked to imported lettuce, the names of the farms supplying the lettuce were not included on the product labels, making it impossible to trace the geographic origin of the produce item (Rosenblum et al., 1990). When combined with issues such as poor patient recall and the extended incubation period for HAV, trace-back of contaminated product is very difficult (Calder et al., 2003; Fiore, 2004).

Items such as green onions, which have recently been implicated in HAV outbreaks in the United States, may become contaminated at any time during the production and processing continuum by contaminated soils, water, or human handling. However, as this particular produce requires extensive human handling during harvesting, it has been suggested that human handling is perhaps the most likely source of virus contamination (Dentinger et al., 2001). Several recent enteric virus outbreaks associated with the consumption of contaminated fresh produce are presented in Table 13.2.

3.1. Preharvest Control Strategies

The Guide to Minimize Microbial Food Safety Hazards for Fresh Fruits and Vegetables (U.S. Food and Drug Administration, 1998) provides a framework

Table 13.2 Epidemiological studies summarizing recent enteric virus outbreaks associated with produce items

Agent	*Food*	*Sample tested*	*Detection Methods*	*Conclusions*	*Reference*
HAV	Blueberries	Clinical and food	RT-PCR, Gel electrophoresis, Dot-blot hybridization, Sequencing	Contamination by infected food handlers or by faecally polluted groundwater	Calder et al., 2003
HAV	Green onions	Clinical	Serological testing	Contamination occurred in the distribution system or during growing, harvest, packing or cooling	CDC, 2003
HAV	Green onions	Clinical	RT-PCR, Sequencing	Contamination probably during harvesting	Dentinger et al., 2001
Norovirus	Lettuce	Clinical	Radioimmunoassay, EM	Unsanitary handling of lettuce or cross-contamination by raw seafood	Griffin et al., 1982
HAV	Strawberries	Clinical	RT-PCR, Sequencing	Contamination occurred during harvest due to unsanitary conditions	Hutin et al., 1999
HAV	Lettuce	Clinical	Serologic testing	Contamination occurred before local distribution	Lisa et al., 1990
HAV	Strawberries	Clinical and food	Immunoselection, RT-PCR, Hybridization	Contamination occurred probably by an infected picker	Niu et al., 1992
Norovirus	Raspberries	Clinical and food	RT-PCR, Hybridization, Sequencing	Contaminated water (irrigation or before packaging)	Ponka et al., 1999
HAV	Raspberries	Clinical	Immunoassay	Contamination at the picking stage	Reid, 1987

for the identification and implementation of practices likely to decrease the risk of pathogen contamination in fresh produce from production, packaging, and transport based on Good Agricultural Practices (GAPs) and Good Manufacturing Practices (GMPs). This document provides guidance for the proper management, handling, and application of animal manure. Emphasis should also be placed on assuring that waters used in production (for irrigation and pesticide application) are of high quality and do not present a human health hazard. There is, however, little conclusive data regarding the efficacy of sewage treatment on virus inactivation or on the degree of virus persistence in treated sewage or sludge. The virucidal efficacy of sewage disinfection can often be limited due to virus aggregation and association with particulate matter, and the occurrence of enteric viruses in sewage is usually sporadic. Estimates of the efficacy of secondary sewage treatment and disinfection on enteric viruses removal range from 1 to 2 orders of magnitude, while chlorination can remove an additional 1–3 orders of magnitude of enteric viruses depending on the dose, temperature, and contact time (Schaub and Oshiro, 2000). Unfortunately, sewage spills, storm-related contamination of surface waters, illicit discharge of waste, and residential septic system failures are widely recognized as the leading sources of surface water and groundwater contamination, which may impact fruit and vegetable production (Suslow et al., 2003). Scientific reports that document the feasibility and performance of various methods of on-farm water treatment (such as chlorination, peroxyacetic acid, UV, and ozone treatment) are also scarce. The risk associated with the reuse of wastewater for irrigation also requires further investigation (Gantzer et al., 2001).

Because many produce items are subjected to extensive human handling during harvesting, preharvest food safety strategies should also focus on food handlers. On-site toilet and hand-washing facilities should be readily accessible, well supplied, and kept clean. All employees (full-time, part-time, and seasonal personnel), including supervisors, should have a good working knowledge of basic sanitation and hygiene principles, including proper hand-washing techniques (FDA, 1998). The employees should be instructed to report any active cases of illness to their supervisors before beginning work (Koopmans et al., 2002). Furthermore, the presence of children at picking sites should be discouraged, and appropriate childcare programs should be available so that workers are not forced to bring their young children into the fields (Fiore, 2004).

3.2. Postharvest Control Strategies

Many produce items are washed before entering the distribution phase of the farm-to-fork chain. Washing fresh produce can reduce overall microbial food safety hazards so long as the water used in such rinses is of adequate quality. Of course, water (and ice) used for rinsing and packaging must originate from a pristine source or be decontaminated with chlorine or by some other disinfection method. However, washing and disinfection may not be sufficient to eliminate viral contamination from vegetables. Surface morphology and physiologic characteristics of the produce item(s) certainly complicate disinfection efficacy; leafy vegetables can be more difficult to decontaminate because of their rough or wrinkled surfaces, and small fruits like raspberries and blackberries have more porous and complex surfaces that can entrap virus particles (Richards, 2001). When fresh produce is cut or damaged, viruses can be sequestered in abrasions. The most commonly used sanitizers for washing fruits and vegetables are chlorine, chlorine dioxide, and organic acids (Seymour and Appleton, 2001). Ozone has been put forth as a potential disinfectant but may be less promising because oxidation of food components may result in discoloration as well as deterioration of flavor. Toxicity and reactivity are other disadvantages associated with ozone.

3.2.1. Water, Produce Washes, and Household Chemicals

Produce items are frequently washed at numerous steps along the postharvest continuum. A general rule of thumb is that water washing alone can remove about $1\log_{10}$ of microbiological contaminants from the surface of produce items, keeping in mind that this estimate varies with factors such as produce type, virus type, degree of viral contamination, and water temperature. For instance, Lukasik et al. (2003) evaluated a variety of simple methods to remove viruses from a model produce commodity. Using strawberries artificially contaminated with poliovirus and bacteriophages MS2, ΦX174, and PRD1, these investigators found that water immersion and hand rubbing of the berries in water held at 22°C or 43°C resulted in removal or inactivation of 41–79% and 60–90% of the input virus, respectively. Overall, hand rubbing in water held at a higher temperature (43°C) facilitated virus removal. These same investigators also evaluated a commercial produce wash called Fit® (Proctor and Gamble) for its ability to reduce virus load in artificially contaminated strawberries and found that virus inactivation ranged between 80% and 90%. Finally, these authors reported that automatic dishwashing detergent (ADWD, 0.05%) and 10% vinegar were more effective than Healthy Harvest®, 0.05% liquid dishwashing detergent, or 2% NaCl for removal of viruses from strawberries immersed in lukewarm water. In this case, supplementing washwater with vinegar and ADWD produced virus reductions ranging from 95% to >99% (Lukasik et al., 2003).

3.2.2. Chlorine, Chlorine Dioxide, and Other Comparative Studies

There is a long history of using chlorine to control microbial contamination in water. Unfortunately, there is a paucity of published data regarding the efficacy of chlorine in the inactivation of viruses from the surface of produce items. In one recent study, Lukasik et al. (2003) reported on the efficacy of chlorine in inactivating viruses from inoculated strawberries. The levels of bacteriophages MS2, ΦX174, and PRD1 and poliovirus type 1 were reduced by 70.4% to 99.5% when strawberries were immersed for 2 min in 43°C water containing 0.3 ppm to 300 ppm of chlorine. Free chlorine at 200 ppm was considered optimal because it gave the same degree of inactivation as did 300 ppm free chlorine.

The HAV was reduced more readily (around 96%) when strawberries were washed in tap water supplemented with 2 ppm chlorine dioxide (ClO_2)

for 30 min, as compared with the same ClO_2 concentration used in wash water for a 30-s exposure period (around 67% inactivation). These results suggest that, under realistic processing conditions, chlorine dioxide washes are not very effective in reducing viral risks associated with this product (Mariam and Cliver, 2000).

In a study comparing the antiviral activity of commonly used antimicrobials (5.25% sodium hypochlorite, quaternary ammonium compounds, and 15% peroxyacetic acid–11% hydrogen peroxide) for rinsing produce, it was found that FCV could survive in strawberries and lettuce after they had been washed for 10 min at room temperature (Gulati et al., 2001). Only peroxyacetic acid–hydrogen peroxide formulations were proved to effectively reduce FCV titers by $3\log_{10}$, although this occurred only at concentrations four times higher than those permitted by the FDA. In general, organic acids are unlikely to cause significant inactivation of enteric viruses, because viruses have mechanisms that facilitate their survival in the low acidity of the stomach. For example, HAV has been demonstrated to be extremely acid stable, remaining infectious after 5 hr of exposure to pH 1 at room temperature (Scholz et al., 1989).

In a recent study, trisodium phosphate (TSP; 1%), a common household cleaner, was found to be as effective as 0.5% hydrogen peroxide for the reduction of representative bacteriophages and poliovirus type 1 from artificially inoculated strawberries immersed in water held at 43°C; the inactivation rates ranged from 97% to >99% (Lukasik et al., 2003). A 0.5% solution of hydrogen peroxide, however, caused bleaching of the product and although this was ameliorated with a 10-fold decrease in concentration, the efficacy of the 0.05% peroxide solution was essentially the same as that of tap water washes alone. Cetylpyridinium chloride (CPC) was less effective at virus inactivation, ranging from 85% to 97% when applied to the surface of the strawberries (Lukasik et al., 2003).

Taken together, the data on chemical disinfection for the inactivation of viruses from food surfaces is not all that promising and is quite variable. For instance, disinfectants incorporated in the wash water may not be effective in removing or inactivating viruses that have penetrated through the skin of the produce or those that might have entered tissues through cuts and abrasions. An additional hurdle is that the surfaces and textures of fruits and vegetables may be rough, wrinkled (leafy vegetables), or porous (strawberries, raspberries, and blackberries), allowing the entrapment of viruses, thereby sequestering them from disinfectants. In general, washing produce items individually rather than in bulk is recommended, as bulk washing may result in the infiltration of viruses into produce items that were not initially contaminated, (Richards, 2001).

3.2.3. Alternative Decontamination Methods

Ultraviolet radiation has been suggested as an alternative to chlorine for water disinfection. FCV and poliovirus type 1 have proved to be highly susceptible to inactivation by UV radiation, with $3\log_{10}$ reductions achieved by doses of 23 and 40 mJ/cm^2, respectively (Gerba et al., 2002; Thurston-Enriquez et al., 2003). Bench-scale ozone disinfection of water using a dose of 0.37 mg of ozone/liter at pH 7 and 5°C resulted in at least a 3 $\log_{10}$ reduction of norovirus and poliovirus type 1 within a contact time of 10 s. This promising technology may some day prove to be an alternative to chlorine for the disinfection of produce wash water. Although many novel disinfectant washes with effective antiviral properties are available at the consumer and processor levels, their use directly on produce surfaces is frequently prohibitive due to unacceptable organoleptic changes in the produce. Likewise, gamma radiation between 2.7 and 3.0 kGy has been shown to reduce HAV on lettuce and strawberry surfaces by 1 $\log_{10}$. However, this dose currently exceeds the U.S. standards for the use of irradiation to control sprouting and pest infestation in produce items (Bidawid et al., 2000).

3.2.4. Temperature

Viruses can survive in contaminated fruits and vegetables under household refrigeration conditions (Kurdziel et al., 2001). The survival of poliovirus on the surface of foods has been demonstrated in many previous studies (Ansari et al., 1988; Mbithi et al., 1992; Abad et al., 1994). For example, poliovirus titers dropped by 1 $\log_{10}$ after 12 days of refrigerated storage in lettuce and white cabbage, whereas on green onions and fresh raspberries its concentration remained unchanged under the same storage conditions. Similarly, a study on the persistence of HAV on fresh produce (lettuce, fennel, and carrot) demonstrated produce-specific variation in the ability of the virus to withstand refrigeration. More specifically, HAV survived on lettuce until the ninth day of refrigeration but decreased to undetectable levels after 7 and 4 days of refrigeration on fennel and carrots, respectively (Croci et al., 2002).

In a multistate HAV outbreak associated with the consumption of frozen strawberries, it was apparent that the virus had survived storage at frozen temperatures for up to 2 years (Niu et al., 1992). In a laboratory-based study of frozen strawberries contaminated with poliovirus, the investigators found only 1 $\log_{10}$ reduction in virus titer within the first 9 days of freezing (Kurdziel et al., 2001). Rotavirus SA-11 survived for almost a month on lettuce, radishes, and carrots when stored at refrigeration temperatures, while survival was significantly less when the produce was stored at room temperature (Badawy et al., 1985). In another study, 93% of the initial rotavirus contamination could still be recovered from lettuce after the inoculum was allowed to dry for 4 hr at room temperature (O'Mahony et al., 2000).

4.0. READY-TO-EAT FOODS

Ready-to-eat (RTE) foods are defined as those products that are edible without washing, cooking, or additional preparation by the consumer or by the food service establishment (Public Health Service, 1999). In general, this means that such foods are not subjected to a terminal heating step prior to consumption. In RTE foods, transmission of enteric viruses through food handlers is widespread. Indeed, recent epidemiological surveillance data (1988 to 1992) indicate that poor personal hygiene of infected food-handlers was the most commonly cited factor contributing to food-borne outbreaks of HAV (96%) and norovirus-associated gastroenteritis (78%) (Bean et al., 1997). The cost of viral disease outbreaks due to infected food-workers can in some instances be very high because they frequently involve secondary transmission, and, especially for HAV, the cost of widespread prophylaxis is very high (Daniels et al., 2000).

4.1. The Epidemiological Significance of RTE Foods

4.1.1. The Role of Fecal-Oral Transmission

Food handlers may transmit viruses to foods from a contaminated surface, from another food, or from contaminated hands. The ultimate source of viral contamination is usually human fecal matter, although vomitus may also contain infectious virus. Because contamination of RTE foods occurs post-processing, no level of upstream food processing will control the problem (Richards, 2001). One of the major hazards for cooked RTE foods arises through handling with bare hands (Bryan, 1995). Technically, any RTE food handled by a symptomatic or asymptomatic virus carrier can become contaminated. However, certain food products have received considerably more attention (e.g., salads, raw fruits and vegetables, bakery products) over the years, probably due to their association with high-profile outbreaks. Recent enteric virus outbreaks associated with RTE foods are shown in Table 13.3.

Nonenveloped viruses, such as rotavirus, noroviruses and HAV, survive better on skin than do enveloped ones such as herpes and influenza viruses (Springthorpe and Sattar, 1998). There is strong evidence suggesting that contaminated hands frequently play a role in virus spread, acting as either virus donors or recipients. Hands can become contaminated by direct contact with any virus-containing fluid from self or others; they may also become individually contaminated by contacting virus-contaminated surfaces or objects (Sattar et al., 2002). The extent of such contamination will vary depending on a variety of factors, including the virus load, the degree of discharge from the host, the hand-washing habits of the infected person, and the efficiency with which virus is transferred and persists. Considerable amounts of HAV (16–30% of the initially recoverable virus) remained infectious on finger pads after 4 hr, even though 68% of viral infectivity was lost within the first 1 hr (Mbithi et al., 1992). In another study, rotavirus remained infectious on human hands for 60 min after its inoculation, and it could be transferred from the contaminated hands to animate and nonporous inanimate surfaces (Ansari et al., 1988). In fact, twice as much virus was transferred by the hand-to-hand route when compared with that transferred between hands and nonporous inanimate surfaces. Transfer studies with PDR-1 phage, used as a surrogate for human viruses, revealed that infectious particles could be transmitted under ordinary circumstances from the surface of fomites, such as

Table 13.3 Epidemiological studies summarizing recent enteric virus outbreaks associated with RTE foods

Agent	*Food*	*Samples tested*	*Detection Methods*	*Conclusions*	*Reference*
Norovirus	Salads	Outbreak samples (clinical, serum and food)	RT-PCR	Ill food handler contaminated the salads	Ansderson et al., 2001
Norovirus	Box lunch	Outbreak samples (clinical)	RT-PCR EM	Person-to-person transmission	Becker et al., 2000
Norovirus	Bakery products	Outbreak samples (clinical and serum)	Immune-EM, Radioimmunoassay	Ill handler during the outbreak	Kuritsky et al., 1984
Norovirus	Turkey salad sandwiches	Outbreak samples (clinical and food)	EM, Immune-EM, bacteriological examination	Mechanical transmission of the virus or pre-symptomatic food handler	Lo et al., 1994
Norovirus	Deli sandwiches	Outbreak samples (clinical and food)	RT-PCR (single and nested) Sequencing Southern hybridization	Food handler slicing the ham had an ill infant	Parashar et al., 1998
Norovirus	Potato salad	Outbreak samples (clinical)	EM, RT-PCR	Kitchen assistant vomited in the sink, where the salad was later prepared	Patterson et al., 1997
HAV	Bread	Outbreak samples (clinical)	Serum and saliva tests for IgM and IgG	Handler with soiled hands contaminated samples when wrapping them	Warburton et al., 1991

telephone receivers, to the hands of a person using the receiver. If there were subsequent fingertip-to-mouth contact, infection might result (Rusin et al., 2002).

Hepatitis A, with an incubation period of 15–50 days, appears more readily transmitted during the latter half of the incubation period, meaning that food workers in retail settings with acute HAV infection can readily contaminate RTE products if they do not practice adequate personal hygiene (Daniels et al., 2000). Although adults are considered infectious only in the first few days of a norovirus infection, it has been shown recently that they can shed viruses in their feces for up to 2 weeks from disease onset (White et al., 1986; Yotsuyanagi et al., 1996; Parashar et al., 1998). Infected infants may be able to shed virus for more than 2 weeks (Daniels et al., 2000). Further complicating the issue is evidence of presymptomatic fecal excretion from food handlers while incubating the disease (Lo et al., 1994). Indeed, an outbreak has been documented in which a food handler was able to transmit calicivirus a few hours before becoming symptomatic (Gaulin et al., 1999).

4.1.2. The Role of Vomitus and Secondary Spread

Although the fecal-oral transmission route is the most important in promoting the spread of noroviruses, the role of vomitus cannot be overlooked. More than 30 million virus particles can be liberated from one vomiting episode, and when compared with an infectious dose of 10–100 particles, this is a significant virus load (Caul, 1994). The importance of this lies in its contribution to secondary spread, because aerosolization of vomitus can result in infection of exposed subjects who inhale and subsequently swallow the aerosolized virus (Marks et al., 2000). Air currents generated by air conditioning or open windows can disperse aerosols widely (Caul, 1994), while ceiling fans can also contribute to the virus spread (Marks et al., 2000). Evidence for respiratory spread has not been documented, and seems unlikely, as replication of noroviruses in respiratory mucosal cells does not occur (Lopman et al., 2002).

4.1.3. The Role of Fomites

Enteric viruses may persist for extended periods of time in foods and on materials and objects that are commonly found in institutions and domestic environments, including paper, cotton cloth, aluminum, china, latex, and polystyrene (Abad et al., 1994). Thus, viruses can be transmitted by mechanical transfer from the contaminated object (Lo et al., 1994). A recent outbreak of norovirus in an elder-care residential hostel in Australia is a case in point. In this case, the vomitus of an infected individual served as the source of virus that contaminated furniture and carpets. The virus remained infectious even after these items were professionally cleaned, serving as an intermediate source of virus transmission (Liu et al., 2003). In general, it is difficult to investigate whether and to what extent fomites assist in the spread of enteric viruses (Abad et al., 1994). Moreover, apart from more predictable surfaces like carpets and toilets seats, other surfaces such as lockers, curtains, and commodes have also been implicated in virus transmission in hospital outbreaks (Green et al., 1998).

4.2. Prevention Strategies

4.2.1. Decontamination of Hands

For hand decontamination to be successful in controlling viral food-borne disease outbreaks, three elements must be in place: (i) an effective disinfecting agent, (ii) adequate use instructions, and (iii) regular compliance (Sattar et al., 2002). Because hands are believed to play an important role in virus spread, the efficiency of several hand-washing agents has been investigated. In the first of such studies, Mbithi et al. (1993) showed that most surface disinfectants, even the alcohol-containing ones, were not able to eliminate poliovirus type 1 and HAV, based on an efficacy criteria of a $3\log_{10}$ reduction in virus titer (99.9% inactivation). A medicated liquid soap was the most effective against both viruses, although there were virus-to-virus differences in inactivation. Disturbing was the fact that as much as 20% of the initial virus inoculum could still be detected on hands after washing and drying, and nearly 2% of the input virus could be readily transferred to other surfaces. These investigators pointed to a need for establishment of new standards in the selection of effective formulations for hand-washing agents with respect to antiviral activities (Mbithi et al., 1993). Moreover, it is generally recognized that more work is needed to establish a standard hand-washing regimen upon which inactivation claims against viruses can be based for labeling purposes (Sattar et al., 2002). In a study of the efficacy of common hand disinfectants against a porcine enterovirus, all of the agents were proved ineffective with the exception of a 1% chlorine bleach solution (Cliver and Kostenbader, 1984). Ethanol-based hand rubs contributed to the reduction of FCV spread, but because they were not as effective as water and soap, the investigators suggested that they are perhaps more useful in the decontamination of hands between hand-washing events (Bidawid et al., 2004). A recent study suggested that contact for 30 s with 1-propanol or ethanol solutions on hands could reduce FCV titer as much as $4\log_{10}$ (Gehrke et al., 2004). The same study indicated that an increase in disinfection effectiveness did not correlate with an increase in alcohol concentration, as alcohol-based solutions of 70% were more effective against FCV than were 90% solutions. This is in agreement with earlier findings reporting that a 70% alcohol formulation was effective for decontaminating rotavirus from hands (Ansari et al., 1989). The same group investigated the efficacy of aqueous solutions of chlorhexidine gluconate (Savlon and Cida-stat) in reducing rotavirus from hands, and the degree of virus removal was the same as that observed with tap water alone (Ansari et al., 1989).

Nearly 46%, 18%, and 13% of infectious FCV was transmitted from experimentally contaminated hands to ham, lettuce, and metal surfaces, respectively (Bidawid et al., 2004). On the contrary, less efficient virus transfer occurred from contaminated ham (6%), lettuce (14%), and metal surfaces (7%) to hands. In both cases, FCV transfer could be significantly interrupted if soiled hands were washed with water or both water and soap before contacting the recipient surface (Bidawid et al., 2004). Hand-washing with water was similarly effective in interrupting HAV transfer from contaminated hands to lettuce (Bidawid et al., 2000). The reduction of rotavirus from finger pads was approximately $3 \log_{10}$ better when using a gel containing 60% ethanol as a hand disinfectant than using a simple hard-water rinse (Sattar et al., 2000). Water rinsing after the application of the hand antiseptic agent followed by immediate drying can provide further reduction of viruses on washed hands (Ansari et al., 1989). Tap water is used in most studies for rinsing hands, although its composition may vary geographically and temporally. Moreover, organic and inorganic compounds in water may facilitate the removal of viruses from hands (Ansari et al., 1989). Residual moisture on hands after hand-washing has been found to play an important role in the transfer of viruses (Springthorpe and Sattar, 1998), meaning that air drying may be a critical step for virus removal, especially if the hand-washing agents are not very effective (Ansari et al., 1991). Hot-air drying of hands contaminated with porcine enterovirus type 3 was found to reduce the virus titer by 92% (Cliver and Kostenbader, 1984). A study by Ansari et al. (1991) found that, regardless of the hand-washing agent used, electric air drying produced the highest reduction in rotavirus when compared with either paper towels or cloth towels. For instance, after washing with soap and water and with no drying step, there was a 77% reduction of rotavirus on hands. On the other hand, a reduction of 92% was observed after warm-air drying compared with 87% and 80% virus removal using paper towel or cloth drying, respectively.

4.2.2. Decontamination of Surfaces

Contaminated surfaces can readily transmit viruses to hands or food upon contact. The survival of human enteric viruses on environmental surfaces depends on several factors, including temperature, relative humidity, type of surface, and virus type. The results of several surface inactivation studies are summarized in Table 13.4. HAV, for example, can remain infectious on nonporous inanimate surfaces for several days, and this survival is influenced by relative humidity and air temperature (Mbithi et al., 1991). In a large study, Mbithi et al. (1991) reported that the half-life of HAV ranged from more than 7 days at relatively low humidity and 5°C to about 2hr at high humidity and 35°C. On the contrary, under the same experimental conditions, poliovirus type 1 survival was proportional to the level of relative humidity and temperature, with longer survival occurring at high relative humidity.

The persistence of human enteric viruses on environmental surfaces in the presence of fecal material has been investigated with results varying by both virus and surface. Abad et al. (1994) found that the survival of human rotavirus and HAV on surfaces was not affected by the presence of fecal material, while enteric adenovirus and poliovirus persistence on nonporous surfaces (aluminum, china, glazed tile, latex, and polystyrene) was enhanced

Table 13.4 Surface inactivation of viruses by common disinfectants

Surface	*Virus*	*Agent/Concentration*	*Contact time*	*Efficiency*	*Comments*	*References*
Polystyrene	HAV, HRV	Sodium chlorite, Ethanol, Chlorhexidine digluconate, Sodium hypochlorite Phenol and Sodium phenate, Diethylentriamine,	1 min, 28°C	~3 log reduction <3 log reduction	Presence of organic matter not increase virus persistence after disinfection	Abad et al., 1997
Food-contact surface	FCV	5.25% sodium hypochlorite, 1.75% iodine and 6.5% phosphoric acid, 3 quaternary ammonium compounds, 15% peroxyacetic acid and 11% hydrogen peroxide, 2 phenolic compounds	1 and 10 min, 22°C	>3 log reduction <3 log reduction >5 log reduction	Sodium hypochlorite: efficient only at 5,000 ppm available chlorine Phenolic compounds: effective at 4× recommended concentration	Gulati et al., 2001
Polyvinyl chlorine, High-density polyethylene, Aluminum, Stainless steel, Copper	HAV	10% quaternary ammonium and 5% glutaraldehyde, 12% sodium hypoclorite, 2.9% dodecylbenzene sulfonic acid and 16% phosphoric acid 10% quaternary ammonium, 2% iodine, 2% stabilized chlorine dioxide	1 or 5 min, 4 and 22°C	2–7 log reduction 3–5 log reduction <3 log reduction	Efficiency increases at 22°C, 5 min contact time and concentration of 3,000 ppm of the active ingredient	Jean et al., 2003

Table 13.4 *Continued*

Surface	*Virus*	*Agent/Concentration*	*Contact time*	*Efficiency*	*Comments*	*References*
Stainless steel disks	HAV	2% gluteraldehyde, quaternary ammonium compound (with 23% HCl and sodium hypochlorite with >5,000 ppm chlorine),	1 min	>3 log reduction	Only 2% gluteraldehyde, a Quat compound with 23% HCl and sodium hypochlorite with >5,000 ppm available chlorine are effective	Mbithi et al., 1990
		phenolics, iodine-based products, alcohols, solutions of acetic, peracetic, citric and phosphoric acids		<3 log reduction		
Stainless steel disks	rotavirus	0.1% o-phenylphenl and 79% ethanol,		>4 log reduction	Only the disinfectant spray reduced virus infectivity more significantly than tap water alone (<1 log reduction)	Sattar et al., 1994
		7.05% quat diluted 1:128 in tap water		<1 log reduction		
		6% sodium hypochlorite diluted to give 800 ppm free chlorine, 14.7% phenol diluted 1:128 in tap water	10 min	<2 log reduction		

by the presence of feces. The persistence of the latter two viruses was unaffected, however, by the presence of fecal matter when deposited on porous (paper and cotton cloth) surfaces. Other important findings from this study focused on the ability of each virus to survive when dried on a surface. For example, the reduction of HAV and human rotavirus infectivity when placed on several fomites and dried for a period of 3–5 hr was not as significant as it was for adenovirus and poliovirus, implying that the former two viruses may be more likely to be transmitted after substantial environmental persistence. In another study, rotavirus SA-11 suspended in a stool preparation could be detected on contaminated environmental surfaces after 60 min of drying, while the same virus survived for only 30 min when suspended in water (Keswick et al., 1983). Similarly, cell culture–adapted human rotavirus was found to be significantly protected from drying when it was in a 10% fecal suspension rather than in distilled water (Ward et al., 1991). The presence of fecal material not only increases virus survival but has also been shown to protect poliovirus from the action of disinfectants (Hejkal et al., 1979). In fact, a fourfold increase in residual chlorine was required to achieve the same degree of inactivation for poliovirus type 1 suspended in feces as compared with a suspension of free virus at pH 8 and 22°C.

Cleaning and disinfection of surfaces are of major importance in the prevention of enteric viral disease. In general, there is a paucity of information on the efficacy of most commercial disinfectants against enteric viruses. Comparing a range of disinfectants used on experimentally contaminated polystyrene surfaces under conditions suggested by the manufacturer, only sodium chlorite proved to be effective against HAV and human rotavirus (Abad et al., 1994, 1997a). Work done with several commercial disinfectants used in the food industry showed that only products containing gluteraldehyde and sodium hypochlorite were effective in HAV inactivation, and their efficacy improved when the compounds were used at high concentrations and for a relatively long time (Jean et al., 2003). A majority of chemical disinfectants used in both institutional and domestic environments do not effectively inactivate HAV (Mbithi et al., 1990). Of the 20 formulations tested, only 2% gluteraldehyde, a quaternary ammonium compound containing 23% HCl, and sodium hypochlorite with free chlorine in excess of 5,000 ppm had demonstrable virucidal efficacy. These results supported the use of sodium hypochlorite for surface disinfection and were validated in a more recent study using FCV as a norovirus surrogate (Gulati et al., 2001). From a variety of disinfectants used at manufacturer-recommended concentrations, only sodium hypochlorite at 5,000 ppm available chlorine (200 ppm of chlorine is the FDA allowable level) was effective in reducing more than $3\log_{10}$ of FCV. In another study, the efficacy of commercially available disinfectants was tested and the most suitable for environmental surfaces was reported to be freshly prepared hypochlorite solution at high concentrations (1,000 ppm) (Doultree et al., 1999). In a transfer study, Sattar et al. (1994) reported a $4\log_{10}$ inactivation of rotavirus transfer from stainless steel disks to fingerpads of volunteers using a disinfectant spray (0.1% *o*-phenyl phenol and 79% ethanol), while domestic bleach (6% sodium hypochlorite diluted to give 800 ppm free chlorine) and a phenol-based product (14.7% phenol diluted 1:128 in tap water) provided reductions of almost $3\log_{10}$ of virus infectivity. No detectable virus was transferred to fingerpads from disks treated with these three agents, but when the disks were cleaned with tap water or a quaternary ammonium–based product, the transfer rates were 5.6% and 7.6%, respectively. In another study, Sattar (2004) tested the anti-FCV activity of various microbicides and discovered that the most effective one at the shortest contact time (1 min) was domestic bleach (5% sodium hypochlorite, 1,000 ppm available chlorine), which reduced the titer of FCV by nearly $4.5\log_{10}$.

4.2.3. Education, Training, and Supervision

Fingers, and particularly fingernails, are thought to be the most important part of the hand in terms of the transfer and spread of pathogenic microflora (Lin et al., 2003). Fingernails are of particular concern because fecal material may be readily deposited in this location and is subsequently difficult to remove. This may be particularly important for those having long or "artificial" fingernails. A recent study demonstrated that the most effective way to remove virus from artificially inoculated fingernails was to scrub them with a nailbrush using soap (regular or antibacterial) and water. Alternatively, employees should be encouraged to maintain short nails, as these are less likely to harbor fecal material than long ones (Lin et al., 2003).

Indeed, food employees should not touch exposed RTE food with their bare hands and should use suitable items such as deli tissue, spatulas, tongs, and single-use gloves (smooth, durable, and nonabsorbent) or dispensing equipment (FDA, 2001). Gloves are the only FDA-approved barrier method allowable to date. Issues impacting the efficacy of gloves in preventing viral disease include the glove material, glove permeability, duration of wearing, and hand-washing techniques prior to and after wearing. Glove leaks are more frequent with vinyl than with latex gloves (Guzewich and Ross, 1999), and frequent replacement of disposable gloves is encouraged, particularly when gloves get damaged or soiled, or when interruptions occur in the operation. Where food contact by handlers is unavoidable, careful practices such as frequent hand-washing and prevention of cross-contamination during handling and preparation are suggested. Apart from the "no bare-hand contact with RTE" policy in the Food Code, there is no direct information on the effectiveness of hand hygiene and gloving regimens in the food industry (Paulson, 2003).

It is of major importance that food-handlers, including seasonal workers, should have appropriate health and hygiene education. Such training should also cover the potential risk of enteric virus transmission due to sick children in the household (Koopmans and Duizer, 2004). If a food preparation staff member reports a diarrhea or vomiting episode while at work, he or she should not be allowed to enter the kitchen again. All of the food handled by that worker, as well as any other food that may have been exposed to aerosolized vomitus, should be destroyed (Lo et al., 1994). Potentially contaminated surfaces in the kitchen should be thoroughly decontaminated with a freshly prepared hypochlorite solution that releases 1,000 ppm of available chlorine. Frequently handled objects such as taps and door knob should also be decontaminated, and bleach-sensitive items should be cleaned with detergent and hot water (Chadwick et al., 2000). Managers of food manufacturing, catering, and food service industries should restrict ill food-handlers from working directly with food or food equipment and provide a sick leave policy that allows workers to stay home while ill (Centers for Disease Control, 2003b). Moreover, as soon as food handlers return to work, they should be instructed that they still may be shedding virus for a period of days to weeks after symptoms have abated and that they should continue practicing stringent personal hygiene (Koopmans and Duizer, 2004).

Increased awareness of the risk of gastrointestinal disease due to virus transmission is encouraged among food safety professionals and the public. However, rapid control of viral gastroenteritis outbreaks can be difficult. For instance, while transmission via contaminated food or water may sometimes be prevented or contained, the potential for person-to-person spread cannot be eliminated. This is especially challenging for settings where close contact inevitably occurs (i.e., university dormitories) and especially where there are a number of susceptible people in confined quarters (hospitals, nursing homes, daycare centers) (Kilgore et al., 1996). The failure to recognize and report HAV among children in daycare facilities is an important and contributing factor in disease propagation (Gingrich et al., 1983). Viruses can be readily transmitted in daycare settings; however, it is the "silent" transmission through poor personal hygiene of food handlers that is becoming increasingly recognized (Sattar et al., 2002).

For gastroenteritis outbreaks in hospitals and nursing homes, efforts to control virus circulation in the environment by immediate isolation of the case(s) should be undertaken. The timing of the last cleaning process should ideally be at least 72 hr after resolution of the last case (Chadwick et al., 2000). Hypochlorite is generally not recommended for the disinfection of carpets; however, steam has been suggested as a viable alternative (Cheesbrough et al., 1997; Chadwick et al., 2000). Indeed, during a viral gastroenteritis outbreak in a hotel, the carpets appeared visually clean after cleaning with detergent and vacuuming, but nonetheless they remained contaminated with infectious norovirus (Cheesbrough et al., 2000).

4.2.4. Vaccination

Immunity to HAV confers complete protection against reinfection. To date, there are three FDA-licensed HAV vaccines on the market. These are generally administered as a single primary immunization, followed by a booster dose 6–12 months later (Lemon, 1997). The vaccine efficacy is 94–100% and protection is likely to last for more than 20 years after vaccination (Fiore, 2004). Hepatitis A vaccination has been limited to high-risk groups and is currently approved for use in the United States in children over 2 years of age. Routine vaccination of all food handlers as a pre-exposure prophylaxis is not recommended because their occupation does not put them at unusually high risk of infection. Furthermore, vaccinating all restaurant employees is unlikely to happen, as the cost to restaurant owners often exceeds the perceived benefits, even during a hepatitis A epidemic (Meltzer et al., 2001). However, some believe that HAV vaccination should be routinely available to people with increased occupational risk such as food handlers, health care workers in infectious disease and pediatrics sections, medical staff in laboratories handling stool samples, staff in daycare centers, and sewage treatment plant workers (Hofmann et al., 1992).

Immunoglobulin (IgG) is not the recommended choice for pre-exposure prophylaxis as it provides only short term (1–2 months) protection from HAV infection (Fiore, 2004). However, postexposure prophylaxis with IgG has been shown to be effective in eliminating or reducing the severity of hepatitis A infection, provided it is administered within 2 weeks of exposure and not within the late incubation period of the disease (Pavia et al., 1990). In the case of food-related exposures, it is recommended that IgG postexposure prophylaxis should be given to all food handlers, even if only one handler in that facility has been diagnosed with HAV infection. Moreover, IgG is recommended for the patrons of that establishment provided all of the following considerations exist: (i) the infected handler was responsible for handling RTE foods and was not wearing gloves, (ii) the infected handler had poor hygiene practices or had diarrheal symptoms, and (iii) the patrons can be identified and treated within 2 weeks of exposure (Committee on Infectious Diseases—American Academy of Pediatrics, 1991). The efficacy of administering HAV vaccine for postexposure prophylaxis remains to be established. It has been demonstrated that anti-HAV seroconversion occurs 14 days after vaccination and the average incubation period for HAV infection is 28–30 days, suggesting that vaccination may be protective if given within a few days of exposure (Koff, 2003).

A safe and effective vaccine against noroviruses would reduce the burden of the disease, which may be of particular importance for controlling gastroenteritis in children of developing countries, partly because repeated diarrheal episodes can cause damage to the intestinal mucosa leading to the development of malnutrition (Kapikian et al., 1996). Recent human challenge studies with the noroviruses have demonstrated both short-term and long-term immunity (Johnson et al., 1990; Parrino et al., 1977). The distinct epidemiological patterns of the noroviruses have introduced some technical difficulties for vaccine development. For instance, there may be multiple antigenically and genetically distinct strains of noroviruses circulating at any one time, and there is evidence indicating that infection with one strain does not provide cross-protection against other strains. This virus genus remains noncultivable, so it is difficult to routinely determine the presence of neutralizing antibodies in an infected individual (Estes et al., 2000; Hale et al., 2000). Recently, specific histological blood group antigens have been identified as putative ligands for the attachment of different norovirus strains to mucosal cell surfaces (Harrington et al.,

2004; Hutson et al., 2004). These observations require further investigation to determine whether there is any correlation between individual differences in susceptibility and specific virus genotype and/or the genetic background of the host (Harrington et al., 2004). The production of recombinant norovirus capsid protein and its formulation as an oral vaccine is an alternative approach to vaccination; however, the efficacy of this type of approach has yet to be established (Estes et al., 2000).

5.0. CONCLUSIONS

For all food commodities, preventing direct contact with human fecal material (and in some instances, direct exposure to vomitus) is obviously the first consideration in controlling virus contamination. Prevention of sewage disposal in harvesting waters is critical in controlling viral contamination of shellfish. However, the lack of correlation between the fecal coliform index and the presence of enteric viruses may at times complicate the ability of regulators to recognize contamination when it occurs. For produce items, adherence to GAPs, including attention to personal hygiene of field workers, is essential to controlling contamination at the preharvest phase. Likewise, proper personal hygiene, including the use of barrier protection and appropriate hand and surface decontamination, provides a first line of defense for preventing viral contamination of RTE foods. A critical consideration is providing food handlers with appropriate and ongoing education in hygienic practices. Designing effective educational programs for food handlers is notoriously difficult, as this itinerant population may have limited English language fluency, generally has lower educational attainment, and turns over quite rapidly.

It is clear from this discussion that low temperatures (refrigeration and freezing) are not reliable means by which to reduce enteric viruses in contaminated foods. High temperature (heating) may be effective in some instances, but recommended time-temperature combinations are virus-specific and will vary with the food commodity. A common theme for the food items discussed here is that most do not undergo a terminal heating step before consumption, so the relevance of heating may be limited simply by virtue of the specific commodity.

The efficacy of other postharvest controls is also somewhat limited, and, in general, these may reduce but will not eliminate viral contamination in foods. This is the case with depuration, relaying, and ionizing radiation as applied to raw, molluskan shellfish. High hydrostatic pressure appears to be a more promising technology, but much more data are needed to definitively establish its efficacy in inactivating viruses in this commodity. Likewise, washing produce with fresh, clean water, both with or without the addition of chemical disinfectants, may also reduce virus load in contaminated items, but it is important to note that the efficacy of such decontamination varies with both virus and produce type.

Widespread vaccination may eventually become an effective control measure for HAV, but it is not likely that norovirus vaccination will be a reality in the near future. It must also be noted that, once an outbreak occurs, strict infection control measures must be instituted to prevent further virus dissemination. Indeed, there are many opportunities for future research in prevention and control. Specifically, research is needed to identify effective intervention and control strategies, to develop improved monitoring and detection methods, to expand immunization options, and to develop successful food-handler educational programs. Working together, scientists can make further inroads in the prevention and control of viral food-borne diseases into the future.

6.0. REFERENCES

Abad, F. X., Pinto, R. M., and Bosch, A., 1994, Survival of enteric viruses on environmental fomites. *Appl. Environ. Microbiol.* 60:3704–3710.

Abad, F. X., Pinto, R. M., and Bosch, A., 1997a, Disinfection of human enteric viruses on fomites. *FEMS Microbiol. Lett.* 156:107–111.

Abad, F. X., Pinto, R. M., Gajardo, R., and Bosch, A., 1997b, Viruses in mussels: public health implications and depuration. *J. Food Prot.* 60:677–681.

Allwood, P. B., Malik, Y. S., Hedberg, C. W., and Goyal, S. M., 2003, Survival of F-specific RNA coliphage, feline calicivirus, and *Escherichia coli* in water: a comparative study. *Appl. Environ. Microbiol.* 69:5707–5710.

Anderson, A. D., Garrett, V. D., Sobel, J., Monroe, S. S., Fankhauser, R. L., Schwab, K. J., Bresee, J. S., Mead, P. S., Higgins, C., Campana, J., and Glass, R., 2001, Multistate outbreak of Norwalk-like virus gastroenteritis associated with a common caterer. *Am. J. Epidemiol.* 154:1013–1019.

Ansari, S. A., Sattar, S. A., Springthorpe, V. S., Wells, G. A., and Tostowaryk, W., 1988, Rotavirus survival on human hands and transfer of infectious virus to animate and nonporous inanimate surfaces. *Appl. Environ. Microbiol.* 26:1513–1518.

Ansari, S. A., Sattar, S. A., Springthorpe, V. S., Wells, G. A., and Tostowaryk, W., 1989, In vivo protocol of hand-washing agents against viruses and bacteria: experiments with rotavirus and *Escherichia coli*. *Appl. Environ. Microbiol.* 55:3113–3118.

Ansari, S. A., Sattar, S. A., Springthorpe, V. S., Tostowaryk, W., and Wells, G. A., 1991, Comparison of cloth, paper and warm air drying in eliminating viruses and bacteria from washed hands. *Am. J. Infect. Control* 19:243–249.

Aristeguieta, C., Koenders, I., Windham, D., Ward, K., Gregos, E., Gorospe, L., Walker, J., and Hammond, R., 1995, Multistate outbreak of viral gastroenetritis associated with consumption of oysters—Apalachicola Bay, Florida, December 1994–January 1995. *Morbid. Mortal. Weekly Rep.* 44:37–39.

Badawy, A. S., Gerba, C. P., and Kelley, L. M., 1985, Survival of rotavirus SA-11 on vegetables. *Food Microbiol.* 2:199–205.

Bean, N. H., Goulding, J. S., Daniels, M. T., and Angulo, F. J., 1997, Surveillance for foodborne disease outbreaks—United States, 1988–1992. *J. Food Prot.* 60: 1265–1286.

Becker, K. M., Moe, C. L., Southwick, K. L., and MacCormack, J. N., 2000, Transmission of norwalk virus during a football game. *N. Engl. J. Med.* 343:1223–1227.

Berg, D. E., Kohn, M. A., Farley, T. A., and McFarland, L. M., 2000, Multi-state outbreaks of acute gastroenteritis traced to fecal-contaminated oysters harvested from Louisiana. *J. Infect. Dis.* 181(Suppl. 2):381–386.

Berlin, D. L., Herson, D. S., Hicks, D. T., and Hoover, D. G., 1999, Response of pathogenic *Vibrio* species to high hydrostatic pressure. *Appl. Environ. Microbiol.* 65: 2776–2780.

Beuret, C., Baumgartner, A., and Schluep, J., 2003, Virus-contaminated oysters: a three-month monitoring of oysters imported to Switzerland. *Appl. Environ. Microbiol.* 69:2292–2297.

Bidawid, S., Farber, J. M., and Sattar, S. A., 2000a, Contamination of foods by food handlers: experiments on hepatitis A virus transfer to food and its interruption. *Appl. Environ. Microbiol.* 66:2759–2763.

Bidawid, S., Farber, J. M., and Sattar, S. A., 2000b, Inactivation of hepatitis A virus (HAV) in fruits and vegetables by gamma irradiation. *Int. J. Food Microbiol.* 57: 91–97.

Bidawid, S., Malik, N., Adegbunrin, O., Sattar, S. A., and Farber, J. M., 2004, Norovirus cross-contamination during food handling and interruption of virus transfer by hand antisepsis: experiments with feline calicivirus as a surrogate. *J. Food Prot.* 67:103–109.

Blogoslawski, W. J., 1991, Enhancing shellfish depuration, in: *Molluscan Shellfish Depuration* (W. S. Otwell, G. E. Rodrick, and R. E. Martin, eds.), CRC Press, Boca Raton, FL.

Bresee, J. S., Widdowson, M.-A., Monroe, S. S., and Glass, R. I., 2002, Foodborne viral gastroenteritis: challenges and opportunities. *Food Safety* 35:748–753.

Bryan, F. L., 1995, Hazard Analysis: the link between epidemiology and microbiology. *J. Food Prot.* 59:102–107.

Burkhardt, W., and Calci, K. R., 2000, Selective accumulation may account for shellfish-associated viral illness. *Appl. Environ. Microbiol.* 66:1375–1378.

Calder, L., Simmons, G., Thornley, C., Taylor, P., Pritchard, K., Greening, G., and Bishop, J., 2003, An outbreak of hepatitis A associated with consumption of raw blueberries. *Epidemiol. Infect.* 131:745–751.

Callahan, K. M., Taylor, D. J., and Sobsey, M. D., 1995, Comparative survival of hepatitis A virus, poliovirus and indicator viruses in geographically diverse seawaters. *Water Sci. Technol.* 31:189–193.

Caul, E. O., 1994, Small round structured viruses: airborne transmission and hospital control. *Lancet* 343:1240–1241.

Centers for Disease Control, 2001, Norwalk-like viruses: public health consequences and outbreak management. *Morbid. Mortal. Weekly Rep.* 50:1–18.

Centers for Disease Control, 2002, Outbreaks of gastroenteritis associated with Noroviruses on cruise ships—United States 2002. *Morbid. Mortal. Weekly Rep.* 51: 1112–1115.

Centers for Disease Control, 2003a, Foodborne transmission of Hepatitis A—Massachusetts, 2001. *Morbid. Mortal. Weekly Rep.* 52:565–567.

Centers for Disease Control, 2003b, Norovirus activity—United States, 2002. *Morbid. Mortal. Weekly Rep.* 52:41–44.

Centers for Disease Control, 2003c, Hepatitis A outbreak associated with green onions at a restaurant—Monaca, Pennsylvania, 2003. *Morbid. Mortal. Weekly Rep.* 52:1–5.

Chadwick, P. R., Beards, G., Brown, D., Caul, E. O., Cheesbrough, J., Clarke, I., Curry, A., O'Brien, S., Quigley, K., Sellwood, J., and Westmoreland, D., 2000, Management of hospital outbreaks of gastroenteritis due to small round structured viruses. *J. Hosp. Infect.* 45:1–10.

Chalmers, J. W. T., and McMillan, J. H., 1995, An outbreak of viral gastroenteritis associated with adequately prepared oysters. *Epidemiol. Infect.* 115:163–167.

Chang, J. C. H., Ossoff, S. F., Lobe, D. C., Dorfman, M. H., Dumais, C. M., Qualls, R. G., and Johnson, J. D., 1985, UV inactivation of pathogenic and indicator microorganisms. *Appl. Environ. Microbiol.* 49:1361–1365.

Cheesbrough, J. S., Barkess-Jones, L., and Brown, D. W., 1997, Possible prolonged environmental survival of small, round structured viruses. *J. Hosp. Infect.* 35:325–326.

Cheesbrough, J. S., Green, J., Gallimore, C. I., Wright, P. A., and Brown, D. W., 2000, Widespread environmental contamination with Norwalk-like viruses (NLV) detected in a prolonged hotel outbreak of gastroenteritis. *Epidemiol. Infect.* 125: 93–98.

Chironna, M., Germinario, G., De Medici, D., Fiore, A., Di Pasquale, S., Quatro M., and Barbuti, S., 2002, Detection of hepatitis A virus in mussels from different sources marketed in Pulgia region (south Italy). *Int. J. Food Microbiol.* 75: 11–18.

Chung, H., and Sobsey, M. D., 1993, Comparative survival of indicator viruses and enteric viruses in seawater and sediment. *Water Sci. Technol.* 27:425–428.

Chung, H., Jaykus, L.-A., Lovelace, G., and Sobsey, M. D., 1998, Bacteriophages and bacteria as indicators of enteric viruses in oysters and their harvest waters. *Water Sci. Technol.* 38:37–44.

Ciocca, M., 2000, Clinical course and consequences of hepatitis A infection. *Vaccine* 18:S71–S74.

Cliver, D. O., and Kostenbader, K. D., 1984, Disinfection of virus on hands for prevention of food-borne disease. *Int. J. Food Microbiol.* 1:75–87.

Cook, D. W., and Ellender, R. D., 1986, Relaying to decrease the concentration of oyster-associated pathogens. *J. Food Prot.* 49:196–202.

Committee on Infectious Diseases, American Academy of Pediatrics, 1991, Hepatitis A, in: *Report of the Committee on Infectious Diseases* (G. Peter, ed.), pp. 234–237.

Croci, L., Ciccozzi, M., De Medici, D., Di Pasquale, S., Fiore, A., Mele, A., and Toti, L., 1999, Inactivation of hepatitis A virus in heat-treated mussels. *J. Appl. Microbiol.* 87:884–888.

Croci, L., De Medici, D., Scalfaro, C., Fiore, A., Divizia, M., Donia, D., Cosentino, A. M., Moretti, P., and Constantini, G., 2000, Determination of enterovirus, hepatitis A virus, bacteriophages and *Escherichia coli* in Adriatic sea mussels. *J. Appl. Microbiol.* 88:293–298.

Croci, L., De Medici, D., Scalfaro, C., Fiore, A., and Toti, L., 2002, The survival of hepatitis A virus in fresh produce. *Int. J. Food Microbiol.* 73:29–34.

Daniels, N. A., Bergmire-Sweat, D. A., Schwab, K. J., Hendricks, K. A., Reddy, S., Rowe, S. M., Fankhauser, R. L., Monroe, S. S., Atmar, R. L., Glass, R. I., and Mead, P., 2000, A foodborne outbreak of gastroenteritis associated with Norwalk-like viruses: first molecular traceback to deli sandwiches contaminated during preparation. *J. Infect. Dis.* 181:1467–1470.

De Leon, R., and Gerba, C. P., 1990, Viral disease transmission by seafood, in: *Food Contamination from Environmental Sources* (J. O. Nriagu and M. S. Simmons, eds.), Joh Wiley & Sons, New York, pp. 639–662.

De Medici, D., Ciccozzi, M., Fiore, A., Di Pasquale, S., Parlato, A., Ricci-Bitti, P., and Croci, L., 2001, Closed-circuit system for the depuration of mussels experimentally contaminated with hepatitis A virus. *J. Food Prot.* 64:877–880.

Dentinger, C. M., Bower, W. A., Nainan, O. V., Cotter, S. M., Myers, G., Dubusky, L. M., Fowler, S., Salehi, E. D. P., and Bell, B. P., 2001, An outbreak of hepatitis A associated with green onions. *J. Infect. Dis.* 183:1273–1276.

DiGirolamo, R., Liston, J., and Matches, J. R., 1970, Survival of virus in chilled, frozen and processed oysters. *Appl. Microbiol.* 20:58–63.

DiGirolamo, R., and Daley, M., 1973, Recovery of bacteriophage from contaminated chilled and frozen samples of edible West coast crabs. *Appl. Microbiol.* 25: 1020–1022.

DiGirolamo, R., Liston, J., and Matches, J., 1977, Ionic bonding, the mechanism of viral uptake by shellfish mucus. *Appl. Environ. Microbiol.* 33:19–25.

Dore, W. J., and Lees, D. N., 1995, Behavior of *Escherichia coli* and male-specific bacteriophage in environmentally contaminated bivalve mollusks before and after depuration. *Appl. Environ. Microbiol.* 61:2830–2834.

Dore, W. J., Henshilwood, K., and Lees, D. N., 2000, Evaluation of F-specific RNA bacteriophage as a candidate human enteric virus indicator for bivalve molluscan shellfish. *Appl. Environ. Microbiol.* 66:1280–1285.

Dore, W. J., Mackie, M., and Lees, D. N., 2003, Levels of male-specific RNA bacteriophage and *Escherichia coli* in molluscan bivalve shellfish from commercial harvesting areas. *Lett. Appl. Microbiol.* 36:92–96.

Doultree, J. C., Druce, J. D., Birch, C. J., Bowden, D. S., and Marshall, J. A., 1999, Inactivation of feline calicivirus, a Norwalk virus surrogate. *J. Hosp. Infect.* 41:51–57.

Doyle, A., Barataud, D., Gallay, A., Thiolet, J. M., Le Guyaguer, S., Kohli, E., and Vaillant, V., 2004, Norovirus foodborne outbreaks associated with the consumption of oysters from the Etang de Thau, France, December 2002. *Euro Surveillance* 9:24–26.

Enriquez, R., Frosner, G. G., Hochstein-Mintzell, V., Riedemann S., and Reinhardt, G., 1992, Accumulation and persistence of hepatitis A virus in mussels. *J. Med. Virol.* 37:174–179.

Estes, M. K., Ball, J. M., Guerrero, R. A., Opekun, A. R., Gilger, M. A., Pacheco, S. S., and Graham, D. Y., 2000, Norwalk virus vaccines: challenges and progress. *J. Infect. Dis.* 181(Suppl. 2):367–373.

FDA (Food and Drug Administration), 1998, Guidance to industry: guide to minimize microbial food safety hazards for fruits and vegetables. Available at http://www.foodsafety.gov/~dms/prodguid.html.

Fiore, A. E., 2004, Hepatitis A transmitted by food. *Clin. Infect. Dis.* 38:705–715.

Food and Drug Administration, 2001, Gloves—use limitation, in: *Food Code*. Available at http://www.cfsan.fda.gov/~dms/fc01-3.html.

Formiga-Cruz, M., Tofino-Quesada, G., Bofill-Mas, S., Lees, D. N., Henshilwood, K., Allard, A. K., Conden-Hansson, A.-C., Hernroth, B. E., Vantarakis, A., Tsibouxi, A., Vantarakis, A., and Girones, R., 2002, Distribution of human virus in contamination in shellfish from different growing areas in Greece, Spain, Sweden, and the United Kingdom. *Appl. Environ. Microbiol.* 68:5990–5998.

Formiga-Cruz, M., Allard, A. K., Conden-Hansson, A. C., Henshilwood, K., Hernroth, B. E., Jofre, J., Lees, D. N., Lucena, F., Papapetropoulou, M., Rangdale, R. E., Tsibouxi, A., Vantarakis, A., and Girones, R., 2003, Evaluation of potential indicators of viral contamination in shellfish and their applicability to diverse geographical areas. *Appl. Environ. Microbiol.* 69:1556–1563.

Furuta, T., Akiyama, M., Kato, Y., and Nishio, O., 2003, A food poisoning outbreak by purple Washington clam contaminated with norovirus (Norwalk-like virus) and hepatitis A virus. *Kansenshogaku Zasshi* 77:89–94.

Gantzer, C., Dubois, E., Crance, J.-M., Billaudel, S., Kopecka, H., Schwartzbrod, L., Pommepuy, M., and Guyader, F. L., 1998, Influence of environmental factors on the survival of enteric viruses in seawater. *Oceanologica Acta* 21:983–992.

Gantzer, C., Gillerman, L., Kuznetsov, M., and Oron, G., 2001, Adsorption and survival of faecal coliforms, somatic coliphages and F-specific RNA phages in soil irrigated with wastewater. *Water Sci. Tech.* 43:117–124.

Garrett, E. S., Jahncke, M. J., and Tennyson, J. M., 1997, Microbiological hazards and emerging food-safety issues associated with seafoods. *J. Food Prot.* 60: 1409–1415.

Gaulin, C., Frigon, M., Poirier D., and Fournier, C., 1999, Transmission of calicivirus by a foodhandler in the pre-symptomatic phase of illness. *Epidemiol. Infect.* 123: 475–478.

Gehrke, C., Steinmann, J., and Goroncy-Bermes, P., 2004, Inactivation of feline calicivirus, a surrogate of norovirus (formerly Norwalk-like viruses), by different types of alcohol in vitro and in vivo. *J. Hosp. Infect.* 56:49–55.

Gerba, C. P., Gramos, D. M., and Nwachuku, N., 2002, Comparative inactivation of enteroviruses and adenovirus 2 by UV light. *Appl. Environ. Microbiol.* 68: 5167–5169.

Gingrich, G. A., Hadler, S. C., Elder, H. A., and Ash, K. O., 1983, Serological investigation of an outbreak of hepatitis A in a rural day-care center. *Am. J. Public. Health* 73:1190–1193.

Green, J., Wright, P. A., Gallimore, C. I., Mitchell, O., Morgan-Capner, P., and Brown, D. W., 1998, The role of environmental contamination with small round structured viruses in a hospital outbreak investigated by reverse-transcriptase polymerase chain reaction assay. *J. Hosp. Infect.* 39:39–45.

Greening, G. E., Dawson, J., and Lewis, G., 2001, Survival of poliovirus in New Zealand green-lipped mussels, *Perna canaliculus*, on refrigerated and frozen storage. *J. Food Prot.* 64:881–884.

Griffin, M. R., Surowiec, J. J., McCloskey, D. I., Capuano, B., Pierzynski, B., Quinn, M., Wojnarski, R., Parkin, W. E., Greenberg, H., and Gary, G. W., 1982, Foodborne Norwalk virus. *Am. J. Epidemiol.* 115:178–184.

Gulati, B. R., Allwood, P. B., Hedberg, C. W., and Goyal, S. M., 2001, Efficacy of commonly used disinfectants for the inactivation of calicivirus on strawberry, lettuce, and a food-contact surface. *J. Food Prot.* 64:1430–1434.

Guzewich, J., and Ross, M. P., 1999, Interventions to prevent or minimize risks associated with bare-hand contact with ready-to-eat foods. Available at http://www.cfsan.fda.gov/~ear/rterisk.html.

Hale, A., Mattick, K., Lewis, D., Estes, M., Jiang, X., and Green, J., 2000, Distinct epidemiological patterns of Norwalk-like virus infection. *J. Med. Virol.* 62: 9–103.

Halliday, M. L., Kng, L.-Y., Zhou, T.-K., Hu, M.-D., Pan, Q.-C., Fu, T.-Y., Huang, Y.-S., and Hu, S.-L., 1991, An epidemic of hepatitis A attributable to the ingestion of raw clams in Shanghai, China. *J. Infect. Dis.* 164:852–859.

Harrington, P. R., Vinje, J., Moe, C. L., and Baric, R. S., 2004, Norovirus capture with histo-blood group antigens reveals novel virus-ligand interactions. *J. Virol.* 78: 3035–3045.

Harris, L. J., Farber, J. N., Beuchat, L. R., Parish, M. E., Suslow, T. V., Garrett, E. H., and Busta, F. F., 2002, Outbreaks associated with fresh produce: incidence, growth, and survival of pathogens in fresh and fresh-cut produce. *Compr. Rev. Food Sci. Food Safety*, pp. 1–64.

Hejkal, T. W., Wellings, F. M., Larock, P. A., and Lewis, A. L., 1979, Survival of poliovirus within organic solids during chlorination. *Appl. Environ. Microbiol.* 38:114–118.

Hofmann, F., Wehrle, G., Berthold, H., and Koster, D., 1992, Hepatitis A as an occupational hazard. *Vaccine* 10:82–84.

Hutin, Y. J. F., Pool, V., Cramer, E. H., Nainan, O. V., Weth, J., Williams, I. T., Goldstein, S. T., Gensheimer, K. F., Bell, B. P., Shapiro, C. N., Alter, M. J., and Margolis, H. S., 1999, A multistate foodborne outbreak of hepatitis A. *New Engl. J. Med.* 340: 595–601.

Hutson, A. M., Atmar, R. L., and Estes, M. K., 2004, Norovirus disease: changing epidemiology and host susceptibility factors. *Trends Microbiol.* 12:279–287.

Jaykus, L., DeLeon, R., and Sobsey, M. D., 1996, A virion concentration method for detection of human enteric viruses in oysters by PCR and oligoprobe hybridization. *Appl. Environ. Microbiol.* 62:2074–2080.

Jaykus, L.-A., 2000a, Detection of human enteric viruses in foods, in: *Foodborne Disease Handbook: Viruses, Parasites and HACCP*, Vol. 2, 2nd ed. (Y. H. Hui, S. A. Sattar, K. D. Murrell, W. K. Nip, and P. S. Stanfield, eds.), Marcel Dekker, New York, pp. 137–163.

Jaykus, L.-A., 2000b, Enteric viruses as "emerging agents" of foodborne disease. *Irish J. Agri. Food Res.* 39:245–255.

Jaykus, L.-A., Hemard, M. T., and Sobsey, M. D., 1994, Human enteric pathogenic viruses, in: *Environmental Indicators and Shellfish Safety* (C. R. Hackney and M. D. Pierson, eds.), Chapman and Hall, New York, pp. 92–153.

Jean, J., Vachon, J.-F., Moroni, O., Darveau, A., Kukavica-Ibrulj, I., and Fliss, I., 2003, Effectiveness of commercial disinfectants for inactivating hepatitis A virus on agri-food surfaces. *J. Food Protect.* 66:115–119.

Johnson, P. C., Mathewson, J. J., DuPont, H. L., and Greenberg, H. B., 1990, Multiple-challenge study of host susceptibility to Norwalk gastroenetritis in US adults. *J. Infect. Dis.* 161:18–21.

Kapikian, A. Z., Estes, M. K., and Chanock, R. M., 1996, Norwalk group of viruses, in *Fields Virology* (B. N. Fields, D. M. Knipe, and P. M. Howley, eds.), Lippincott-Raven, Philadelphia, pp. 783–810.

Keswick, B., Pickering, L. K., DuPont, H. L., and Woodward, W. E., 1983, Survival and detection of rotaviruses on environmental surfaces in day care centers. *Appl. Environ. Microbiol.* 46:813–816.

Kilgore, P. E., Belay, E. D., Hamlin, D. M., Noel, J. S., Humphrey, C. D., Gary, H. E., Ando, T., Monroe, S. S., Kludt, P. E., Rosenthal, D. S., Freeman, J., and Glass, R. I., 1996, A university outbreak of gastroenteritis due to a small round-structured virus: application of molecular diagnostics to identify the etiolgical agent and patterns of transmission. *J. Infect. Dis.* 173:787–793.

Kingsley, D. H., and Richards, G. P., 2001, Rapid and efficient extraction method for reverse transcription-PCR detection of hepatitis A and Norwalk-like viruses in shellfish. *Appl. Environ. Microbiol.* 67:4152–4157.

Kingsley, D. H., and Richards, G. P., 2003, Persistence of hepatitis A virus in oysters. *J. Food Prot.* 66:331–334.

Kingsley, D. H., Meade, G. K., and Richards, G. P., 2002a, Detection of both hepatitis A virus and Norwalk-like virus in imported clams associated with food-borne illness. *J. Food Prot.* 68:3914–3918.

Kingsley, D. H., Hoover, D. G., Papafragkou, E., and Richards, G. P., 2002b, Inactivation of hepatitis A virus and a calicivirus by high hydrostatic pressure. *J. Food Prot.* 65:1605–1609.

Koff, R. S., 2003, Hepatitis vaccines: recent advances. *Int. J. Parasitol.* 33:517–523.

Koff, R. S., and Sear, H. S., 1967, Internal temperature of steamed clams. *N. Engl. J. Med.* 276:737–739.

Kohn, M. A., Farley, T. A., Ando, T., Curtis, M., Wilson, S. A., Jin, Q., Monroe, S. S., Baron, R. C., McFarland, L. M., and Glass, R. I., 1995, An outbreak of Norwalk virus gastroenteritis associated with eating raw oysters. *J. Am. Med. Assoc.* 273: 466–471.

Koopmans, M., and Duizer, E., 2004, Foodborne viruses: an emerging problem. *Int. J. Food Microbiol.* 90:23–41.

Koopmans, M., von Bonsdorff, C.-H., Vinge, J., De Medici, D., and Monroe, S., 2002, Foodborne viruses. *FEMS Microbiol. Rev.* 26:187–205.

Kurdziel, A. S., Wilkinson, N., Langton, S., and Cook, N., 2001, Survival of poliovirus on soft fruit and salad vegetables. *J. Food Prot.* 64:706–709.

Kuritsky, J. N., Osterholm, M. T., Greenberg, H. B., Korlath, J. A., Godes, J. R., Hedberg, C. W., Forfang, J. C., Kapikian, A. Z., McCullough, J. C., and White, K. E., 1984, Norwalk gastroenteritis: a community outbreak associated with bakery product consumption. *Ann. Intern. Med.* 100:519–521.

LeBaron, C. W., Furutan, N. P., Lew, J. F., Allen, J. R., Gouvea, V., and Moe, C., 1990, Viral agents of gastroenteritis. *Morbid. Mortal. Weekly Rep.* 39:1–24.

Lemon, S. M., 1997, Type A hepatitis: epidemiology, diagnosis, and prevention. *Clin. Chem.* 43:1494–1499.

Lees, D., 2000, Viruses and bivalve shellfish. *Int. J. Food Microbiol.* 59:81–116.

Legnani, P., Leoni, E., Lev, D., Rossi, R., Villa, G. C., and Bisbini, P., 1998, Distribution of indicator bacteria and bacteriophages in shellfish and shellfish-growing waters. *J. Appl. Microbiol.* 85:790–798.

Leoni, E., Bevini, C., Esposti, S. D., and Graziano, A., 1998, An outbreak of intrafamiliar hepatitis A associated with clam consumption: epidemic transmission to a school community. *Eur. J. Epidemiol.* 14:187–192.

Lin, C.-M., Wu, F.-M., Kim, H.-K., Doyle, M. P., Michaels, B. S., and Williams, L. K., 2003, A comparison of hand washing techniques to remove *Escherichia coli* and caliciviruses under natural or artificial fingernails. *J. Food Prot.* 66:2296–2301.

Liu, B., Maywood, P., Gupta, L., and Campbell, B., 2003, An outbreak of Norwalk-like virus gastroenteritis in a aged-care residential hostel. *NSW Public Health Bull.* 14:105–109.

Lo, S., Gilbert, J., and Hetrick, F., 1976, Stability of human enteroviruses in estuarine and marine waters. *Appl. Environ. Microbiol.* 32:245–249.

Lo, S. V., Connolly, A. M., Palmer, S. R., Wright, D., Thomas, P. D., and Joynson, D., 1994, The role of pre-symptomatic food handler in a common source outbreak of food borne SRVS gastroenteritis in a group of hospitals. *Epidemiol. Infect.* 113: 513–521.

Lopman, B. A., Brown, D. W., and Koopmans, M., 2002, Human caliciviruses in Europe. *J. Clin. Virol.* 4:137–160.

Lukasik, J., Bradley, M. L., Scott, T. M., Dea, M., Koo, A., Hsu, W.-Y., Bartz, J. A., and Farrah, S. R., 2003, Reduction of poliovirus 1, bacteriophages, *Salmonella montevideo*, and *Escherichia coli* 0157:H7 on strawberries by physical and disinfectant washes. *J. Food Prot.* 66:188–193.

Mallet, J. C., Beghian, L. E., Metcalf, T. G., and Kaylor, J. D., 1991, Potential of irradiation technology for improving shellfish sanitation. *J. Food Safety.* 11:231–245.

Mariam, T. W., and Cliver, D. O., 2000, Hepatitis A virus control in strawberry products. *Dairy Food Environ. Sanit.* 20:612–616.

Marks, P. J., Vipond, I. B., Carlisle, D., Deakin, D., Fey, R. E., and Caul, E. O., 2000, Evidence for airborne transmission of Norwalk-like virus (NLV) in a hotel restaurant. *Epidemiol. Infect.* 124:481–487.

Mbithi, J. N., Springthorpe, V. S., and Sattar, S. A., 1990, Chemical disinfection of hepatitis A virus on environmental surfaces. *Appl. Environ. Microbiol.* 56:3601–3604.

Mbithi, J. N., Springthorpe, V. S., and Sattar, S. A., 1991, Effect of relative humidity and air temperature on survival of hepatitis A virus on environmental surfaces. *Appl. Environ. Microbiol.* 57:1394–1399.

Mbithi, J. N., Springthorpe, V. S., Boulet, J. R., and Sattar, S. A., 1992, Survival of hepatitis A virus on human hands and its transfer on contact with animate and inanimate surfaces. *J. Clin. Microbiol.* 30:757–763.

Mbithi, J. N., Springthorpe, V. S., and Sattar, S. A., 1993, Comparative in vivo efficiences of hand-washing agents against hepatitis A virus (HM-175) and poliovirus Type 1 (Sabin). *Appl. Environ. Microbiol.* 59:3463–3469.

McCaustland, K. A., Bond, W. W., Bradley, D. W., Ebert, J. W., and Maynard, J. E., 1982, Survival of hepatitis A in feces after drying and storage for 1 month. *J. Clin. Microbiol.* 16:957–958.

McDonnell, S., Kirkland, K. B., Hlady, W. G., Aristeguieta, C., Hopcins, R. S., Monroe, S. S., and Glass, R. I., 1997, Failure of cooking to prevent shellfish-associated viral gastroenteritis. *Arch. Intern. Med.* 157:111–116.

Mead, P. S., Slutsker, L., Dietz, V., McCaig, L. F., Bresee, J. S., Shapiro, C., Griffin, P. M., and Tauxe, R. V., 1999, Food-related illness and death in the United States. *Emerg. Infect. Dis.* 5:607–625.

Meltzer, M. I., Shapiro, C. N., Mast, E. E., and Arcari, C., 2001, The economics of vaccinating restaurant workers against hepatitis A. *Vaccine* 19:2138–2145.

Metcalf, T. G., Melnick, J. L., and Estes, M. K., 1995, Environmental virology: from detection of virus in sewage and water by isolation to identification by molecular biology—a trip of over 50 years. *Ann. Rev. Microbiol.* 49:461–487.

Millard, J., Appleton, H., and Parry, J. V., 1987, Studies on heat inactivation of hepatitis A virus with special reference to shellfish. *Epidemiol. Infect.* 98:397–414.

Mishu, B., Handler, S. C., Boaz, V.A., Hutcheson, R. H., Horan, J. M., and Shaffner, W., 1990, Foodborne hepatitis A: evidence that microwaving reduces risk? *J. Infect. Dis.* 162:655–658.

Mounts, A. W., Ando, T., Koopmans, M., Bresee, J. S., Noel, J., and Glass, R. I., 2001, Cold weather seasonality of gastroenteritis associated with Norwalk-like viruses. *J. Infect. Dis.* 181(Suppl. 2):284–287.

Muniain-Mujika, I., Girones, R., and Lucena, F., 2000, Viral contamination of shellfish: evaluation of methods and analysis of bacteriophages and human viruses. *J. Virol. Methods* 89:109–118.

Muniain-Mujika, I., Girones, R., Tofino-Quesada, G., Calvo, M., and Lucena, F., 2002, Depuration dynamics of viruses in shellfish. *Int. J. Food Microbiol.* 77:125–133.

Muniain-Mujika, I., Calvo, M., Lucena, F. and Girones, R., 2003, Comparative analysis of viral pathogens and potential indicators in shellfish. *Int. J. Food Microbiol.* 83:75–85.

National Shellfish Sanitation Program, 2000, Guide for the Control of Molluscan Shellfish, Model Ordinance. U. S. Food and Drug Administration, Center for Food Safety and Applied Nutrition, Office of Seafood. Available at http://vm.cfsan.fda.gov/~ear/nsspotoc.html.

Nishida, T., Kimura, H., Saitoh, M., Shinohara, M., Kato, M., Fukuda, S., Munemura, T., Mikami, T., Kawamoto, A., Akiyama, M., Kato, Y., Nishi, K., Kozawa, K., and

Nishio, O., 2003, Detection, quantitation and phylogenetic analysis of noroviruses in Japanese oysters. *Appl. Environ. Microbiol.* 69:5782–5786.

Niu, M. T., Polish, L. B., Robertson, B. H., Khanna, B. K., Woodruff, B. A., Shapiro, C. N., Miller, M. A., Smith, J. D., Gedrose, J. K., Alter, M. J., and Margolis, H. S., 1992, Multistate outbreak of hepatitis A associated with frozen strawberries. *J. Infect. Dis.* 166:518–524.

O'Brien, S. J., Mitchell, R. T., Gillespie, I. A., and Adak, G. K., 2000, The microbiological status of ready-to-eat fruit and vegetables. Discussion paper: Advisory Committee on the Microbiological Safety of Food ACM/476, pp. 1–34.

O'Mahony, J., O'Donoghue, M., Morgan, J. G., and Hill, C., 2000, Rotavirus survival and stability in foods as determined by an optimised plaque assay procedure. *Int. J. Food Microbiol.* 61:177–185.

Parashar, U. D., Dow, L., Fankhauser, R. L., Humphrey, C. D., Miller, I. J., Ando, T., Williams, K. S., Eddy, C. R., Noel, J. S., Ingram, T., Bresee, J. S., Monroe, S. S., and Glass, R. I., 1998, An outbreak of viral gastroenteritis associated with consumption of sandwiches: implications for the control of transmission by food handlers. *Epidemiol. Infect.* 121:615–621.

Parrino, T. A., Schreiber, D. S., Trier, J. S., Kapikian, A. Z., and Blacklow, N. R., 1977, Clinical immunity in acute gastroenetroenetritis caused by Norwalk agent. *N. Engl. J. Med.* 297:86–89.

Patterson, W., Haswell, P., Fryers, P. T., and Green, J., 1997, Outbreak of small round structured virus gastroenteritis arose after kitchen assistant vomited. *Commun. Dis. Rep.* 7:101–103.

Paulson, D. S., 2003, Handwashing, gloving, and disease transmission by the food preparer, in: *Handbook of Topical Antimicrobial: Industrial Applications in Consumer Products and Pharmaceuticals* (D. S. Paulson, ed.), Marcel Dekker, New York, pp. 255–270.

Pavia, A. T., Nielsen, L. Armington, Thurman, D. J., Tierney, E., and Nichols, C. R., 1990, A community-wide outbreak of hepatitis A in a religious community: impact of mass administration of immune globulin. *Am. J. Epidemiol.* 131:1085–1093.

Pina, S., Puig, M., Lucena, F., Jofre, J., and Girones, R., 1998, Viral pollution in the environment and in shellfish: human adenovirus detection by PCR as an index of human viruses. *Appl. Environ. Microbiol.* 64:3376–3382.

Ponka, A., Maunula, L., von-Bonsdorff, C.-H., and Lyytikainen, O., 1999, An outbreak of calicivirus associated with consumption of frozen raspberries. *Epidemiol. Infect.* 123:469–474.

Potasman, I., Paz, A., and Odeh, M., 2002, Infectious outbreaks associated with bivalve shellfish consumption: a worldwide perspective. *Clin. Infect. Dis.* 35:921–928.

Power, U. F., and Collins, J. K., 1989, Differential depuration of poliovirus, *Escherichia coli*, and a coliphage by the common mussel, *Mytilus edulis*. *Appl. Environ. Microbiol.* 55:1386–1390.

Public Health Service, 1999, *Food Code*. National Technology Information Service, U.S. Public Health Service, Springfield, VA.

Reid, T. M. S., 1987, Frozen rasberries and hepatitis A. *Epidemiol. Infect.* 98:109–112.

Richards, G. P., 2001, Food-borne pathogens: enteric virus contamination of foods through industrial practices: a primer on intervention strategies. *J. Indust. Microbiol. Biotechnol.* 27:117–125.

Roderick, G. E., and Schneider, K. R., 1994, Depuration and relaying of molluscan shellfish, in: *Environmental Indicators and Shellfish Safety* (C. R. Hackney and M. D. Pierson, eds.), Chapman and Hall, New York.

Romalde, J. L., Area, E., Sanchez, E., Ribao, C., Torrado, I., Abad, X., Pinto, R. M., Barja, J. L., and Bosch, A., 2002, Prevalence of enterovirus and hepatitis A virus in bivalve molluscs from Galicia (NW Spain): inadequacy of the EU standards of microbiological quality. *Int. J. Food Microbiol.* 74:119–130.

Rosenblum, L. S., Mirkin, I. R., Allen, D. T., Sufford, S., and Hadler, S. C., 1990, A multifocal outbreak of hepatitis A traced to commercially distributed lettuce. *Am. J. Public. Health* 80:1075–1079.

Rusin, P., Maxwell, S., and Gerba, C. P., 2002, Comparative surface-to-hand and fingertip-to-mouth transfer efficiency of gram-positive bacteria, gram-negative bacteria, and phage. *J. Appl. Microbiol.* 93:585–592.

Sair, A. I., D'Souza, D. H., and Jaykus, L.-A., 2002, Human enteric viruses as causes of foodborne disease. *Compr. Rev. Food Sci. Food Safety* 1:73–89.

Sattar, S. A., 2004, Microbicides and the environmental control of nosocomial viral infections. *J. Hosp. Infect.* 56:564–569.

Sattar, S. A., Jacobsen, H., Rahman, H., Cusack, T. M., and Rubino, J. R., 1994, Interruption of rotavirus spread through chemical disinfection. *Infect. Control. Hosp. Epidemiol.* 15:751–756.

Sattar, S. A., Abebe, M., Bueti, A. J., Jampani, H., Newman, J., and Hua, S., 2000, Activity of an alcohol-based hand gel against human adeno-, rhino-, and rotaviruses using the fingerpad method. *Infect. Control. Hosp. Epidemiol.* 21:516–519.

Sattar, S. A., Springthorpe, V. S., Tetro, J., Vashon, R., and Keswick, B., 2002, Hygienic hand antiseptics: should they not have activity and label claims against viruses? *Am. J. Infect. Control* 30:355–372.

Schaub, S. A., and Oshiro, R. K., 2000, Public health concerns about caliciviruses as waterborne contaminants. *J. Infect. Dis.* 181(Suppl. 2):374–380.

Scholz, E., Heinricy, U., and Flehmig, B., 1989, Acid stability of hepatitis A virus. *J. Gen. Virol.* 70:2481–2485.

Seymour, I. J., and Appleton, H., 2001, Foodborne viruses and fresh produce. *J. Appl. Microbiol.* 91:759–773.

Shieh, Y.-S. C., Monroe, S. S., Fankhauser, R. L., Langlois, G. W., Burkhardt, W., III, and Baric, R. S., 2000, Detection of Norwalk-like virus in shellfish implicated in illness. *J. Infect. Dis.* 181(Suppl. 2):360–366.

Sinton, L. W., Hall, C. H., Lynch. P. A., and Davies-Colley, R. J., 2002, Sunlight inactivation of fecal indicator bacteria and bacteriophages from waste stabilization pond effluent in fresh and saline waters. *Appl. Environ. Microbiol.* 68: 1122–1131.

Slomka, M. J., and Appleton, H., 1998, Feline calicivirus as a model system for heat inactivation studies of small round structured viruses in shellfish. *Epidemiol. Infect.* 121:401–407.

Sobsey, M. D., and Jaykus, L.-A., 1991, Human enteric viruses and depuration of bivalve molluscs, in: *Molluscan Shellfish Depuration* (W. S. Otwell, G. E. Rodrick, and R. E. Martin, eds.), CRC Press, Boca Raton, FL, pp. 71–114.

Somerset, I. J., 1991, Current U.S. commercial shellfish depuration, in: *Molluscan Shellfish Depuration* (W. S. Otwell, G. E. Rodrick, and R. E. Martin, eds.), CRC Press, Boca Raton, FL.

Sorber, C. A., 1983, Removal of viruses from wastewater and effluent by treatment processes, in: *Viral Pollution of the Environment* (G. Berg, ed.), CRC Press, Boca Raton, FL, pp. 39–52.

Springthorpe, S., and Sattar, S., 1998, Handwashing: what can we learn from recent research? *Infect. Control Today* 2:20–28.

Suslow, T. V., Oria, M. P., Beuchat, L. R., Garrett, E. H., Parish, M. E., Harris, L. J., Farber, J. N., and Busta, F. F., 2003, Production practices as risk factors in microbial food safety of fresh and fresh-cut produce. *Compr. Rev. Food Sci. Food Safety* 2(Suppl.):1–40.

Thurston-Enriquez, J. A., Haas, C. N., Jacangelo, J., Riley, K., and Gerba, C. P., 2003, Inactivation of feline calicivirus and adenovirus type 40 by UV radiation. *Appl. Environ. Microbiol.* 69:577–582.

Tierney, J. T., Sullivan, R., Peeler, J. T., and Larkin, E. P., 1982, Persistence of poliovirus in shellstock and shucked oysters stored at refrigeration temperature. *J. Food Protect.* 45:1135–1137.

Warburton, A. R. E., Wreghitt, T. G., Rampling, A., Buttery, R., Ward, K. N., Perry, K. R., and Perry, J. V., 1991, Hepatitis A outbreak involving bread. *Epidemiol. Infect.* 106:199–202.

Ward, B. K., Chenoweth, C. M., and Irving, L. G., 1982, Recovery of viruses from vegetable surfaces. *Appl. Environ. Microbiol.* 44:1389–1394.

Ward, R. L., Bernstein, D. I., Knowlton, D. R., Sherwood, J. R., Young, E. C., Cusack, T. M., Rubino, J. R., and Schiff, G. M., 1991, Prevention of surface-to-human transmission of rotaviruses by treatment with disinfectant spray. *J. Clin. Microbiol.* 29:1991–1996.

White, K. E., Osterholm, M. T., Mariotti, J. A., Korlath, J. A., Lawrence, D. H., Ristinen, T. L., and Greenberg, H. B., 1986, A foodborne outbreak of Norwalk virus gastroenteritis: evidence for post recovery transmission. *Am. J. Epidemiol.* 124:120–126.

Wilkinson, N., Kurdziel, A. S., Langton, S., Needs, E., and Cook, N., 2001, Resistance of poliovirus to inactivation by high hydrostatic pressures. *Innov. Food Sci. Emerg. Techol.* 2:95–98.

Yates, M. V., Gerba, C. P., and Kelley, L. M., 1985, Virus persistence in groundwater. *Appl. Environ. Microbiol.* 49:778–781.

Yotsuyanagi, H., Koike, K., Yasuada, K., Moriya, K., Shintani, Y., Fujie, H., Kurokawa, K., and Iino, S., 1996, Prolonged fecal excretion of hepatitis A virus in adult patients with hepatitis A as determined by polymerase chain reaction. *Hepatology* 24:10–13.

Index

D

Q

R

S